D0780980

ADVANCED NUMBER THEORY

Harvey Cohn

Distinguished Professor of Mathematics
City University of New York

Dover Publications, Inc.
New York

dedicated to TONY and SUSAN

Published in Canada by General Publishing Company, Ltd., 30 Lesmill Road, Don Mills, Toronto, Ontario.
Published in the United Kingdom by Constable and Company, Ltd., 10 Orange Street, London WC2H 7EG.

This Dover edition, first published in 1980, is an unabridged and corrected republication of the work first published in 1962 by John Wiley & Sons, Inc., under the title *A Second Course in Number Theory*.

International Standard Book Number: 0-486-64023-X
Library of Congress Catalog Card Number: 80-65862

Manufactured in the United States of America
Dover Publications, Inc.
180 Varick Street
New York, N.Y. 10014

PREFACE

The prerequisites for this book are the "standard" first-semester course in number theory (with incidental elementary algebra) and elementary calculus. There is no lack of suitable texts for these prerequisites (for example, *An Introduction to the Theory of Numbers*, by I. Niven and H. S. Zuckerman, John Wiley and Sons, 1960, can be cited as a book that introduces the necessary algebra as part of number theory). Usually, very little else can be managed in that first semester beyond the transition from improvised combinatorial amusements of antiquity to the coherently organized background for quadratic reciprocity, which was achieved in the eighteenth century.

The present text constitutes slightly more than enough for a second-semester course, carrying the student on to the twentieth century by motivating some heroic nineteenth-century developments in algebra and analysis. The relation of this textbook to the great treatises will necessarily be like that of a historical novel to chronicles. We hope that once the student knows what to seek he will find "chronicles" to be as exciting as a "historical novel."

The problems in the text play a significant role and are intended to stimulate the spirit of experimentation which has traditionally ruled number theory and which has indeed become resurgent with the realization of the modern computer. A student completing this course should acquire an appreciation for the historical origins of linear algebra, for the zeta-function tradition, for ideal class structure, and for genus theory. These ideas, although relatively old, still make their influence felt on the frontiers of modern mathematics. Fermat's last theorem and complex multiplication are unfortunate omissions, but the motive was not to depress the degree of difficulty so much as it was to make the most efficient usage of one semester.

My acknowledgments are many and are difficult to list. I enjoyed the benefits of courses under Bennington P. Gill at City College and Saunders MacLane at Harvard. The book profited directly from suggestions by my students and from the incidental advice of many readers, particularly Burton W. Jones and Louis J. Mordell. I owe a special debt to Herbert S. Zuckerman for a careful reading, to Gordon Pall for major improvements, and to the staff of John Wiley and Sons for their cooperation.

HARVEY COHN

Tucson, Arizona
October 1961

CONTENTS

Note: The sections marked with * or ** might be omitted in class use if there is a lack of time. (Here the ** sections are considered more truly optional.)

PART 2. IDEAL THEORY IN QUADRATIC FIELDS

PART 3. APPLICATIONS OF IDEAL THEORY

INTRODUCTORY SURVEY

DIOPHANTINE EQUATIONS

The most generally enduring problem of number theory is probably that of diophantine equations. Greek mathematicians were quite adept at solving in integers x and y the equation

$$ax + by = c,$$

where a, b, and c are any given integers. The close relation with the greatest common divisor algorithm indicated the necessity of treating *unique factorization* as a primary tool in the solution of diophantine equations.

The Greek mathematicians gave some sporadic attention to forms of the more general equation

$$f(x, y) = Ax^2 + Bxy + Cy^2 + Dx + Ey + F = 0, \tag{1}$$

but achieved no sweeping results. They probably did not know that every equation of this kind can be solved "completely" by characterizing all solutions in a finite number of steps, although they had success with special cases such as $x^2 - 3y^2 = 1$. In fact, they used continued fraction techniques in both linear and quadratic problems, indicating at least esthetically a sense of unity. About 1750 Euler and his contemporaries became aware

---

This section presupposes some familiarity with elementary concepts of group, congruence, Euclidean algorithm, and quadratic reciprocity (which are reviewed in Chapter I).

of the systematic solvability in a finite number of steps. Yet it was not until 1800 that Gauss gave in his famous *Disquisitiones Arithmeticae* the solution that still remains a model of perfection.

Now a very intimate connection developed between Gauss's solution and *quadratic reciprocity*, making unique factorization (in the linear case) and quadratic reciprocity (in the quadratic case) parallel tools. Finally, about 1896, Hilbert achieved the reorganization of the quadratic theory, making full use of this coincidence and thus completing the picture.

MOTIVATING PROBLEM IN QUADRATIC FORMS

The first step in a general theory of quadratic diophantine equations was probably the famous theorem of Fermat (1640) relating to a (homogeneous) quadratic *form* in x, y.

A prime number p is representable in an essentially unique manner by the form $x^2 + y^2$ for integral x and y if and only if $p \equiv 1$ modulo 4 (or $p = 2$).

It is easily verified that $2 = 1^2 + 1^2$, $5 = 2^2 + 1^2$, $13 = 3^2 + 2^2$, $17 = 4^2 + 1^2$, $29 = 5^2 + 2^2$, etc., whereas the primes 3, 7, 11, 19, etc., have no such representation. The proof of Fermat's theorem is far from simple and is achieved later on as part of a larger result.

At the same time, Fermat used an identity from antiquity:

$$(x^2 + y^2)(x'^2 + y'^2) = (xx' - yy')^2 + (xy' + x'y)^2,$$

easily verifiable, since both sides equal $x^2x'^2 + y^2y'^2 + x'^2y^2 + x^2y'^2$. He used this formula to build up solutions to the equation

$$x^2 + y^2 = m \tag{2}$$

for values of m which are not necessarily prime. For example, from the results

$$3^2 + 2^2 = 13, \qquad (x = 3,\ y = 2),$$
$$2^2 + 1^2 = 5, \qquad (x' = 2,\ y' = 1),$$

we obtain

$$7^2 + 4^2 = 65, \qquad (xx' - yy' = 4,\ xy' + x'y = 7).$$

If we interpret the representation for 13 as

$$(-3)^2 + 2^2 = 13 \qquad (x = -3,\ y = 2),$$

whereas

$$2^2 + 1^2 = 5, \qquad (x' = 2,\ y' = 1),$$

then we obtain

$$(-8)^2 + 1^2 = 65, \qquad (xx' - yy' = -8,\ xy' + x'y = 1);$$

but the reader can verify that $65 = 7^2 + 4^2 = 8^2 + 1^2$ are the only representations obtainable for 65 in the form $x^2 + y^2$, to within rearrangements of summands or changes of sign. If we allow the trivial additional operation of using (x, y), which are not relatively prime ($(kx)^2 + (ky)^2 = k^2m$), we can build up all solutions to (2), from those for prime m.

Thus Fermat's result, stated more compactly, is the following:

Let $$Q(x, y) = x^2 + y^2.$$

Then all relatively prime solutions (x, y) *to the problem of representing*

$$Q(x, y) = m$$

for m any integer are achieved by means of the successive application of two results called genus and composition theorems.

GENUS THEOREM

$$Q(x, y) = p \tag{3}$$

can be solved in integral x, y *for* p *a prime if and only if* $p \equiv 1 \pmod 4$, *or* $p = 2$. *The representation is unique, except for obvious changes of sign or rearrangements of* x *and* y.

COMPOSITION THEOREM

$$Q(x, y)\,Q(x', y') = Q(xx' - yy', x'y + xy'). \tag{4}$$

In the intervening years until about 1800, Euler, Lagrange, Legendre, and others invented analogous results for a variety of quadratic forms. Gauss (1800) was the first one to see the larger problem and to achieve a complete generalization of the genus and composition theorems. The main result is too involved even to state here, but a slightly more difficult special result will give the reader an idea of what to expect. (See Chapter XIII.)

Let $$Q_1(x, y) = x^2 + 5y^2,$$
$$Q_2(x, y) = 2x^2 + 2xy + 3y^2.$$

Then all relatively prime solutions (x, y) *to the problem of representing*

$$Q_1(x, y) = m$$

or

$$Q_2(x, y) = m$$

for m any integer are achieved by means of the successive application of the following two results.

GENUS THEOREM

$$(5)\qquad \begin{bmatrix} Q_1(x, y) \\ Q_2(x, y) \end{bmatrix} = p, \text{ a prime, if and only if } p \equiv \begin{pmatrix} 1, 9 \\ 3, 7 \end{pmatrix} \pmod{20},$$

in an essentially unique fashion. (The only special exceptions are, $Q_1(0, 1) = 5$, $Q_2(1, 0) = 2$.)

COMPOSITION THEOREM

$$(6a)\qquad \begin{cases} Q_1(x, y)\, Q_1(x', y') = Q_1(xx' - 5yy', x'y + xy') \\ Q_1(x, y)\, Q_2(x', y') = Q_2(xx' - x'y - 3yy', xy' + 2x'y + yy') \\ Q_2(x, y)\, Q_2(x', y') = Q_1(2xx' + xy' + x'y - 2yy', xy' + x'y + yy'). \end{cases}$$

One may protest (in vain) that he is interested only in $Q_1(x, y)$, but it is impossible to separate $Q_1(x, y)$ and $Q_2(x, y)$ in the composition process For instance,

$$Q_2(1, 1) = 7, \qquad (x = 1,\ y = 1),$$

$$Q_2(0, 1) = 3, \qquad (x' = 0, y' = 1),$$

and, from the last of the composition formulas,

$$Q_1(-1, 2) = 21, \quad (2xx' + xy' + x'y - 2yy' = -1, xy' + x'y + yy' = 2).$$

Thus, to represent 21 by Q_1, we are forced to consider possible representations of factors of 21 by Q_2. The reader may find the following exercise instructive along these lines:

Find a solution to $Q_1(x, y) = 29$ by trial and error and build from the preceding results solutions to $Q_1(x, y) = 841$ and $Q_2(x, y) = 203$.

Those readers who are familiar with the concept of a group will recognize system (6*a*) symbolically as

$$(6b)\qquad \begin{cases} \mathbf{Q}_1^{\,2} = \mathbf{Q}_1 \text{ (identity)}, \\ \mathbf{Q}_1\mathbf{Q}_2 = \mathbf{Q}_2, \\ \mathbf{Q}_2^{\,2} = \mathbf{Q}_1. \end{cases}$$

In this manner we are led from quadratic forms into algebra!

USE OF ALGEBRAIC NUMBERS

The reader will probably note that the decomposition theorem resembles the method of multiplication of complex numbers:

$$(7a) \qquad (x + iy)(x' + iy') = (xx' - yy') + i(xy' + yx'),$$

where, of course, $i = \sqrt{-1}$. The composition theorems for $Q_1(x, y)$ and $Q_2(x, y)$ can be similarly explained by use of $\sqrt{-5}$ if we solve for x'' and y'' in each of the following equations:

$$(7b) \quad \begin{cases} (x + \sqrt{-5}y)(x' + \sqrt{-5}y') = (x'' + \sqrt{-5}y''), \\ (x + \sqrt{-5}y)(2x' + y' + \sqrt{-5}y') = (2x'' + y'' + \sqrt{-5}y'') \\ (2x + y + \sqrt{-5}y)(2x' + y' + \sqrt{-5}y') = 2(x'' + \sqrt{-5}y''), \end{cases}$$

but we shall defer all details to Chapter XIII.

The important point, historically, is that before the time of Gauss mathematicians strongly feared the possibility of developing a contradiction if reliance was placed on such numbers as $\sqrt{-1}$, $\sqrt{-5}$, and they would use these numbers "experimentally," although their final proofs were couched in the immaculate language of traditional integral arithmetic; yet eventually they had to accept radicals as a necessary simplifying device.

A second guiding influence in the introduction of radicals was the famous conjecture known as *Fermat's last theorem*:

If n is an integer ≥ 3, the equation

$$x^n + y^n = z^n$$

has no solution in integers (x, y, z), except for the trivial case in which $x y z = 0$. The result is still not proved for all n, nor is it contradicted. Here Cauchy, Kummer, and others achieved, for special n, remarkable results by factoring the left-hand side. We shall ignore this very important development in order to unify the material, but we cannot fail to see its relevance (say) for $n = 3$, if we write

$$x^3 + y^3 = (x + y)(y + \rho y)(x + \rho^2 y),$$
$$\rho = (-1 + \sqrt{-3})/2, \rho^2 = (-1 - \sqrt{-3})/2.$$

The introduction of such numbers as ρ, $\sqrt{-1}$, $\sqrt{-5}$ resulted in a further development by Dedekind (1870) of a systematic theory of algebraic numbers. These are quantities α defined by equations, for instance, of degree k,

$$A_0\alpha^k + A_1\alpha^{k-1} + \cdots + A_k = 0$$

with integral coefficients. It turned out that quadratic surds ($k = 2$) were an extremely significant special case whose properties to this very day are not fully generalized to $k > 2$. Thus the importance of this special (quadratic) case cannot be overestimated in the theory of algebraic numbers of arbitrary degree k.

In this book we try to get the best of both worlds: we use quadratic forms with *integral* coefficients or factor the forms (using *algebraic* number theory), depending on which is more convenient.

PRIMES IN ARITHMETIC PROGRESSION

If we examine $Q_1(x, y)$ and $Q_2(x, y)$ more carefully, we find that in both cases the discriminant is -20, (the discriminant is the usual value, $d = B^2 - 4AC$ for the form $Ax^2 + Bxy + Cy^2$). Actually, the number of forms required for a complete composition theorem associated with a discriminant is (essentially) a very important integer called the *class number*, written $h(d)$. Thus, referring to $Q(x, y)$, we find $h(-4) = 1$; and referring to $Q_1(x, y)$, $Q_2(x, y)$, we find $h(-20) = 2$. The value of the class number is one of the most *irregular* functions in number theory. Gauss (1800) and Dirichlet (1840), however, did obtain "exact" formulas for the class number. They used continuous variables and the limiting processes of calculus, or the tools of analysis.

One of the most startling results in number theory developed when Dirichlet used this class-number formula to show the following result:

There is an infinitude of primes in any arithmetic progression

$$a, a + d, a + 2d, a + 3d, \cdots,$$

provided $(a, d) = 1$, *and* $d > 0$.

The fact that *quadratic* forms had originally provided the clue to a problem involving the *linear* form $a + xd$ has not been completely assimilated even today. Despite the occurrence of "direct" demonstrations of the result of Dirichlet, the importance of the original ideas is manifest in the wealth of unsolved related problems in algebraic number theory.

We are thus concerned with the remarkable interrelation between the theory of integers and analysis. The role of number theory as a fountainhead of algebra and analysis is the central idea of this book.

PART 1

BACKGROUND MATERIAL

chapter I

Review of elementary number theory and group theory

NUMBER THEORETIC CONCEPTS

1. Congruence

We begin with the concept of divisibility. We say[1] a divides b if there is an integer c such that $b = ac$. If a divides b, we write $a \mid b$, and if a does not divide b we write $a \nmid b$. If $k \geq 0$ is an integer for which $a^k \mid b$ but $a^{k+1} \nmid b$, we write $a^k \parallel b$, which we read as "a^k divides b exactly."

If $m \mid (x - y)$, we write

$$x \equiv y \pmod{m} \tag{1}$$

and say that x is *congruent* to y *modulo* m. The quantity m is called the modulus, and all numbers congruent (or equivalent) to $x \pmod{m}$ are said to constitute a *congruence* (or equivalence) *class*. Congruence classes are preserved under the rational integral operations, addition, subtraction, and multiplication; or, more generally, from the congruence (1) we have

$$f(x) \equiv f(y) \pmod{m} \tag{2}$$

where $f(x)$ is any polynomial with integral coefficients.

[1] Lower case italic letters denote integers (positive, negative, or zero), unless otherwise stated.

2. Unique Factorization

It can be shown that any two integers a and b not both 0 have a *greatest common divisor* $d(>0)$ such that if $t \mid a$ and $t \mid b$ then $t \mid d$, and conversely, if t is any integer (including d) that divides d, then $t \mid a$ and $t \mid b$. We write $d = \gcd(a, b)$ or $d = (a, b)$. It is more important that for any a and b there exist two integers x and y such that

$$ax + by = d. \tag{1}$$

If $d = (a, b) = 1$, we say a and b are *relatively prime*.

One procedure for finding such integers x, y is known as the *Euclidean algorithm*. (This algorithm is referred to in Chapter VI in another connection, but it is not used directly in this book.)

We make more frequent use of the *division algorithm*, on which the Euclidean algorithm is based: if a and b are two integers ($b \neq 0$), there exists a quotient q and a remainder r such that

$$a = qb + r \tag{2}$$

and, most important, $a \equiv r \pmod{b}$ where

$$0 \leq r < |b|. \tag{3}$$

The congruence classes are accordingly called *residue* (remainder) classes.

From the foregoing procedure it follows that if $(a, m) = 1$ then an integer x exists such that $(x, m) = 1$ and $ax \equiv b \pmod{m}$. From this it also follows that the symbol $b/a \pmod{m}$ has integral meaning and may be written as x if $(a, m) = 1$.

An integer p greater than 1 is said to be a *prime* if it has no positive divisors except p and 1. The most important result of the Euclidean algorithm is the theorem that if the prime p is such that $p \mid ab$ then $p \mid a$ or $p \mid b$. Thus, by an elementary proof, any nonzero integer m is representable in the form

$$m = \pm p_1^{a_1} p_2^{a_2} \cdots p_t^{a_t}, \tag{4}$$

where the p_i are distinct primes. The representation is unique within rearrangement of factors. Each factor $p_i^{a_i}$ is called *primary*.

EXERCISE 1. Observe that

$$\tfrac{1}{2} + \tfrac{1}{3} = \tfrac{5}{6}$$

$$\tfrac{1}{2} \equiv 4, \qquad \tfrac{1}{3} \equiv 5. \qquad \tfrac{5}{6} \equiv 2 \pmod{7},$$

$$4 + 5 \equiv 2 \pmod{7}.$$

Write down and prove a general theorem enabling us to use ordinary arithmetic to work with fractions modulo m (if the denominators are prime to m).

EXERCISE 2. Prove $\sum_{x=1}^{p-1} 1/x \equiv 0 \pmod{p}$, ($p$ odd).

EXERCISE 3. From the remarkable coincidence $2^4 + 5^4 = 2^7 \cdot 5 + 1 = 641$ show $2^{32} + 1 \equiv 0 \pmod{641}$. *Hint.* Eliminate y between the pair of equations $x^4 + y^4 = x^7 y + 1 = 0$ and carry the operations over to integers (mod 641).

EXERCISE 4. Write down and prove the theorem for the solvability or non-solvability of $ax \equiv b \pmod{m}$ when $(a, m) > 1$.

3. The Chinese Remainder Theorem

If $m = rs$ where $r > 0$, $s > 0$, then every congruence class modulo m corresponds to a unique pair of classes in a simple way, i.e., if $x \equiv y \pmod{m}$, then $x \equiv y \pmod{r}$ and $x \equiv y \pmod{s}$. If $(r, s) = 1$, the converse is also true; every pair of residue (congruence) classes modulo r and modulo s corresponds to a single residue class modulo rs. This is called the Chinese remainder theorem.[1] One procedure for defining an x such that $x \equiv a \pmod{r}$ and $x \equiv b \pmod{s}$ uses the Euclidean algorithm, since $(x =)a + rt = b + su$ constitutes an equation in the unknowns t and u, as in (1) of §2.

As a result of this theorem, if we want to solve the equation

$$f(x) \equiv 0 \pmod{m}, \tag{1}$$

all we need do is factor $m = p_1^{a_1} p_2^{a_2} \cdots p_s^{a_s}$ and then solve each of the equations

$$f(x) \equiv 0 \pmod{p_i^{a_i}} \tag{2}$$

for as many roots as occur (possibly none). If x_i is a solution to (2), we apply the Chinese remainder theorem step-by-step to solve simultaneously the equations

$$x \equiv x_i \pmod{p_i^{a_i}}, \qquad (i = 1, 2, \cdots, s), \tag{3}$$

to obtain a solution to (1). If r_i is the number of incongruent solutions to (2), there will be $\prod_{i=1}^{s} r_i$ incongruent solutions. (The result is true even if one or more $r_i = 0$.)

EXERCISE 5. In a game for guessing a person's age x, one discreetly requests three remainders: r_1 when x is divided by 3, r_2 when x is divided by 4, and r_3 when x is divided by 5. Then

$$x \equiv 40r_1 + 45r_2 + 36r_3 \pmod{60}.$$

Discuss the process for the determination of the integers 40, 45, 36.

[1] The theorem was not handed down from China but was found to have also been known there since antiquity.

4. Structure of Reduced Residue Classes

A residue class modulo m will be called a *reduced residue class* (mod m) if each of its members is relatively prime to m. If $m = p_1^{a_1}p_2^{a_2}\cdots p_s^{a_s}$ (prime factorization), then any number x relatively prime to m may be determined modulo m by equations of the form

$$(1) \qquad x \equiv x_i \pmod{p_i^{a_i}}, \qquad (x_i, p_i) = 1, \qquad (i = 1, 2, \cdots, s).$$

The number of reduced residue classes modulo p^a is given by the *Euler ϕ function*:

$$(2) \qquad \phi(p^a) = p^a\left(1 - \frac{1}{p}\right).$$

By the Chinese remainder theorem the number of reduced residue classes modulo m is $\phi(m)$, where

$$(3) \quad \phi(m) = \prod_{i=1}^{s} \phi(p_i^{a_i}) = m[1 - (1/p_1)][1 - (1/p_2)]\cdots[1 - (1/p_s)].$$

By the Fermat-Euler theorem, if $(b, m) = 1$, then

$$(4) \qquad b^{\phi(m)} \equiv 1 \pmod{m}.$$

A number g is a primitive root of m if

$$(5) \qquad g^k \not\equiv 1 \pmod{m} \quad \text{for} \quad 0 < k < \phi(m).$$

Only the numbers $m = p^a$, $2p^a$, 2, and 4 have primitive roots (where p is an odd prime). But then, for such a value of m, all y relatively prime to p are representable as

$$(6) \qquad y \equiv g^t \pmod{m},$$

where t takes on all $\phi(m)$ values; $t = 0, 1, 2, \cdots, \phi(m) - 1$.

The accompanying tables (see appendix) give the minimum primitive root g for such prime $p < 100$ and represent y in terms of t and t in terms of y modulo p. Generally, t is called the *index* (abbr. I in the tables) and y is the *number* (abbr. N). Of course, the index is a value modulo $\phi(m)$, and the operation of the index recalls to mind elementary logarithms.

EXERCISE 6. Verify the index table modulo 19 and solve

$$2^{10}y^{60} \equiv 14^{70} \pmod{19}$$

by writing

$$10 \text{ ind } 2 + 60 \text{ ind } y \equiv 70 \text{ ind } 14 \pmod{18}$$

(and using Exercise 4, etc.).

5. Residue Classes for Prime Powers[1]

In the case of an *odd* prime power p^a, for a fixed base p, a single value g can be found that will serve as a primitive root for all exponents $a > 1$. In fact, g need be selected to serve only as the primitive root of p^2, or, even more simply, as shown in elementary texts, g can be any primitive root of p with just the further property $g^{p-1} \not\equiv 1 \pmod{p^2}$. We then take (6) of §4 to represent an arbitrary reduced residue class $y \pmod{p^a}$, using the minimum positive g for definiteness.

In the case of powers of 2, the situation is much more complicated. The easy results are (taking odd y) for different powers of 2

$$y = 1 \pmod{2}, \quad \text{trivially,} \tag{1}$$

$$y \equiv (-1)^{t_0} \pmod{4}, \qquad t_0 = 0, 1; \tag{2}$$

but for odd y, modulo 8, we find there is no primitive root. Thus there is no way of writing *all* odd $y \equiv g^t \pmod{8}$ for $t = 0, 1, 2, 3$. We must write

$$y \equiv (-1)^{t_0} 5^{t_1} \pmod{8}, \qquad t_0 = 0, 1, \qquad t_1 = 0, 1, \tag{3}$$

yielding the following table of all odd y modulo 8.

TABLE 1

y	1	3	5	7
t_0	0	1	0	1
t_1	0	1	1	0

More generally, if we consider residues modulo 2^a, $a \geqq 3$, we find the odd y are accounted for by

$$y \equiv (-1)^{t_0} 5^{t_1} \pmod{2^a}; \qquad t_0 = 0, 1; \qquad t_1 = 0, 1, \cdots, (2^a/4) - 1. \tag{4}$$

This result makes 5 a kind of "half-way" primitive root modulo 2^a for each $a \geq 3$. For instance $5^{t_1} \equiv 1 \pmod{2^a}$ when $t_1 = \phi(2^a)/2 = 2^a/4$ but for no smaller positive value of t_1. Let us collect these remarks:

If we factor $m = p_1^{a_1} p_2^{a_2} \cdots p_s^{a_s}$, and if $(y, m) = 1$, then y is uniquely determined by a set of exponents as follows: for odd primes p_i with primitive root $g_i \pmod{p_i^2}$.

$$y \equiv g_i^{t_i} \pmod{p_i^{a_i}}, \qquad 0 \leq t_i < \phi(p_i^{a_i}). \tag{5a}$$

[1] In this section and those that follow the proofs are less elementary than before. The reader should not hesitate to consult some elementary text in the bibliography if the desired conclusion does not sound familiar.

If there is an even prime present call it $p_1(=2)$. *Then if* $a_1 = 1$ *all* y *are congruent to one another* (*mod* 2), *if* $a_1 = 2$,

(5b) $$y \equiv (-1)^{t_0} \pmod 4, \qquad 0 \le t_0 < \phi(4) = 2;$$

and if $a_1 \ge 3$

(5c) $$y \equiv (-1)^{t_0} 5^{t_1} \pmod{2^{a_1}}, \qquad 0 \le t_0 < 2; \qquad 0 \le t_1 < \phi(2^{a_1})/2.$$

The index of y in general is not an exponent but an ordered n-tuple[1] of exponents or a vector.[2] If we assume the primitive roots in (5*a*) are fixed for each odd p_i as the minimum positive value, we can write

(6) $$\mathbf{ind}\,(y) = [t_0, t_1, t_2, \cdots, t_s],$$

where *each* t_i *is taken modulo the value* $\phi(p_i^{a_i})$, *or* 2, *or* $\frac{1}{2}\phi(2^{a_1})$, as required by (5*a*), (5*b*), and (5*c*).

Thus, if $m = 17$, we represent $y \equiv 3^{t_1} \pmod{17}$, and

(7a) $$\mathbf{ind}\,(y) = [t_1], \qquad (t_1 \text{ determined modulo } 16).$$

Here the vector is merely the index. On the other hand, if $m = 24 = 2^3 \cdot 3$, we write

(7b) $$y \equiv 2^{t_2} \pmod 3, \qquad y \equiv (-1)^{t_0} 5^{t_1} \pmod 8,$$

(7c) $$\mathbf{ind}\,(y) = [t_0, t_1, t_2], \qquad (t_0, t_1, t_2 \text{ determined modulo } 2).$$

We can easily see the vectors corresponding to 5, 7, and 11 ($\equiv$ 35 modulo 24);

$$\mathbf{ind}\,(5) = [0, 1, 1], \qquad \mathbf{ind}\,(7) = [1, 0, 0], \qquad \mathbf{ind}\,(11) = [1, 1, 1].$$

In accordance with the usual vector laws, we define addition [with each t_i determined modulo $\phi(p_i^{a_i})$, 2, or $\phi(2^{a_1})/2$, according to (5*a*), (5*b*), or (5*c*)]. Let

(8) $$\mathbf{ind}\,(y) = [t_0, t_1, \cdots, t_2], \qquad \mathbf{ind}\,(y') = [t_0', t_1', \cdots t_s'].$$

Then

(9) $$\mathbf{ind}\,(y) + \mathbf{ind}\,(y') = [t_0 + t_0', t_1 + t_1', \cdots, t_s + t_s'].$$

We then have an obvious theorem

(10) $$\mathbf{ind}\,(yy') = \mathbf{ind}\,(y) + \mathbf{ind}\,(y').$$

[1] The statements such as (6) must be suitably modified in case an entry such as t_0 is absent (when m is odd) as well as when t_1 is absent (or $2^2 \parallel m$). (Effectively, $n = s$ or $s \pm 1$.)

[2] We use the term "vector" intuitively as an "ordered n-tuple of components with addition and subtraction defined for two vectors by a component-by-component operation."

EXERCISE 7. From representations (7*b*) and (7*c*) draw the conclusion that $\mathbf{ind}\,(y^2) = [0, 0, 0]$ for all y, for which $(y, 24) = 1$. (In other words all such y are solutions to $y^2 \equiv 1$, modulo 24).

EXERCISE 8. Find all m for which, whenever $(y, m) = 1$, then $y^4 \equiv 1 \pmod{m}$, using the index vector notation as in Exercise 7.

GROUP THEORETIC CONCEPTS

6. Abelian Groups and Subgroups

In the development of number theory, structurally similar proofs had been repeated for centuries before it was realized that a great convenience could be achieved by the use of groups.

We shall ultimately repeat the earlier results (§5) in group theoretic language. We need consider only finite commutative groups in this book.

A finite *commutative* (or *abelian*) *group* **G** is a set of objects:

$$\mathbf{G}: \{\mathbf{a}_1, \mathbf{a}_2, \cdots, \mathbf{a}_h\}, \tag{1a}$$

with a well-defined binary operation (symbolized by $\otimes$) and subject to the following rules:

$$\mathbf{a}_i \otimes \mathbf{a}_j = \mathbf{a}_j \otimes \mathbf{a}_i \qquad \text{(Commutative law)} \tag{1b}$$

$$\mathbf{a}_i \otimes (\mathbf{a}_j \otimes \mathbf{a}_k) = (\mathbf{a}_i \otimes \mathbf{a}_j) \otimes \mathbf{a}_k, \qquad \text{(Associative law)} \tag{1c}$$

for every $\mathbf{a}_i$ and $\mathbf{a}_j$ an $\mathbf{a}_k$ exists such that

$$\mathbf{a}_j \otimes \mathbf{a}_k = \mathbf{a}_i. \qquad \text{(Division law)} \tag{1d}$$

From these axioms it follows that a *unique* element, called the identity and written **e**, exists for which $\mathbf{a}_i \otimes \mathbf{e} = \mathbf{a}_i$. The number of elements h of the group is called the *order of the group*. The powers of a are written with exponents $\mathbf{a} \otimes \mathbf{a} = \mathbf{a}^2$, etc. The axiom ($d$) can be interpreted as meaning that the set

$$\mathbf{a}_j \otimes \mathbf{a}_1, \mathbf{a}_j \otimes \mathbf{a}_2, \cdots, \mathbf{a}_j \otimes \mathbf{a}_h$$

constitutes a rearrangement of the group elements (1) for any choice of $\mathbf{a}_j$.

A *subgroup* is a subset of elements of the group which under the operation $\otimes$, themselves form a group. It can be verified that the subgroup contains the same identity **e** as **G**. A well-known result, that of Lagrange, is that *the order of a subgroup divides the order of the group.*[1]

The groups that are involved modulo m are of two types, additive and multiplicative.

[1] Gauss in his *Disquisitiones* was particularly blind to groups and repeated the proof everytime he used this result (implicitly). Modern books on number theory, at long last, take greater cognizance of groups than did Gauss. Despite this fact, his results on quadratic forms were a stimulus to the group concept.

The *additive* group modulo m has as elements all m residue classes (both those relatively prime to m and those not relatively prime to m). In accordance with our earlier notation, we would write the residue class merely as $\mathbf{x}$. The group operation $\otimes$ is addition modulo m, and for convenience we represent it by $+$, or $\mathbf{x} + \mathbf{y} = \mathbf{x + y}$. This statement is exceedingly transparent and we see that (1d) calls for subtraction, i.e.,

$$(2) \qquad \mathbf{x}_j + \mathbf{x}_k = \mathbf{x}_i \quad \text{means} \quad \mathbf{x}_k = \mathbf{x}_i - \mathbf{x}_j$$

and $\mathbf{e} = \mathbf{0}$ in the usual way.

The *multiplicative* group modulo m, $\mathbf{M}(m)$ has as elements those $\phi(m)$ residue classes relatively prime to m. The operation $\otimes$ is multiplication modulo m and (1d) is less trivial; indeed, it is equivalent to the fact that a_i/a_j represents an integer (mod m) relatively prime to m if $(a_i, m) = (a_j, m) = 1$. We again represent residue classes by $\mathbf{x}$.

EXERCISE 9. With a convenient numbering of elements, let $\mathbf{a}_1 = \mathbf{e}$ and let $\mathbf{K} = \{\mathbf{a}_1, \mathbf{a}_2, \cdots, \mathbf{a}_t\}$ be a subgroup of order t in $\mathbf{G}$ [given by (1a)]. Let $\mathbf{K}_i$ denote the so-called *coset* $\{\mathbf{a}_i \otimes \mathbf{a}_1, \mathbf{a}_i \otimes \mathbf{a}_2, \cdots, \mathbf{a}_i \otimes \mathbf{a}_t\}$ for $i = 1, 2, \cdots, h$. Show that either $\mathbf{K}_i$ and $\mathbf{K}_j$ have no element in common or that they agree completely (permitting rearrangement of elements in each coset). From this result show $t \mid h$ (Lagrange's lemma) and that there are h/t different cosets.

EXERCISE 10. Show that the Fermat-Euler theorem [(4), §4] is a consequence of Lagrange's lemma by establishing the subgroup of $\mathbf{M}(m)$ generated by powers of b modulo m where $(b, m) = 1$.

7. Decomposition into Cyclic Groups

A *cyclic group* is one that consists of powers of a single element called the *generator*. Two simple examples immediately come to mind.

The additive group modulo m is generated by "powers" of $\mathbf{1}$. Here, of course, the operation $\otimes$ is addition, so the powers are $\mathbf{1}$, $\mathbf{1} + \mathbf{1} = \mathbf{2}$, $\mathbf{1} + \mathbf{1} + \mathbf{1} = \mathbf{3}$, etc., and, of course, $\mathbf{m}$ can be written as $\mathbf{0}$.

If m has a primitive root g, the multiplicative group modulo m has $\phi(m)$ elements and is generated by powers of g (under multiplication) namely $g^1, g^2, \cdots, g^{\phi(m)} (\equiv 1) \pmod{m}$.

The *order of a group element* is defined accordingly as the order of the cyclic group which it generates. By Lagrange's lemma, the order of a group element divides the order of the group.

We use the notation Z or $Z(m)$ to denote a cyclic group of order m, whether it is multiplicative or additive. Thus the multiplicative group modulo m is cyclic, or, symbolically,

$$(1) \qquad \mathbf{M}(m) = \mathbf{Z}(\phi(m))$$

if and only if a primitive root exists modulo m.

Not every abelian group is cyclic, as we shall see, but for every abelian group $\mathbf{G}$ we can find a set of *generators* $\mathbf{g}_0, \mathbf{g}_1, \cdots, \mathbf{g}_s$ such that $\mathbf{g}_i$ is of order h_i and an arbitrary group element of $\mathbf{G}$ is representable *uniquely* as

$$\mathbf{g} = \mathbf{g}_0^{t_0} \otimes \mathbf{g}_1^{t_1} \otimes \cdots \otimes \mathbf{g}_s^{t_s} \tag{2}$$

(meaning that the t_i are determined modulo h_i by the element $\mathbf{g}$). This result is called the Kronecker decomposition theorem (1877). We shall prove it under lattice point theory in Chapter V, but no harm can be done by using it in the meantime. We write this decomposition, purely symbolically, as

$$\mathbf{G} = \mathbf{Z}(h_0) \times \mathbf{Z}(h_1) \times \cdots \times \mathbf{Z}(h_s). \tag{3}$$

The order of $\mathbf{G}$ must be $h_0 h_1 \cdots h_s$ (by the uniqueness of the representation (2) of $\mathbf{g}$ through exponents modulo h_i).

For the time being we note that Kronecker's result holds easily for $\mathbf{M}(m)$, the multiplicative group modulo m for each m. This is a simple reinterpretation of the representation for the reduced residue class modulo m given in (6), §5. We represented the multiplicative $\mathbf{M}(m)$ by the additive group on

$$\mathbf{ind}\,(y) = [t_0, t_1, \cdots, t_s], \tag{4}$$

where t_i is represented modulo h_i, as in §5. Then, for instance, the generators are $\mathbf{g}_0 = [1, 0, \cdots, 0]$ $\mathbf{g}_1 = [0, 1, \cdots, 0], \cdots, \mathbf{g}_s = [0, 0, \cdots, 1]$ and

$$\mathbf{M}(m) = \mathbf{Z}(h_0) \times \mathbf{Z}(h_i) \times \cdots \times \mathbf{Z}(h_s). \tag{5}$$

Here $h_t = \phi(p_t^{a_t})$, with the usual provisions that when $8 \mid m$, $h_0 = 2$, $h_1 = \phi(2^{a_1})/2$; when $2^2 \parallel m$, $h_0 = 2$ and the h_1 term is missing, as provided in (5*a*), (5*b*), (5*c*) of §5.

We note, in conclusion, that *the group* $\mathbf{G}$ given in (3) *is cyclic if and only if* $(h_0, h_1) = (h_0, h_2) = (h_1, h_2) = \cdots = 1$. (We recall that in the group $\mathbf{M}(m)$, $2 \mid h_i$ so that $\mathbf{M}(m)$ is seen to be generally noncyclic and thus no primitive root exists modulo m generally). To review the method of proof, let us take $\mathbf{G} = \mathbf{Z}(h_0) \times \mathbf{Z}(h_1)$ of order $h_0 h_1$. First, we verify that $\mathbf{g}_0 \otimes \mathbf{g}_1$ is of order $h_0 h_1$ if $(h_0, h_1) = 1$; hence it generates $\mathbf{G}$. For if

$$(\mathbf{g}_0 \otimes \mathbf{g}_1)^x = \mathbf{e},$$

then

$$\mathbf{g}_0^x \otimes \mathbf{g}_1^x = \mathbf{e}.$$

By the uniqueness of representation of element $\mathbf{e}$, $x \equiv 0 \pmod{h_0}$ and $x \equiv 0 \pmod{h_1}$, whence $h_0 h_1 \mid x$. Second, we note that if $(h_0, h_1) = d > 1$

no element $\mathbf{g}$ of $\mathbf{G}$ can be of order h_0h_1. Indeed, the order of $\mathbf{g}_0^{t_0} \otimes \mathbf{g}_1^{t_1}$ cannot exceed h_0h_1/d. For

$$\begin{aligned}(\mathbf{g}_0^{t_0} \otimes \mathbf{g}_1^{t_1})^{h_0h_1/d} &= \mathbf{g}_0^{t_0h_0h_1/d} \otimes \mathbf{g}^{t_1h_0h_1/d} \\ &= (\mathbf{g}_0^{h_0})^{t_0h_1/d} \otimes (\mathbf{g}_1^{h_1})^{t_1h_0/d} \\ &= \mathbf{e} \otimes \mathbf{e} = \mathbf{e}.\end{aligned}$$

Q.E.D.

EXERCISE 11. If $m = p_1p_2$, where p_1 and p_2 are different odd primes, does the statement $\mathbf{M}(m) = \mathbf{Z}(p_1 - 1) \times \mathbf{Z}(p_2 - 1)$ mean that every reduced residue class $x \pmod m$ has a unique representation as $x \equiv g_1^{t_1}g_2^{t_2} \pmod m$ where $0 \le t_j < p_j - 1, (j = 1, 2)$? *Hint.* Take $m = 15$.

EXERCISE 12. Show that in a cyclic group of even order half the elements are perfect squares and in a cyclic group of odd order all the elements are perfect squares. Square all elements of $\mathbf{Z}(6)$ and $\mathbf{Z}(5)$ as illustrations.

EXERCISE 13. Do the statements of Exercise 12 apply to noncyclic groups?

QUADRATIC CONGRUENCES

8. Quadratic Residues

The values of a for which the congruence in x,

$$x^2 \equiv a \pmod p$$

is solvable are called *quadratic residues* of the odd prime p. The quadratic residue character is denoted[1] by the Legendre symbol $\left(\frac{a}{p}\right)$ [also written (a/p)], where

$$(1)\quad \begin{cases} \left(\dfrac{a}{p}\right) = 1 & \text{if } x^2 \equiv a \pmod p \text{ solvable and } (a, p) = 1, \\ \left(\dfrac{a}{p}\right) = 0 & \text{if } (a, p) = p, \\ \left(\dfrac{a}{p}\right) = -1 & \text{if } x^2 \equiv a \pmod p \text{ unsolvable.} \end{cases}$$

Thus $[1 + (a/p)]$ is the number of solutions modulo p to the equation $x^2 \equiv a$ modulo p for any a. Easily

$$\left(\frac{a_1}{p}\right) = \left(\frac{a_2}{p}\right) \quad \text{if } a_1 \equiv a_2 \pmod p,$$

and

$$\left(\frac{a_1}{p}\right) \cdot \left(\frac{a_2}{p}\right) = \left(\frac{a_1a_2}{p}\right).$$

[1] We refer to a and p as "numerator" and "denominator" (for want of more suitable universally established terms).

Thus the evaluation of the symbol (a/p) reduces to the evaluation of the symbols $(-1/p)$, $(2/p)$, and (q/p), where q is any odd prime.

The famous *quadratic reciprocity* relations are

$$\left(\frac{-1}{p}\right) = (-1)^{(p-1)/2}, \tag{2a}$$

$$\left(\frac{2}{p}\right) = (-1)^{(p^2-1)/8} \tag{2b}$$

$$\left(\frac{q}{p}\right) = \left(\frac{p}{q}\right)(-1)^{(p-1)/2\cdot(q-1)/2}, \tag{2c}$$

where p and q are odd positive primes. These relations enable us to evaluate (q/p) by continued inversion and division in a manner described in elementary texts. To avoid the factor $(-1)^{(p-1)/2\cdot(q-1)/2}$, we could write $(q/p) = (p^*/q)$ where $p^* = p(-1/p)$. For example, $3^* = -3$, $5^* = 5$; thus $(q/3) = (-3/q)$, whereas $(q/5) = (5/q)$.

A very useful relation due to Euler is

$$\left(\frac{a}{p}\right) \equiv a^{(p-1)/2} \pmod p \tag{3}$$

for p an odd prime and $(a, p) = 1$.

The equation

$$x^2 \equiv a \pmod{p^s} \tag{4}$$

can also be shown to present no greater difficulty for $s > 1$ than for $s = 1$. The fundamental case is where $(a, p) = 1$. There we can show, if p is odd, that the solvability of

$$x^2 \equiv a \pmod p, \qquad (a, p) = 1, \tag{5a}$$

leads to the solvability of

$$x^2 \equiv a \pmod{p^s}, \qquad s \geq 1. \tag{5b}$$

Correspondingly, if

$$a \equiv 1 \pmod 8, \tag{6a}$$

then we can solve

$$x^2 \equiv a \pmod{2^s}, \qquad s \geq 3. \tag{6b}$$

The details are illustrated in Exercises 14 and 15.

EXERCISE 14. Show that if

$$x_s^2 \equiv a \pmod{p^s}, \qquad (p \text{ odd}), (a, p) = 1,$$

we can find a value $k \pmod{p}$ for which

$$x_{s+1} = x_s + kp^s \equiv x_s \pmod{p^s}$$

and

$$x_{s+1}^2 \equiv a \pmod{p^{s+1}}.$$

Construct the sequence x_1, x_2, x_3, x_4, starting with $x_1 = 2$, $a = -1$, $p = 5$, $x_1^2 \equiv -1 \pmod{5}$.

EXERCISE 15. Show that if

$$x_s^2 \equiv a \pmod{2^s}, \qquad s \geqq 3, \qquad a \equiv 1 \pmod{8},$$

we can find a value k such that

$$x_{s+1} = x_s + k2^{s-1} \equiv x_s \pmod{2^{s-1}}, \qquad (k = 0 \text{ or } 1),$$

and

$$x_{s+1}^2 \equiv a \pmod{2^{s+1}}.$$

Construct the sequence $(1 =)x_3$, x_4, x_5, x_6 for

$$x_s^2 \equiv 17 \pmod{2^s}.$$

9. Jacobi Symbol

As an aid in evaluating the symbol (a/p) numerically, we introduce a generalized symbol for greater flexibility, namely (a/b). For $b = \pm \prod^{(i)} p_i^{a_i}$ we define

$$(1) \qquad \left(\frac{a}{b}\right) = \prod^{(i)} \left(\frac{a}{p_i}\right)^{a_i} \qquad \begin{cases} a \text{ positive, negative, or zero,} \\ b \text{ odd, nonzero.} \end{cases}$$

For $b = \pm 1$ we define the symbol as 1.

Then it can be shown that for a, b, positive and odd;

$$\begin{cases} (2a) & \left(\frac{-1}{b}\right) = (-1)^{(b-1)/2}, \\ (2b) & \left(\frac{2}{b}\right) = (-1)^{(b^2-1)/8}, \\ (2c) & \left(\frac{a}{b}\right) = \left(\frac{b}{a}\right)(-1)^{(b-1)/2 \cdot (a-1)/2}. \end{cases}$$

A necessary and sufficient condition that

$$(3) \qquad x^2 \equiv a \pmod{pq}$$

be solvable for p, q distinct primes not dividing a is that the individual Legendre symbols (a/p), (a/q) all be $+1$. If the Jacobi symbol (a/pq) is -1, (3) is unsolvable.

There are many cases in which the evaluation of (a/p) (Legendre symbol) can be facilitated by treating it as a Jacobi symbol in order to invert. The answer is the same, as both symbols must agree for (a/p). We shall ultimately see that the introduction of the Jacobi symbol is more than a convenience; it is a critical step in the theory of quadratic forms.

Thus we conclude the review of elementary number theory. The deepest result is, of course, quadratic reciprocity, which we shall prove anew in Chapter XI from an advanced standpoint.

EXERCISE 16 (Dirichlet). Evaluate $(365/1847)$ as a strict Legendre symbol and (inverting) as a Jacobi symbol. (1847 is a prime.)

EXERCISE 17. Show that even when a is negative, if $|a| > 1$, $b > 1$ and a and b are odd, then

$$\left(\frac{a}{b}\right) = \left(\frac{b}{a}\right)(-1)^{(b-1)/2 \cdot (a-1)/2}. \tag{4}$$

EXERCISE 18. If $|a| > 1$, $|b| > 1$, with a and b both negative and odd, show that

$$\left(\frac{a}{b}\right) = -\left(\frac{b}{a}\right)(-1)^{(b-1)/2 \cdot (a-1)/2}. \tag{5}$$

EXERCISE 19. Find an expression for $(-1/b)$ for b odd and negative and show Exercises 17 and 18 to be valid when $|a|$ or $|b| = 1$.

*chapter II

Characters

1. Definitions

An important question which we develop here is the manner of distinguishing by analytic means a residue class modulo m, which is really an abstract concept. In our case this means we are trying to represent a whole residue class,

$$y \equiv a \pmod{m}, \tag{1}$$

by a set of ordinary (real or complex) numbers $\chi(a)$, called *characters.*

We start more generally by defining characters for a finite abelian group[1] $\mathbf{G}$ of order h with elements $\mathbf{a}_1, \mathbf{a}_2, \cdots, \mathbf{a}_h$ (where $\mathbf{a}_1 = \mathbf{e}$, the unit element). We call the character χ a function over all group elements, or

$$\chi(\mathbf{a}_1), \chi(\mathbf{a}_2), \cdots, \chi(\mathbf{a}_h)_2 \tag{2}$$

with the properties

$$\chi(\mathbf{a}_i) \neq 0, \tag{3}$$

$$\chi(\mathbf{a}_i)\chi(\mathbf{a}_j) = \chi(\mathbf{a}_i\mathbf{a}_j). \tag{4}$$

It is easily seen that $\chi(\mathbf{e}) = 1$, using $\mathbf{a}_1\mathbf{e} = \mathbf{a}_1$ in (4). Furthermore, if h is the order of the group, then $\mathbf{a}^h = \mathbf{e}$, for any element $\mathbf{a}$ of the group. Thus $[\chi(\mathbf{a})]^h = \chi(\mathbf{a}^h) = 1$ and $\chi(\mathbf{a})$ is an h-root of unity, i.e.,

$$\chi(\mathbf{a}) = \exp(2\pi it/h) = \cos 2\pi t/h + i \sin 2\pi t/h \tag{5}$$

[1] Henceforth the group operation will be written $\mathbf{ab}$ instead of $\mathbf{a} \otimes \mathbf{b}$.

for an appropriate t. There are h such roots of unity for the h values $t = 0, 1, 2, \cdots, h-1$. Thus the number of possible characters χ determined by the values (2) is at most h^h. Actually, there are precisely h characters, as we shall soon see. (Naturally, two characters are different if and only if they differ for one or more group elements.)

The *product* of two different characters, denoted by $\{\chi_1\chi_2\}$, is a character if we define for an arbitrary group element $\mathbf{a}$,

$$\{\chi_1\chi_2\}(\mathbf{a}) = \chi_1(\mathbf{a})\chi_2(\mathbf{a}) \tag{6}$$

using ordinary multiplication. Then easily $\{\chi_1\chi_2\}(\mathbf{a})$ is never zero. Furthermore, for group elements $\mathbf{a}_i$ and $\mathbf{a}_j$

$$\begin{aligned}\{\chi_1\chi_2\}(\mathbf{a}_i\mathbf{a}_j) &= \chi_1(\mathbf{a}_i\mathbf{a}_j)\chi_2(\mathbf{a}_i\mathbf{a}_j) = \chi_1(\mathbf{a}_i)\chi_1(\mathbf{a}_j)\chi_2(\mathbf{a}_i)\chi_2(\mathbf{a}_j)\\ &= \{\chi_1\chi_2\}(\mathbf{a}_i)\{\chi_1\chi_2\}(\mathbf{a}_j).\end{aligned}$$

Hence $\chi_1\chi_2$ has the properties of a character in (3) and (4). We define χ_i^n in like fashion. We can now have a *group of characters* $\mathbf{X}$. We call $\{\chi_1/\chi_2\}$ the (obvious) *quotient character* using ordinary division:

$$\{\chi_1/\chi_2\}(\mathbf{a}_i) = \chi_1(\mathbf{a}_i)/\chi_2(\mathbf{a}_i). \tag{7}$$

In the same spirit we define the unit character by

$$\chi_0(\mathbf{a}_i) = 1 \text{ for all } \mathbf{a}_i. \tag{8}$$

In the case of residue classes under multiplication, we can identify $\mathbf{a}$, the group element, with $y \equiv a$, the residue class, and use $\chi(\mathbf{a})$ and $\chi(a)$ interchangeably, with "modulo m" and $(a, m) = 1$ understood. We can also write $\chi(a) = 0$ if $(a, m) \neq 1$ without contradicting the multiplication law (5).

As an example, consider first the reduced residues modulo 5. Clearly $y^4 \equiv 1 \pmod 5$ so that from (5) $\chi(\mathbf{a})$ have only one of the four values $i, -1, -i, +1$. The reader can verify the following characters (four in number):

TABLE 1

Reduced Residues Modulo 5

$\mathbf{a} = \mathbf{e}$ $y \equiv 1$	$\mathbf{a}_2$ 2	$\mathbf{a}_3$ 3	$\mathbf{a}_4$ 4
$\chi_0(\mathbf{a}) = 1$	1	1	1
$\chi_1(\mathbf{a}) = 1$	i	$-i$	-1
$\chi_3(\mathbf{a}) = 1$	$-i$	i	-1
$\chi_2(\mathbf{a}) = 1$	-1	-1	1

To illustrate the group property for residue classes, we note the relations

(9a) $$2 \cdot 3 \equiv 1 \pmod 5,$$

(10a) $$\chi_j(2)\chi_j(3) = \chi_j(1),$$

(11a) $$(-i)(i) = 1, \qquad \text{(e.g., taking } j = 3).$$

To illustrate the group property for characters, we note the relations

(9b) $$\chi_2(\mathbf{a}) = \chi_1(\mathbf{a})^2, \quad \chi_3(\mathbf{a}) = \chi_1(\mathbf{a})^3, \quad \chi_0(\mathbf{a}) = \chi_1(\mathbf{a})^4,$$

(10b) $$\chi_3(\mathbf{a}_j) = \chi_1(\mathbf{a}_j)^3,$$

(11b) $$i = (-i)^3, \qquad \text{(e.g., taking } j = 3).$$

We observe that the cyclic structure of **M**(5) somehow carries over to the characters.

The reader can verify the following scheme modulo 8:

TABLE 2
Reduced Residues Modulo 8

$\mathbf{a} = \mathbf{e}$	$\mathbf{a}_2$	$\mathbf{a}_3$	$\mathbf{a}_4$
$y \equiv 1$	3	5	7
$\chi_0(\mathbf{a}) = 1$	1	1	1
$\chi_1(\mathbf{a}) = 1$	−1	−1	1
$\chi_2(\mathbf{a}) = 1$	1	−1	−1
$\chi_3(\mathbf{a}) = 1$	−1	1	−1

and the properties

(9c) $$\begin{cases} \chi_1^{\,2} = \chi_2^{\,2} = \chi_3^{\,2} = \chi_0, \\ \chi_1\chi_2 = \chi_3, \quad \chi_2\chi_3 = \chi_1, \quad \chi_3\chi_1 = \chi_2. \end{cases}$$

These properties are capable of generalization, as we shall now see.

2. Total Number of Characters

The main result[1] is that an abelian group of order h has precisely h different characters.

To see this result, first consider a single cycle of order h.

$$\mathbf{G} = \mathbf{Z}(h).$$

[1] For this chapter we assume Kronecker's theorem on cyclic structure, proved in Chapter V.

The elements of **G** are expressed in terms of a generator **a** as

$$\mathbf{a}, \mathbf{a}^2, \cdots, \mathbf{a}^h(= \mathbf{e}).$$

Then, clearly, the values of $\chi(\mathbf{a})$ are of the form $\exp 2\pi iu/h$, and

$$\chi(\mathbf{a}^t) = \exp 2\pi itu/h, \qquad \text{for } 0 \leq t < h. \tag{1}$$

Thus we obtain h different characters as u varies:

$$\chi_u(\mathbf{a}^t) = \exp 2\pi itu/h, \qquad 0 \leq u < h. \tag{2}$$

For each character (or for each fixed u) the h group elements are generated at t varies to 0 to $h - 1$. The properties (3), (4) of §1 are easily verified for the characters in (1) and (2).

The comparison with Table 1 in §1 (above) for reduced residues modulo 5 should be altogether clear if we note that $\mathbf{a} \equiv 2$ and that $\mathbf{a}^t$ has the successive values $2^1 \equiv 2$, $2^2 \equiv 4$, $2^3 \equiv 3$, $2^0 \equiv 1$ (all statements holding modulo 5, of course).

Next, let $\mathbf{G} = \mathbf{Z}(h_1) \times \mathbf{Z}(h_2)$, an abelian group of order h_1h_2 which is decomposed into two cyclic groups of order h_1 and h_2:

$$\mathbf{a}_1, \mathbf{a}_1^2, \cdots, \mathbf{a}_1^{h_1} (= \mathbf{e}),$$
$$\mathbf{a}_2, \mathbf{a}_2^2, \cdots, \mathbf{a}_2^{h_2} (= \mathbf{e}).$$

Thus the general element of **G** is

$$\mathbf{g} = \mathbf{a}_1^{t_1}\mathbf{a}_2^{t_2}, \qquad 0 \leq t_1 < h_1, \qquad 0 \leq t_2 < h_2. \tag{3}$$

Then we write

$$\chi_{u_1u_2}(\mathbf{g}) = \exp 2\pi i(t_1u_1/h_1 + t_2u_2/h_2). \tag{4}$$

As u_1 and u_2 vary, they are reduced modulo h_1 and h_2, respectively, so that

$$0 \leq u_1 < h_1, \qquad 0 \leq u_2 < h_2. \tag{5}$$

We generate all h_1h_2 (different) characters, as we verify below. The reader can refer to residue classes modulo 8 in Table 2, §1 (above), Here $h_1 = h_2 = 2$, χ_0, χ_1, χ_2, χ_3 can easily be identified with the 4 characters $\chi_{u_1u_2}$ in (4), where $u_1 = 0, 1$ and $u_2 = 0, 1$.

Thus, when we have an abelian group of order h (using cyclic structure), *we can show that there are h characters*. We can even see that *the character group has the same cyclic structure*. (These proofs occupy the rest of the section.) Specifically, let

$$\mathbf{G} = \mathbf{Z}(h_0) \times \mathbf{Z}(h_1) \times \cdots \times \mathbf{Z}(h_s) \tag{6}$$

so that an arbitrary element of **G** is

$$\mathbf{a} = \mathbf{a}_0^{t_0}\mathbf{a}_1^{t_1} \cdots \mathbf{a}_s^{t_s}, \qquad (t_i \bmod h_i), \tag{7}$$

Then, as we have shown for $s = 1$, here are $h = h_0 \cdots h_s$ possible characters:

(8a) $$\chi_{u_0 u_1 \cdots u_s}(\mathbf{a}) = \exp 2\pi i\left(\frac{t_0 u_0}{h_0} + \cdots + \frac{t_s u_s}{h_s}\right), \qquad 0 \le u_i < h_i.$$

A convenient notation is

(8b) $$\chi_{u_0 u_1 \cdots u_s}(\mathbf{a}) = e(t_0/h_0)^{u_0} \cdots e(t_s/h_s)^{u_s},$$
$$0 \le u_i < h_i, \qquad 0 \le t_i < h_i,$$

using the function

(9) $$e(\xi) = \exp 2\pi i \xi.$$

This function has the obvious period 1, e.g., $e(\xi + 1) = e(\xi)$, as well as the exponential property $e(\xi + \eta) = e(\xi)e(\eta)$. Also, $e(\xi) = 1$ if and only if ξ is an integer.

It is not obvious that the h characters listed in (8a) are different. For instance, if $u_0 \not\equiv v_0 (\mathrm{mod}\ h_0)$, we must verify that for some a

(10a) $$\chi_{u_0 \cdots u_s}(\mathbf{a}) \ne \chi_{v_0 \cdots v_s}(\mathbf{a})$$

But we need only take $\mathbf{a} = \mathbf{a}_0$ in (7), then the relation (10a) follows from the obvious result that (with $t_0 = 1,\ t_1 = t_2 = \cdots t_s = 0$),

(10b) $$\exp 2\pi i u_0/h_0 \ne \exp 2\pi i v_0/h_0.$$

In similar fashion if $\mathbf{a} \ne \mathbf{e}$ we can find some χ in the set (8a) for which $\chi(\mathbf{a}) \ne 1$. For example, if $\mathbf{a} \ne \mathbf{e}$, then in the decomposition,

(11) $$\mathbf{a} = \mathbf{a}_0^{t_0} \cdots \mathbf{a}_s^{t_s}, \qquad 0 \le t_i < b_i,$$

we note one exponent (say) $t_0 \not\equiv 0\ (\mathrm{mod}\ h_0)$. Thus,

(12) $$\chi_{100\cdots 0}(\mathbf{a}) = \exp 2\pi i t_0/h_0 \ne 1.$$

Now let us assume there are c characters in the character group $\mathbf{X}$; we know $c \ge h$. We shall show $c = h$. First note the results:

(13) $$\sum_{\mathbf{a} \text{ in } \mathbf{G}} \chi(\mathbf{a}) = \begin{cases} h \text{ if } \chi = \chi_0, \\ 0 \text{ if } \chi = \text{any other (fixed) character;} \end{cases}$$

(14) $$\sum_{\chi \text{ in } \mathbf{X}} \chi(\mathbf{a}) = \begin{cases} c \text{ if } \mathbf{a} = \mathbf{e}, \\ 0 \text{ if } \mathbf{a} = \text{any other (fixed) group element.} \end{cases}$$

First take (13). For $\chi = \chi_0$ the conclusion is obvious. For $\chi \ne \chi_0$ write $\sum_{\mathbf{a} \text{ in } \mathbf{G}} \chi(\mathbf{a}) = S$. We have for each fixed χ an element $\mathbf{a}^*$ such that $\chi(\mathbf{a}^*) \ne 1$ (by definition, since $\chi \ne \chi_0$); whence

$$S = \sum_{\mathbf{a} \text{ in } \mathbf{G}} \chi(\mathbf{a}\,\mathbf{a}^*) = \sum_{\mathbf{a} \text{ in } \mathbf{G}} \chi(\mathbf{a})\chi(\mathbf{a}^*) = S\chi(\mathbf{a}^*),$$

since as $\mathbf{a}$ varies $\mathbf{a}^*\mathbf{a}$ is a rearrangement of the group elements denoted by $\mathbf{a}$. The conclusion follows from the fact that $\chi(\mathbf{a}^*) \neq 1$; hence $S = 0$.

Likewise in (14), for $\mathbf{a} = \mathbf{e}$ the conclusion is obvious. For $\mathbf{a} \neq \mathbf{e}$ write $\sum_{\chi \text{ in } \mathbf{X}} \chi(\mathbf{a}) = S$. But we have a special character χ^* for which $\chi^*(\mathbf{a}) \neq 1$, by (12). Therefore,

$$S = \sum_{\chi \text{ in } \mathbf{X}} \{\chi^*\chi\}(a) = \sum_{\chi \text{ in } \mathbf{X}} \chi^*(\mathbf{a}) \cdot \chi(\mathbf{a}) = S\chi^*(\mathbf{a}),$$

since, once more, with χ^* fixed, $\{\chi^*\chi\}$ is a rearrangement of the characters χ of $\mathbf{X}$. Thus, since $\chi^*(\mathbf{a}) \neq 1$, $S = 0$.

The final result now follows:

$$h = c. \tag{15}$$

For proof, set

$$\sum_{\chi}\sum_{\mathbf{a}} \chi(\mathbf{a}) = \sum_{\mathbf{a}}\sum_{\chi} \chi(\mathbf{a}),$$

using relation (13) on the left and relation (14) on the right!

From (13) we find that, since $\chi_1/\chi_2 = \chi_0$ exactly when $\chi_1 = \chi_2$,

$$\sum_{\mathbf{a} \text{ in } \mathbf{G}} \chi_1(\mathbf{a})\chi_2^{-1}(\mathbf{a}) = \begin{cases} h \text{ when } \chi_1 = \chi_2, \\ 0 \text{ when } \chi_1 \neq \chi_2. \end{cases} \tag{16}$$

Such results would be more laborious to prove by using the cyclic structure. The "dual" result is

$$\sum_{\chi \text{ in } \mathbf{X}} \chi(\mathbf{a}_1)\chi(\mathbf{a}_2^{-1}) = \begin{cases} h \text{ when } \mathbf{a}_1 = \mathbf{a}_2, \\ 0 \text{ when } \mathbf{a}_1 \neq \mathbf{a}_2. \end{cases} \tag{17}$$

Again we note $\chi(\mathbf{a}_1/\mathbf{a}_2) = \chi(\mathbf{a}_1)/\chi(\mathbf{a}_2)$, and $\mathbf{a}_1/\mathbf{a}_2 = e$ exactly when $\mathbf{a}_1 = \mathbf{a}_2$.

The relations (16) and (17) are called "orthogonality" conditions, by analogy with the perpendicularity of two (ordinary) geometric vectors

$$[A_1, \cdots, A_m], [B_1, \cdots, B_m], \text{ namely } \sum_{i=1}^{m} A_iB_i = 0.$$

EXERCISE 1. Show that there are no other characters than those listed in (8*a*) *directly* by considering value of $\chi(\mathbf{a}_0), \chi(\mathbf{a}_1), \cdots, \chi(\mathbf{a}_s)$ for the generating group elements in (7).

In this exercise and the next take $s = 2$ for convenience.

EXERCISE 2. Show (16) and (17) directly from the explicit form of the characters in (8*a*).

3. Residue Classes

We noticed in Chapter I, §5, that the residue classes $y \equiv a$ modulo m, relatively prime to m, can be represented additively by the vector index.

$$\textbf{ind } y = [t_0, t_1, \cdots, t_s], \tag{1}$$

where t_i are the exponents used in (5*a*, *b*, *c*) of Chapter I, §5, with

$$m = p_1^{a_1} p_2^{a_2} \cdots p_s^{a_s}.$$

The congruence classes of t_i are somewhat varied but can be represented symbolically by

$$(2) \qquad 0 \leq t_i < h_i,$$

where $h_i = \phi(p_i^{a_i})$, 2, or $\frac{1}{2}\phi(2^{a_1})$, as the case may require (for $i > 1$, $i = 0$, $i = 1$). Of course, $\phi(m) = h_0 h_1 \cdots h_s$, the order of the group of reduced residue classes $\mathbf{M}(m)$, making the usual allowances for missing components if $8 \nmid m$.

The additive group of indices may be represented symbolically in the usual form of (6) in §2 (above) or of (5) in §7, Chapter I;

$$(3) \qquad \mathbf{M}(m) = \mathbf{Z}(h_0) \times \mathbf{Z}(h_1) \times \cdots \times \mathbf{Z}(h_s),$$

with h_i defined as the moduli of (5*a*, *b*, *c*) in Chapter I, §5.

Its characters are seen to be in the form of (8*b*) and (9) in §2 (above). We think of the arguments as integers y rather than group elements, written as

$$(4) \qquad \chi_{u_0 u_1 \cdots u_s}(y) = e(t_0/h_0)^{u_0}\, e(t_1/h_1)^{u_1} \cdots e(t_s/h_s)^{u_s}.$$

For more convenient symbolism, we use the new symbol with a single subscript:

$$(5) \qquad \chi_{p_i^{a_i}}(y) = e(t_i/h_i) = \chi_{0\cdots 010 \cdots 0}(y)$$

where the subscript $0 \cdots 010 \cdots 0$ symbolically denotes $u_i = 1$ and $u_j = 0$ for all other u_j, $(j \neq i)$.

When $p_1^{a_1} = 2^{a_1}$, $a_1 \geq 3$), there are two symbols corresponding to $e(t_0/h_0)$ and $e(t_1/h_1)$, which we denote by $\chi_4(y)$ and $\chi_{2^{a_1}}(y)$. We recognize, of course, $\chi_4(y) = (-1/y) = (-1)^{(y-1)/2}$ if $y > 0$ and odd and $\chi_8(y) = (2/y) = (-1)^{(y^2-1)/8}$, for example, when $a_1 = 3$.

From now on, $\chi_1(y)$ (not $\chi_0(y)$) will denote the unit character.

The illustrations we now give are self-explanatory, except for the right-hand marginal notations of type "(a/b)" and "$M = \cdots$" which are explained in §4 and §7 (below). The reader should identify the characters of (4) with those on pages 29 and 30.

EXERCISE 3. Construct a similar table modulo 9, 15, 16, and 24. (See $m = 12$ in §6, below.)

4. Resolution Modulus

We next consider extending the domain of values over which χ is defined. To begin with, we might set $\chi = 0$ where it has been previously undefined,

TABLE 3

$m = 3, \quad \phi(3) = 2.$

$y \equiv$	1 2^0	2 2^1	$\equiv 2^{t_1}$	
$t_1 =$	0	1		
$\chi_3(y) = \chi_3(2^{t_1}) = \exp 2\pi i t_1/2 =$	1	-1	$(M = 3)$,	$(-3/y)$
$\chi_1 = \chi_3^2 =$	1	1	$(M = 1)$,	$(9/y)$

TABLE 4

$m = 4 = 2^2, \quad \phi(4) = 2$

$y \equiv$	1 $(-1)^0$	3 $(-1)^1$	$\equiv (-1)^{t_0}$	
$t_0 =$	0	1		
$\chi_4(y) = (-1)^{t_0} =$	1	-1	$(M = 4)$,	$(-4/y)$
$\chi_1 = \chi_4^2 =$	1	1	$(M = 1)$,	$(4/y)$

TABLE 5

$m = 5, \quad \phi(5) = 4$

$y \equiv$	1 2^0	2 2^1	3 2^3	4 2^2	$= 2^{t_1}$	
$t_1 =$	0	1	3	2		
$\chi_5(y) = \chi_5(2^{t_1}) =$ $\exp 2\pi i t_1/4 = \chi_5 =$	1	i	$-i$	-1	$(M = 5)$	
$\chi_5^2 =$	1	-1	-1	1	$(M = 5)$,	$(5/y)$
$\chi_5^3 =$	1	$-i$	i	-1	$(M = 5)$	
$\chi_5^4 = \chi_1 =$	1	1	1	1	$(M = 1)$,	$(25/y)$

TABLE 6

$m = 7, \quad \phi(7) = 6$

$y \equiv$	1 3^0	2 3^2	3 3^1	4 3^4	5 3^5	6 3^3	3^{t_1}
$t_1 =$	0	2	1	4	5	3	
$\chi_7(y) = \exp 2\pi i t_1/6 = \chi_7 =$	1	$\frac{-1+i\sqrt{3}}{2}$	$\frac{1+i\sqrt{3}}{2}$	$\frac{-1-i\sqrt{3}}{2}$	$\frac{1-i\sqrt{3}}{2}$	-1	
$\chi_7^2 =$	1	$\frac{-1-i\sqrt{3}}{2}$	$\frac{-1+i\sqrt{3}}{2}$	$\frac{-1+i\sqrt{3}}{2}$	$\frac{-1-i\sqrt{3}}{2}$	1	$(M = 7)$
$\chi_7^3 =$	1	1	-1	1	-1	-1	$(M = 7)$, $(-7/y)$
$\chi_7^4 =$	1	$\frac{-1+i\sqrt{3}}{2}$	$\frac{-1-i\sqrt{3}}{2}$	$\frac{-1-i\sqrt{3}}{2}$	$\frac{-1+i\sqrt{3}}{2}$	1	$(M = 7)$
$\chi_7^5 =$	1	$\frac{-1-i\sqrt{3}}{2}$	$\frac{1-i\sqrt{3}}{2}$	$\frac{-1+i\sqrt{3}}{2}$	$\frac{1+i\sqrt{3}}{2}$	-1	$(M = 7)$
$\chi_1 = \chi_7^6 =$	1	1	1	1	1	1	$(M = 1)$, $(49/y)$

TABLE 7

$m = 8 = 2^3, \quad \phi(8) = 4$

$y \equiv$	1, 1,	3, -5,	5, 5,	7 -1,	$\equiv (-1)^{t_0}5^{t_1}$	
$t_0 =$	0,	1,	0,	1		
$t_1 =$	0,	1,	1,	0		
$\chi_4(y) = (-1)^{t_0} =$	1,	-1,	1,	-1	$(M = 4)$,	$(-4/y)$
$\chi_8(y) = (-1)^{t_1} =$	1,	-1,	-1,	1	$(M = 8)$,	$(8/y)$
$\chi_4\chi_8 =$	1,	1,	-1,	-1	$(M = 8)$,	$(-8/y)$
$\chi_4^2 = \chi_8^2 = \chi_1 =$	1,	1,	1,	1	$(M = 1)$,	$(4/y)$

knowing that the multiplication rules of §1 (above) still apply, although division is restricted to the original range. We, of course, want less trivial extensions.

When a character $\chi(y)$ is defined for y modulo m, with $(y, m) = 1$, it might still happen that

$$\chi(y_1) = \chi(y_2) \tag{1}$$

whenever

$$(y_1, m) = (y_2, m) = 1$$

and

$$y_1 \equiv y_2 \pmod{M'}, \tag{2}$$

for some other M' in addition to m. The smallest M' (>0) for which the property holds is called the *resolution modulus M*.

For example, the unit character modulo m is defined by

$$\begin{cases} \chi_1(y) = 1. & (y, m) = 1, \\ \chi_1(y) = 0, & (y, m) > 1, \end{cases} \tag{3}$$

but it might just as well have been defined modulo 1 by $\chi^*(y) = 1$ and then $\chi_1 = \chi^*(y)$ specialized to only those y where $(y, m) = 1$.

A less trivial example is $\chi_4(y)$ modulo 8 (see Table 7 above), for which $M = 4$, e.g.,

$$\begin{cases} \chi_4(y) = 1, & \text{if } y \equiv 1 \pmod 4, \\ \chi_4(y) = -1, & \text{if } y \equiv -1 \pmod 4). \end{cases} \tag{4}$$

The resolution modulus is indicated in the margin of the table by $(M = \cdots)$.

LEMMA 1. An equivalent definition of the resolution modulus M is the least value of the positive integer M^* with the property that

$$\chi(y) = 1 \tag{5}$$

whenever $y \equiv 1 \pmod{M^*}$ and $(y, m) = 1$. For proof see Exercise 4 (below).

We next define the natural extension of a character $\chi(y)$ modulo m to a character modulo M where M is the resolution modulus of $\chi(y)$. The process is trivial, of course, unless the values of y for which $(y, M) = 1$ include more values than those for which $(y, m) = 1$. As a nontrivial case, for example, if $m = 15$ and $M = 3$, we might have a character $\chi^\dagger(y)$ modulo 15 which is none other than $\chi_3(y)$ but limited in domain of definition to $(y, 15) = 1$ and trivially extended to 0 when $(y, 15) > 1$. (Compare the following table with Table 3.)

TABLE 8

$y \pmod{15}$	0	1	2	3	4	5	6	7	8	9	10	11	12	13	14
$\chi^\dagger(y)$	0	1	−1	0	1	0	0	1	−1	0	0	−1	0	1	−1
$\chi_3(y)$	0	1	−1	0	1	−1	0	1	−1	0	1	−1	0	1	−1

We would like to know how to retrieve $\chi_3(y)$ from $\chi^\dagger(y)$ by the "natural" process of noticing that $\chi^\dagger(y)$ is determined modulo 3 as long as $(y, 5) = 1$.

The basic method is contained in the following lemma:

LEMMA 2. If $(y, M) = 1$, then for given y, M, and m we can find an x such that

(6) $$(y + Mx, m) = 1.$$

Proof. Let $p_i^{a_i}$ be a general primary divisor of m. Then an x_i exists such that $(y + Mx_i, p_i^{a_i}) = 1$. For, if $p_i \mid M$, then $p_i \nmid y$ and $x_i = 0$ suffices. If $p_i \nmid M$, any choice of x_i with $x_i \not\equiv -y/M \pmod{p_i}$ will suffice. By the Chinese remainder theorem, an x exists which satisfies $x \equiv x_i \pmod{p_i}$ and $(y + Mx, m) = 1$. Q.E.D.

Thus, if $\chi(y)$ modulo m has resolution modulus M, we define $\chi^*(y)$, the *natural extension of* χ modulo M, by using $Y = y + Mx$ of the lemma.

(7) $$\begin{cases} \chi^*(y) = \chi(Y) \text{ for } (y, M) = 1 \text{ (even if } (y, m) > 1) \text{ and} \\ \chi^*(y) = 0 \text{ for } (y, M) \neq 1 \end{cases}$$

We can see that the value $\chi(y)$ is unique, despite the latitude in the choice of x by definition of the resolution modulus (since all Y are congruent to one another modulo M). Furthermore, we can see that $\chi^*(y)$ has resolution modulus M (by showing that if $\chi^*(y) = 1$ whenever $y \equiv 1 \pmod{M'}$, and $(y, m) = 1$, then $\chi(y) = 1$).

EXERCISE 4. Prove Lemma 1. (*Hint.* Call M'' the resolution modulus as defined in the lemma. Show trivially that $M \geq M''$, and, using the residue class y_1/y_2, show $M'' \geq M$.)

EXERCISE 5. Show that the resolution modulus is given by the subscript in $\chi_4(y)$ and $\chi_{2^{a_1}}(y)$.

EXERCISE 6. Show that the resolution modulus of a character modulo m is a divisor of every M' for which (2) leads to (1).

EXERCISE 7. The reduced residue class group modulo m has a character whose resolution modulus is m unless $2 \parallel m$. (Show that one such character is $\chi_{11\cdots 1}(y)$ by using each primary factor of m, giving particular care to 2^{a_1}.)

5. Quadratic Residue Characters

We recall that Jacobi's symbol (a/b) had the property that the denominator could vary over *odd* positive or negative integers. The reciprocity law is slightly encumbered if we permit negative signs (see Exercises 17 and 18 of Chapter I, §9), yet even restricting ourselves to odd b we can consider the denominator "more arbitrary" than when b is prime.

We pursue the opposite of the original viewpoint of Legendre's symbol by writing

(1) $$\chi(y) = (a/y),$$

as a function of the *denominator* for odd y. Clearly,

(2) $$\chi(y_1y_2) = \chi(y_1)\chi(y_2)$$

and

(3) $$\chi(y) \neq 0$$

for y odd and relatively prime to a. To see that $\chi(y)$ is a character in the sense of this chapter, we must find a modulus m to which it belongs in the sense that

(4) $$\chi(y) = 1, \quad \text{if } y \equiv 1 \pmod{m}, \qquad (y, m) = 1.$$

It is easily seen that we can take $m = 4a$ if we *take* $y > 0$ *only*, for then the reciprocity law yields for a odd (by Exercise 17, Chapter I)

(5) $$\chi(y) = (y/a)(-1)^{(a-1)/2\cdot(y-1)/2}$$

and easily

(6) $$\chi(y + 4a) = \chi(y).$$

We now ask the vital question: what is the resolution modulus of $\chi(y)$ as a character? To answer this, first we define a *square-free integer* as one which has no perfect square divisor greater than 1. Then we define $k(a)$, the *square-free kernel* of a, as follows: if $a = AB^2$ and A is square-free, then $k(a) = A$.

THEOREM. The character $\chi(y) = (a/y)$, restricted to $y > 0$ and y odd, is a character with resolution modulus $|k(a)|$ if $k(a) \equiv 1$ modulo 4 and $4\,|k(a)|$ otherwise.

Proof. If $(y, a) = 1$, $(a/y) = (k(a)/y)$ directly from definition. Suppose $2 \nmid k(a)$; then, since $y > 0$ and y odd,

(5a) $$(k(a)/y) = (y/k(a))(-1)^{(y-1)/2\cdot(k(a)-1)/2}$$

If $k(a) \equiv 1 \pmod 4$, $(k(a)/y) = (y/k(a))$. Thus the Jacobi symbol (with y odd), (a/y), is determined by y modulo $k(a)$. If $k(a) \not\equiv 1 \pmod 4$, $(k(a)/y)$ is determined by y modulo $4k(a)$, i.e.,

(6a) $$\left(\frac{k(a)}{y}\right) = \left(\frac{k(a)}{y + 4k(a)}\right)$$

as is easily seen.

Next suppose $2 \mid k(a)$. Then $k(a) = 2a^*$, a^* odd.

$$(k(a)/y) = (2/y)(a^*/y).$$

But (a^*/y) is determined by y modulo a^* or $4a^*$, hence modulo $4a^*$ (easily), whereas $(2/y)$ is determined by y modulo 8, hence $(k(a)/y)$ is determined by y modulo $8a^* = 4k(a)$.

To prove that the quantities stated in the theorem *are* resolution moduli in the respective cases, we must show a prime factor p *cannot* be removed from $k(a)$ or $4k(a)$, as the case may be. Thus the integer $k(a)/p$ or $4k(a)/p$ would not be suitable as resolution modulus in the respective cases $k(a) \equiv 1$ or $k(a) \not\equiv 1 \pmod 4$.

First of all let p be odd; we wish to show that if $y \equiv 1 \pmod{k(a)/p}$ [or even $(\text{mod } 4k(a)/p)$] this would be *insufficient* to imply that $\chi(y) = (a/y) = 1$. For proof take y^* such that $(p/y^*) = -1$. [To do this, note that if $y^* \equiv 1 \pmod 4$ from reciprocity $(y^*/p) = (p/y^*)$ and y^* can be chosen simply congruent to a nonresidue modulo p.] Then, if

$$y \equiv 1 \pmod 2, \tag{7a}$$

with

$$y \equiv 1 \pmod{k(a)/p}, \tag{7b}$$

and

$$y \equiv y^* \pmod p, \tag{7c}$$

we can still have $\chi(y) = 0$ if $(y, a) > 1$, whereas, otherwise

$$\chi(y) = (4k(a)/y) = (k(a)/y) = ([k(a)/p]/y)(p/y) = -1. \tag{8}$$

The more difficult case is $p = 2$. For this case $k(a) \not\equiv 1 \pmod 4$.

Then there are two alternatives. If $2 \nmid k(a)$, then $k(a) \equiv -1 \pmod 4$. We choose $y \equiv 1 \pmod{k(a)}$, $y \equiv 3 \pmod 4$, and $y \equiv 1 \bmod (4k(a)/2)$. Thus $\chi(y) = 0$, if $(a, y) \neq 1$, otherwise

$$\chi(y) = (a/y) = (k(a)/y) = (y/k(a))(-1)^{(y-1)/2 \cdot (k(a)-1)/2} = -1, \tag{9}$$

completing the proof for the alternative $2 \nmid k(a)$.

Now, if $2 \mid k(a)$, then $4 \nmid k(a)$, and we can take $y \equiv 1 \pmod{k(a)/2}$ at the same time as $y \equiv 5 \pmod 8$. Thus $y \equiv 1 \pmod 4$ and $y \equiv 1 \pmod{k(a)/2}$, which yields

$$y \equiv 1 \pmod{4k(a)/2}; \tag{10}$$

but, using reciprocity [formulas (4) and (2b) of Chapter I, §9], we see

$$(4k(a)/y) = (\tfrac{1}{2}k(a)/y)(2/y) = (y/\tfrac{1}{2}k(a))(2/y) = -1. \qquad \text{Q.E.D.} \tag{11}$$

The reader may wonder why the symbol (a/y) was not treated as a function of a, which seems to lead to a simpler theory (see Exercise 8 below). The reason will be clear as we find that the symbol (a/y) as a function of y is just right for an important application (Dirichlet's lemma, §7).

EXERCISE 8. Show that $\chi^*(a) = (a/b)$ has resolution modulus $k(b)$ with no restriction that a be positive.

6. Kronecker's Symbol and Hasse's Congruence

The theorem of the preceding section makes the introduction of further concepts mandatory, first as a matter of convenience and then as an essential part of the theory!

Note first of all that if $a \equiv 1 \pmod 4$ then the character $\chi(y) = (a/y)$ must have resolution modulus $|k(a)|$, which is odd and for which $k(a) \equiv 1 \pmod 4$, all odd squares being $\equiv 1 \pmod 4$. Since $k(a) \mid a$, we can define the residue symbol $(a/2)$ as follows:

$$\left(\frac{a}{2}\right) = \left(\frac{a}{2+|a|}\right) = \left(\frac{2+|a|}{a}\right) = \left(\frac{2}{a}\right). \tag{1}$$

Thus we have *Kronecker's extension of Jacobi's symbol*:

$$\left(\frac{a}{2}\right) = \begin{cases} 0 \text{ if } 4 \text{ divides } a, \\ 1 \text{ if } a \equiv 1 \pmod 8, \\ -1 \text{ if } a \equiv 5 \pmod 8, \\ \text{undefined for all other } a. \end{cases} \tag{2}$$

The general symbol (a/b) can be defined by prime decomposition as in Chapter I, §9, to accommodate any resolution modulus of the element a, since any resolution modulus is in one of the "definable" categories for a. Of course, a may have square factors other than 4.

Second, we note that in the theorem in §5 (above) on the resolution modulus of (a/y) the condition that $y > 0$ is required only to ensure the validity of the reciprocity law (5*a*) when $a < 0$. (When $a > 0$, the condition on y may be removed.) It is therefore clear that the sign should be a part of the resolution modulus. We define

$$f(a) = \begin{cases} k(a), & \text{if } k(a) \equiv 1 \pmod 4, \\ 4k(a), & \text{otherwise,} \end{cases} \tag{3}$$

the so-called *conductor* (or *Führer* in German). Then Hasse, for instance, restricts the meaning of a congruence modulo $f(a)$ by saying $y_1 \equiv y_2$ modulo$^+$ $f(a)$ when $y_1 - y_2$ is divisible by $f(a)$ and *when* y_1 *and* y_2 *agree in sign* if $f(a) < 0$ (or with no further restriction if $f(a) > 0$). Thus, embodying the earlier remark on Kronecker's extension, we can prove the following more final improvement of the theorem in §5.

HASSE'S RESOLUTION MODULUS THEOREM

The Jacobi symbol $(a/y) = \chi(y)$, *as a character, can be extended to the Kronecker symbol* $(f(a)/y) = \chi^*(y)$, *so that* $\chi^*(y) = \chi(y)$ *whenever* $\chi(y) \neq 0$. *For this new character,* $\chi^*(y) \neq 0$ *when* y *is relatively prime to*

$f(a)$; and for nonzero values $\chi^*(y_1) = \chi^*(y_2)$ if and only if $y_1 \equiv y_2$ modulo$^+$ *$f(a)$. Also $|f(a)|$ is the minimum value for which the latter congruence property holds in any extension symbol for $\chi(y)$.*

In the rest of the book we consider $\chi(y)$ mainly for $y > 0$, obviating the need for the new congruence symbol. [The symbol $\chi^*(y)$ for negative y is the subject of Exercise 9 (below).]

Consider the further example:

The real characters modulo 12 can be listed in terms of their generators χ_4, χ_3. We note the resolution moduli:

TABLE 9

$m = 12 = 4 \cdot 3, \quad \phi(12) = 4$

$y \equiv$	1	5	7	11		
$\chi_4(y) =$	1	1	-1	-1	$(M = 4)$,	$(-36/y)$
$\chi_3(y) =$	1	-1	1	-1	$(M = 3)$,	$(-12/y)$
$\chi_4(y)\chi_3(y) =$	1	-1	-1	1	$(M = 12)$,	$(12/y)$
$\chi_1(y) =$	1	1	1	1	$(M = 1)$,	$(36/y)$.

The characters $\chi_1(y)$ and $\chi_3(y)$ alone are definable for y even. For $\chi_3(a)$, since $f = -3$, we must restrict y to be positive.

EXERCISE 9. Show for $\chi(b) = (f(a)/b)$, $\chi(-b) = \chi(b)$ but

$$\begin{aligned} \chi(b) &= \chi(f(a) - b) \quad \text{if } f(a) > 0 \text{ and } 0 < b < f(a) \\ &= -\chi(|f(a)| - b) \quad \text{if } f(a) < 0 \text{ and } 0 < b < |f(a)|. \end{aligned}$$

EXERCISE 10. Show that it is impossible to extend (a/y) to $y = 2$, when $a \equiv -1 \pmod 4$. *Hint*. Note $(3/2) = (3/(2 + 12))$ and $(3/2)^2 = (9/4)$ lead to a contradiction with earlier rules for (a/y).

7. Dirichlet's Lemma on Real Characters

We now turn our attention to the *real characters*, i.e., those characters modulo m which take on only real values. If we inspect the tables in §3 (above), we find that except for a few cases in which $m = 7$ and $m = 5$ the imaginary element $i = \sqrt{-1}$ is absent. We then observe that the real characters can be characterized by a Kronecker symbol (g/y) for some suitable g, not always square-free. (This accounts for the notation (g/y) on the right-hand side of the tables.)

DIRICHLET'S LEMMA

Any real character $\chi(y)$ *modulo m can be expressed in the form*

$$\chi(y) = (g/y), \quad y > 0, \tag{1}$$

using Kronecker's extension of the Jacobi symbol. The value of g will be $\equiv 0$ *or* 1 (*mod* 4) *and will depend on the character* χ *as well as m.*

The proof of this theorem is tedious but not difficult. We shall illustrate the proof in the case

$$m = 2^{a_1}p_2^{a_2}p_3^{a_3}, \qquad a_1 \geq 3, \qquad a_2 \geq 1, \qquad a_3 \geq 1, \tag{2}$$

where p_2 and p_3 are distinct odd primes. The most general character modulo m in the notation of §3 is

$$\chi_{u_0u_1u_2u_3}(y) = e(t_0/h_0)^{u_0}e(t_1/h_1)^{u_1}e(t_2/h_2)^{u_2}e(t_3/h_3)^{u_3}, \tag{3}$$

where, of course, the t_i are given by (5*a*, *b*, *c*) in Chapter I, §5, and u_i and t_i are determined modulo h_i, where

$$h_0 = 2, \qquad h_1 = \phi(2^{a_1})/2, \qquad h_2 = \phi(p_2^{a_2}), \qquad h_3 = \phi(p_3^{a_3}) \tag{4}$$

and

$$\phi(m) = h_0h_1h_2h_3. \tag{5}$$

First of all, $\chi_{u_0u_1u_2u_3}(y)$ is a real character if and only if each u_i is a multiple of $h_i/2$. This is fairly elementary, since we can, for instance, choose y^* so that (say) $t_1 = 1$ but the other $t_i = 0$, $(i = 0, 2, 3)$. Then $\chi_{u_0u_1u_2u_3}(y^*) = e(1/h_1)^{u_1} = \exp\ 2\pi i u_1/h_1 = \cos\ 2\pi u_1/h_1 + i \sin\ 2\pi u_1/h_1$. Naturally the imaginary term is absent only if $\sin 2\pi u_1/h_1 = 0$ or $u_1/(\frac{1}{2}h_1)$ is an integer. Likewise, $u_i/(\frac{1}{2}h_i)$ would have to be an integer.

Second, we verify (for $y > 0$)

$$e(t_0/h_0) = \chi_4(y) = (-1/y), \tag{6a}$$

$$e(t_1/h_1)^{h_1/2} = \chi_{2^{a_1}}(y)^{h_1/2} = (2/y), \tag{6b}$$

$$e(t_2/h_2)^{h_2/2} = \chi_{p_2^{a_2}}(y)^{h_2/2} = (y/p_2), \text{ etc.} \tag{6c}$$

This is, of course, the most important and least trivial step. Equation 6*a* follows directly from the fact that $e(t_0/h_0) = (-1)^{t_0}$ and $(-1/y) \equiv y \pmod 4$, whereas in (5*c*), Chapter I, §5,

$$y \equiv (-1)^{t_0}5^{t_1} \pmod{2^{a_1}}. \tag{7}$$

Hence $y \equiv (-1)^{t_0} \pmod 4$. Similarly $e(t_1/h_1)^{h_1/2} = (-1)^{t_1} = 1$ if and only if t_1 is even, or, easily, if[1] and only if $y \equiv \pm(25)^{t_1/2} \equiv \pm 1 \pmod 8$. If we

[1] See Exercise 12 below.

recognize this as the condition that $(2/y) = 1$, we see (6*b*). To see (6*c*), we consider the representation (5*a*), Chapter I, §5,

$$y \equiv g_2^{t_2} \pmod{p_2^{a_2}}, \qquad (y, p_2) = 1. \tag{8}$$

We observe that t_2 *is even if and only if* $(y/p_2) = +1$, for in elementary number theory (Chapter I, §8) we recall $(y/p_2) = 1$ if and only if the congruence

$$y \equiv x^2 \pmod{p_2^{a_2}}, \qquad (y, p_2) = 1$$

is solvable for each a_2. Such values of y occur if[1] and only if t_2 is even in (8), always assuming $(y, p_2) = 1$. Now

$$e(t_2/h_2)^{h_2/2} = (-1)^{t_2} = (y/p_2).$$

Thus, finally, any real character has the form

$$\chi(y) = (-1/y)^{w_0}(2/y)^{w_1}(y/p_2)^{w_2}(y/p_3)^{w_3}, \tag{9}$$

where $w_i = 1$ or 2. We use the notation $p^* = p$ if $p \equiv 1 \pmod 4$ and $p^* = -p$ if $p \equiv -1 \pmod 4$ and find by reciprocity $(y/p) = (p^*/y)$ for $y > 0$; hence we may satisfy (9) by

$$\chi(y) = (g/y), \qquad g = (-1)^{w_0}2^{4-w_1}(p_2^*)^{w_2} \quad (p_3^*)^{w_3}. \tag{10}$$

Note that when the factors 2^{w_1}, $p_2^{w_2}$, or $p_3^{w_2}$ become squares they conveniently make $\chi(y) = 0$ when y is divisible by 2, p_2, or p_3, respectively. Of course, (10) is Dirichlet's lemma for m in the convenient form (2).

EXERCISE 11. Show that the only m for which all characters are real are 1, 2 (trivial), and 3, 4, 6, 8, 12, 24. In fact, this is equivalent to the statement that all $h_i = 2$ in (2) of §3 (above). (Compare Exercise 7, Chapter I.)

EXERCISE 12. Justify the "if and only if" statements in the proof of (6*b*) and (6*c*). (The "only if" is easy but not the "if.")

EXERCISE 13. Justify the factor $2^{(4-w_1)}$ in (10) and write out the similar equation if m is odd or if $2 \parallel m$ and $2^2 \parallel m$.

[1] See Exercise 12 below.

chapter III

Some algebraic concepts

1. Representation by Quadratic Forms

The basic problem is the representation of an integer m by the quadratic form in the integral variables x and y

$$Q(x, y) = Ax^2 + Bxy + Cy^2 = m.$$

The problem is twofold. First of all, we must decide if such a representation is possible or if the Diophantine equation in two unknowns

$$Q(x, y) = m \tag{1}$$

is solvable, and then we must find out how to characterize all solutions, i.e., how to write the general (x, y) satisfying (1).

In this chapter we shall indicate how the problem, in principle, leads to the study of special algebraic systems. A satisfactory solution is not achieved until Chapter XII, and indeed not until we use algebraic numbers, in this case, quadratic surds $(a + b\sqrt{D})/c$, that would arise if we were to solve the equation $Q(x, y) = 0$ by "completing the square."

Specifically, we write

$$4AQ(x, y) = (2Ax + By)^2 - Dy^2, \tag{2}$$

where $D = B^2 - 4AC$ is the discriminant. We assume D is not a perfect square (hence $A \neq 0$), although D may have square divisors.

In the case in which $D < 0$, it follows that $4AQ(x, y) \geq 0$ from (2); hence $Q(x, y)$ is either zero or it agrees in sign with A. (In fact $Q(x, y) = 0$ only when the integers x, y are both zero, as is easily verified.) For $D < 0$

the form is therefore called *positive* (or *negative*) *definite*, according to whether the sign of A is positive (or negative). We note that A cannot be zero, for then $D = B^2$ would be a perfect square.

In the case in which $D > 0$ the form is called *indefinite* since the values of Q have no definite sign. Thus, if $A \neq 0$, values of x and y exist that make $AQ > 0$ (e.g., $y = 0$, $x = 1$) or which make $AQ < 0$ (e.g., $x = -B$, $y = 2A$). If $A = 0$, it can be seen that Q can also be made positive or negative.

The distinction between indefinite and definite forms carries over to the factors as the question whether $\sqrt{D}$ is real or imaginary.

EXERCISE 1. Show that if D is a perfect square the equation $Q(x, y) = m$ has only a finite number of solutions and indicate how they would be formed.

2. Use of Surds

We excluded the simple case in which D is a perfect square, and in the other case we shall introduce the symbol $\sqrt{D}$ to accomplish the factorization of (2) in the last section.

We introduce N, the *norm* symbol for a *fixed* D, not a perfect square: if a, b, c are integers, positive, negative, or zero (but $c \neq 0$), then we define[1] the *conjugate* surds

$$\lambda = \frac{a + b\sqrt{D}}{c}, \qquad \lambda' = \frac{a - b\sqrt{D}}{c}.$$

Thus $\lambda = \lambda'$ if and only if λ is rational ($b = 0$). We can see $(\lambda')' = \lambda$. The *norm* is defined as

$$N(\lambda) = \frac{a^2 - b^2 D}{c^2} = \lambda\lambda'.$$

Thus $N(\lambda') = N(\lambda)$, $N(a/c) = (a^2/c^2)$, $N(\lambda_1\lambda_2) = N(\lambda_1)\,N(\lambda_2)$. (The latter follows from the identity $(\lambda_1\lambda_2)' = \lambda_1'\lambda_2'$ if λ_1 and λ_2 are two surds.) Although we may use several surds in a problem, they will all have the same square-free kernel as D (or the same "reduced" radical) so that the norm symbol N will always have a clear reference. When $D < 0$, the norm, $N(\xi) = |\xi|^2$ (the usual absolute value squared). In this new symbol the factorization (2) of §1 (above) becomes

$$(1) \qquad 4AQ(x, y) = N(2Ax + By + \sqrt{D}\,y).$$

We now assume $A \neq 0$ for convenience.

[1] Unless otherwise specified, Greek letters denote variables which can become irrational.

The set of numbers $\xi = 2Ax + (B + \sqrt{D})y$ is generated by giving integral values to the integers x, y. This set really leads to a set of couples (ξ, ξ') like the cartesian coordinates of analytic geometry:

$$\text{(2a)} \qquad \begin{cases} \xi = 2Ax + (B + \sqrt{D})y, \\ \xi' = 2Ax + (B - \sqrt{D})y. \end{cases}$$

Then the vector $\mathbf{V} = (\xi, \xi')$ is generated by the two vectors $\mathbf{V}_1 = (2A, 2A)$ and $\mathbf{V}_2 = (B + \sqrt{D}, B - \sqrt{D})$ by use of integral coefficients

$$\text{(2b)} \qquad \mathbf{V} = x\mathbf{V}_1 + y\mathbf{V}_2.$$

The problem of (1) is to find a vector $\mathbf{V} = (\xi, \xi')$ of this type, for which

$$\text{(3)} \qquad N(\xi) = 4Am.$$

3. Modules

We define a *module* as a set of quantities closed under addition and subtraction. Thus, when a module contains an element ξ, it contains $0(= \xi - \xi)$ as well as negatives $-\xi(= 0 - \xi)$ and integral multiples ($\xi + \xi$ written 2ξ, $\xi + \xi + \xi$ written 3ξ, etc.) We shall use gothic capital letters $\mathfrak{M}$, $\mathfrak{N}$, $\mathfrak{O}$, etc., to denote modules.

The various vector sets used earlier [such as **ind** y in Chapter I and (ξ, ξ') of §2 above] clearly satisfy the definition of module. For the most important applications we generalize (2*b*) of the preceding section:

We consider combinations of a finite set of vectors[1] $\mathbf{V}_i$,

$$\text{(1)} \qquad \mathbf{u} = x_1\mathbf{V}_1 + x_2\mathbf{V}_2 + \cdots + x_s\mathbf{V}_s,$$

where the x_i range over all integers. The set of these $\mathbf{u}$ forms a module $\mathfrak{M}$ and the vectors $\mathbf{V}_1, \mathbf{V}_2, \cdots, \mathbf{V}_s$ are called a *basis* of the module, written

$$\mathfrak{M} = [\mathbf{V}_1, \mathbf{V}_2, \cdots, \mathbf{V}_s].$$

If the further condition is satisfied that no element $\mathbf{u}$ has two distinct representations of type (1), or in other words, the s-tuple $(x_1, \cdots, x_s)$ is uniquely determined by $\mathbf{u}$, we call the basis *minimal*.[2]

A module $\mathfrak{N}$ consisting of elements from a module $\mathfrak{M}$ is called a *submodule* of $\mathfrak{M}$.

[1] From now on we can consider that vector means "element of module" without contradicting its previous intuitive meaning of "ordered n-tuple."

[2] Beginning with Chapter V, we shall take "basis" to mean "minimal basis" in reference to modules, to simplify terminology. (But compare the *ideal basis* defined in Chapter VII, §4.)

LEMMA 1. Any submodule $\mathfrak{N}$ of the module $\mathfrak{M} = [\mathbf{V}]$ is precisely $[n\mathbf{V}]$ for a properly chosen integer n.

Proof. Assume $\mathfrak{N}$ does not consist only of $\mathbf{0}$. The submodule consists of some of the integral multiples of $\mathbf{V}$. We consider the smallest $|n| > 0$ for which $n\mathbf{V}$ lies in the submodule $\mathfrak{N}$. For any element $m\mathbf{V}$ of $\mathfrak{N}$, $n \mid m$. Otherwise by the division algorithm $m/n = q + r/n$ where $0 < r < |n|$ and the vector

$$r\mathbf{V} = (m\mathbf{V}) - q(n\mathbf{V})$$

belongs to $\mathfrak{N}$, since both $n\mathbf{V}$ and $m\mathbf{V}$ belong to $\mathfrak{N}$. This contradicts the definition of n as the smallest element of its kind. Q.E.D.

Note that we have used the symbol ξ both as the vector $\mathbf{V}$ and the component of $\mathbf{V} = (\xi, \xi')$, depending on which is more convenient.

4. Quadratic Integers

It is clear in some sense that a surd with integral coefficients is a generalization of an integer. Gauss, in fact, defined as "integers" the numbers $a + b\sqrt{-1}$, where a, b are ordinary integers positive, negative, or zero. More generally, one can see that if $\xi = a + b\sqrt{D}$ then $\xi^2 - 2a\xi + (a^2 - b^2D) = 0$, which corresponds to the quadratic form $x^2 - 2axy + (a^2 - b^2D)y^2$. Gauss did insist, however, that the standard form *must have an even middle coefficient*, so that he did not regard $x^2 + xy + y^2$ as "integral" but rather worked with $2x^2 + 2xy + 2y^2$ as the basic form. The relevance to surds becomes apparent if we note that

$$x^2 + xy + y^2 = N\left(x + \frac{1 + \sqrt{-3}}{2}\,y\right),$$

and Gauss in essence rejected[1] $(1 + \sqrt{-3})/2$ as an integer because it "had a denominator" or it did not arise from a form $x^2 + Bxy + Cy^2$ with B *even*. Yet at a later point we shall see that the whole development of algebraic number theory hinges on the use of certain numbers of type $(a + b\sqrt{D})/2$ as integers, as perceived by Dedekind (1871). This new type of integer enabled unique factorization to be extended to quadratic fields in a manner analogous to composition of forms (as in the introductory survey).

If we proceed by generalization of rational numbers, we can say that the rational number $\xi = p/q$ is a root of the equation

$$q\xi - p = 0$$

[1] This surd was in fact treated as an integer by Eisenstein (1844), a pupil of Gauss, but the significance of the proper definition of algebraic integer was not then appreciated.

and the integer is a root of an equation with first coefficient q equal to 1. An equation with coefficient of term of highest degree equal to 1 is called a *monic* equation; more generally it has the form

$$\xi^n + a_1\xi^{n-1} + \cdots + a_n = 0, \qquad (a_i \text{ integral}). \tag{1a}$$

Now it is clear that a general quadratic surd $\xi = (a + b\sqrt{D})/c$ is a root of the equation

$$c^2\xi^2 - 2ac\xi + (a^2 - b^2D) = 0. \tag{2}$$

Without any regard for the middle coefficient, we now define a *quadratic integer* as a solution to any monic quadratic[1] equation:

$$\xi^2 + B\xi + C = 0, \tag{1b}$$

where B and C are integers. Hence for $B = C = 1$, $\xi = (1 \pm \sqrt{-3})/2$ becomes an integer.

From now on, for the sake of clarity, an ordinary integer will often be called a *rational* integer.

5. Hilbert's Example

Hilbert gave a very famous example of an associative and multiplicatively closed set (called a semigroup), which fails to display unique factorization because of the "scarcity" of integers. Consider all positive integers congruent to 1 modulo 4:

$$\mathfrak{H}: 1, 5, 9, 13, 17, 21, 25, 29, \cdots, 441, \cdots.$$

If we define a "prime" number as a number *indecomposable into factors lying in* $\mathfrak{H}$, we find that numbers like

$$5, 9, 13, 17, 21, 29, \cdots, 49 \cdots$$

are "prime" but $25 = 5 \cdot 5$, $45 = 5 \cdot 9, \cdots$, and $441 = 21 \cdot 21 = 9 \cdot 49$ are not. We observe that these last two factorizations are irreconcilable. The most convenient way to resolve the difficulty is to introduce new integers, i.e., to "discover" 3, 7, 11, etc., so that we may write $441 = 3^2 \cdot 7^2$.

Actually, 3 can be "discovered" as the "greatest common divisor" of 9 and 21. Likewise, in algebraic number theory we shall discover that the "greatest common divisor" can even serve as a factor. This is a result which requires a greater development of modules, even leading to a composition theory for forms.

[1] It can be shown that a quadratic surd which fails to be a solution to a monic quadratic equation cannot be the solution to a monic equation of higher degree (Gauss's lemma).

EXERCISE 2. Consider the set $\mathfrak{H}'$ of positive integers which are quadratic residues modulo some fixed m. Show that it has properties like those of $\mathfrak{H}$. Do the same for the set $\mathfrak{H}''$ of positive integers $\equiv 1 \pmod{m}$ for m a fixed integer.

EXERCISE 3. Generalize Hilbert's set to cover all of the foregoing illustrations.

6. Fields

The role of algebraic integers can be seen best by starting with the concept of *field*. A field is a set of quantities taken from the complex numbers closed under the rational operations, namely addition, subtraction, multiplication, and division (excluding division by zero). In elementary number theory the field of rational numbers was introduced.

It is often convenient to extend the definition to quantities consisting of *sets* of real or complex numbers. Thus another type of field, introduced in Chapter I §2, is exemplified by all residue classes modulo p for p, a prime. This is a set of p sets written $\mathbf{0}, \mathbf{1}, \mathbf{2}, \cdots, (\mathbf{p} - \mathbf{1})$. They are clearly seen to be closed under the operations of addition, subtraction, and multiplication, whereas the existence of b/a modulo p for $a \not\equiv 0$ takes care of division. These sets form a *finite* field. The residue classes modulo m are not a field if m is $\neq 0$ and composite. For if $m = ab$, $(|a| \neq 1, |b| \neq 1)$, then $x \equiv 1/a \pmod{m}$ cannot exist (as $1 \equiv ax$ leads to $b \equiv abx \equiv 0$, which is false.)

In quadratic number theory the field we consider is taken to be the set of surds $(a + b\sqrt{D})/c$ for a, b, c integral, D fixed and not a perfect square, and $c \neq 0$. It can be seen that addition, subtraction, multiplication, and division of such quantities lead to quantities of the same form. (This is done in elementary algebra.) This field can be written symbolically as $R(\sqrt{D})$, meaning that the set of surds is *generated* by adjoining $\sqrt{D}$ to the rationals. The field $R(\sqrt{D})$ is called a field over the rationals.

We can state that our problem on "integers" is to characterize all elements of the field $R(\sqrt{D})$ which are also quadratic integers.

The concept behind "field" is due to Riemann (1857), who noted, in regard to function theory, that the difficulties involved in defining $w = \sqrt{z}$ (such as the usual difficulty in sign in the radical) are no worse for w than for any rational function of z and w such as $zw^3 = z^2\sqrt{z}$. This is a gross simplification of Riemann's contribution, but we merely emphasize the peculiar closeness to algebraic number theory where, say, the field $R(\sqrt{2})$ has the same problem. There the important choice (sign of $+\sqrt{2}$ versus sign of $-\sqrt{2}$) is made only once, and this choice of sign distinguishes all elements from conjugates henceforth. Riemann introduced no term, but Dedekind introduced "Körper" (1871), in the sense of "body" or "embodiment" of elements arising from rational operations, which for awhile was rendered in Latin "corpus" by British mathematicians, whereas French mathematicians used the cognate "corps," meaning body.

The word "field" seems to have been introduced by American algebraists who also used "realm" in the interim. Strangely enough, now, in both English and Russian,[1] field and the cognate *polye* mean field in two senses, the algebraic sense discussed here and also the sense of vector field from physics.

EXERCISE 4. Show that any field over the rationals containing $(a + b\sqrt{D})/c$ contains $\sqrt{D}$ (if $b \neq 0$).

EXERCISE 5. Show that if D_0 is square-free then the field generated by $\sqrt{D_0}$ can contain no other reduced radical than $\sqrt{D_0}$. *Hint.* Show first that the field generated by $\sqrt{2}$ will not contain $\sqrt{3}$.

7. Basis of Quadratic Integers

Consider next the problem of deciding when the arbitrary surd ξ of the field generated by $\sqrt{D}$,

$$\xi = (a + b\sqrt{D})/c, \tag{1}$$

is a quadratic integer. First of all we extract from D its (positive or negative) square-free kernel D_0, so that $D = m^2D_0$. Then we can cancel any factor of c which divides both a and b and make $c > 0$ for convenience. Replacing b by b/m, we write

$$\xi = \frac{a + b\sqrt{D_0}}{c}, \qquad \xi' = \frac{a - b\sqrt{D_0}}{c}.$$

Thus $\sqrt{D}$ and $\sqrt{D_0}$ generate the same field. We see that for ξ to be an integer the coefficients in (1*b*) of §4 for ξ, namely,

$$\xi + \xi' = -B \quad \text{and} \quad \xi\xi' = C$$

must be integers. Thus we must restrict a, b, c (relatively prime) so that

$$\frac{2a}{c} = -B \quad \text{and} \quad \frac{a^2 - b^2D_0}{c^2} = C$$

are integers.

First we observe $(a, c) = 1$; otherwise, if for some prime p, $p \mid a$ and $p \mid c$, then for C to be an integer p^2 (which divides the denominator) must

[1] The agreement of English and Russian on the same stem for two uses of field is remarkable, since there is a separate word in almost every other language for the physicists' vector field (*Feld, champs, campo,* etc.).

divide $a^2 - b^2D_0$. From $a^2 - b^2D_0 \equiv 0 \pmod{p^2}$ and $a^2 \equiv 0 \pmod{p^2}$ it follows that $b^2D_0 \equiv 0 \pmod{p^2}$; but since D_0 had no square factor $p \mid b$ and thus $p \mid a, p \mid b, p \mid c$, contradicting our assumption that the fraction (1) was reduced.

Thus since B is integral, if $c \neq 1$, $c = 2$ necessarily, and

$$a^2 - b^2D_0 = c^2C \equiv 0 \pmod 4. \tag{2}$$

We now consider in detail all possibilities concerning the parity of a and b if $c = 2$. We can see that unless a and b are both odd then either a, b, and c are all even or $4 \mid D_0$, leading (either way) to a contradiction. Then $a^2 \equiv b^2 \equiv 1 \pmod 4$ and, from congruence (2), $D_0 \equiv 1 \pmod 4$. Hence, easily, if $D_0 \not\equiv 1 \pmod 4$, c = 1.

Conversely, if $D_0 \equiv 1 \pmod 4$, and if a and b are both odd, we can take $c = 2$, since $a^2 - b^2D_0 \equiv 0 \pmod 4$, making B and C integral and making $\xi = (a + b\sqrt{D_0})/2$ a quadratic integer. Likewise, trivially, if $D_0 \equiv 1 \pmod 4$, ξ is an integer if $c = 2$ and both a and b are even (although the fraction (1) will not be reduced). We cannot make $c = 2$, however, if a and b are of mixed parity, (i.e., one odd and one even). Thus the most general quadratic integer is

$$\xi = \begin{cases} \dfrac{a + b\sqrt{D_0}}{2}, & a \equiv b \pmod 2 \text{ if } D_0 \equiv 1 \pmod 4, \\ a + b\sqrt{D_0}, & \text{all } a, b \text{ if } D_0 \not\equiv 1 \pmod 4. \end{cases} \tag{3}$$

There is nothing in the discussion to exclude $b = 0$. Here $c = 2$ only if $D_0 \equiv 1 \pmod 4$, and a is then also even, so that the only rational numbers that are quadratic integers are ordinary integers.

Note that we have another way of stating the result in (3) if we observe $(a + b\sqrt{D_0})/2 = (a - b)/2 + b(1 + \sqrt{D_0})/2$. Thus we let $(a - b)/2 = a'$ $b = b'$ and under the condition $a \equiv b \pmod 2$ we can set

$$(a + b\sqrt{D_0})/2 = a' + b'(1 + \sqrt{D_0}/2),$$

where a' and b' are *arbitrary* rational integers. Thus we define

$$\omega_0 = \begin{cases} \dfrac{1 + \sqrt{D_0}}{2} & \text{if } D_0 \equiv 1 \pmod 4, \\ \sqrt{D_0} & \text{if } D_0 \not\equiv 1 \pmod 4. \end{cases} \tag{4}$$

Then in both cases a basis of quadratic integers in $R(\sqrt{D})$ is $[1, \omega_0]$. This module is designated by the symbol

$$\mathfrak{O} = [1, \omega_0]. \tag{5}$$

Hence the most general integer in

$R(\sqrt{-1})$ is $x + y\sqrt{-1}$, (basis = $[1, \sqrt{-1}]$),
$R(\sqrt{2})$ is $x + y\sqrt{2}$, (basis = $[1, \sqrt{2}]$),
$R(\sqrt{-2})$ is $x + y\sqrt{-2}$, (basis = $[1, \sqrt{-2}]$),
$R(\sqrt{3})$ is $x + y\sqrt{3}$, (basis = $[1, \sqrt{3}]$),
$R(\sqrt{-3})$ is $x + y(1 + \sqrt{-3})/2$, (basis = $[1, (1 + \sqrt{-3})/2]$),
$R(\sqrt{5})$ is $x + y(1 + \sqrt{5})/2$, (basis = $[1, (1 + \sqrt{5})/2]$),
$R(\sqrt{8})$ is the same as for $R(\sqrt{2})$ (same basis),
$R(\sqrt{-12})$ is the same as for $R(\sqrt{-3})$ (same basis),
$R(\sqrt{20})$ is the same as for $R(\sqrt{5})$ (same basis).

The field $R(\sqrt{m^2D_0})$ is independent of m, and so is $\mathfrak{D}$ and its basis.

8. Integral Domains[1]

A set of quantities taken from the complex numbers which is closed under addition, subtraction, and multiplication (ignoring division) is called a *ring*. If a ring contains the rational integers, it is called an *integral domain*. The quadratic integers of a fixed field $R(\sqrt{D_0})$ form an integral domain which we call $\mathfrak{D}$. For addition and subtraction closure is obvious, and for multiplication it suffices to work with the basis elements: to take the hard case, let $D_0 \equiv 1 \pmod 4$. To establish the closure under multiplication, we note in this case $\omega_0 = (1 + \sqrt{D_0})/2$ and $\omega_0^2 = \omega_0 + (D_0 - 1)/4$. Thus $(a + b\omega_0)(a' + b'\omega_0) = aa' + (ab' + a'b)\omega_0 + bb'\omega_0^2 = (aa' + bb'(D_0 - 1)/4) + (ab' + a'b + bb')\omega_0$, clearly a member of the module $[1, \omega_0] = \mathfrak{D}$.

The closure of $\mathfrak{D}$ makes it possible for us to discuss congruences within $\mathfrak{D}$, i.e., $\xi_1 \equiv \xi_2 \pmod{\eta}$ if $(\xi_1 - \xi_2)/\eta$ is in $\mathfrak{D}$. The congruences then are clearly additive and multiplicative as in rational number theory. Thus, if $f(\xi)$ is a polynomial, with quadratic integers (elements of $\mathfrak{D}$) as coefficients, $\xi_1 \equiv \xi_2 \pmod{\eta}$ implies $f(\xi_1) \equiv f(\xi_2) \pmod{\eta}$. The properties extend to all rings.

Note that $\sqrt{-3} \equiv 1 \pmod 2$ and that $\sqrt{3} \not\equiv 1 \pmod 2$ on the basis of the fact that $(\sqrt{-3} - 1)/2$ is an integer but $(\sqrt{3} - 1)/2$ is *not* an integer in each respective field.

[1] The definitions of *ring* and *integral domain* are restricted to the context of subsets of the complex numbers. Definitions of *integral domain* vary widely in the literature, but we follow the spirit of the original efforts to generalize rational integers.

THEOREM I. If the integral domain $\mathfrak{O}$ of all quadratic integers of $R(\sqrt{D})$ contains an integral domain $\mathfrak{O}^*$ which does not consist wholly of rationals, then $\mathfrak{O}^*$ is characterized by some fixed positive rational integer n as the set of integers of $\mathfrak{O}$ which are congruent to a (variable) rational integer modulo n.

Proof. Clearly the aggregate of quadratic integers $\mathfrak{O}$, which are in $\mathfrak{O}$ and congruent to a rational integer modulo n, is closed under addition, subtraction, and multiplication by means of the ring property of rational integers.

The converse is less immediate. Consider the terms $x + y\omega_0$ of the arbitrary integral domain $\mathfrak{O}^*$. To avoid a triviality (the case in which no irrationals occur), we note $y \neq 0$ for some terms. For every element $x + y\omega_0$ which occurs in $\mathfrak{O}^*$, $y\omega_0$ must occur in $\mathfrak{O}^*$, since $\mathfrak{O}^*$ contains all integers x. We consider the smallest such $|y|$; call it n. Then, for $\mathfrak{O}^*$, all terms $y\omega_0$ (in $x + y\omega_0$) must be multiples of $n\omega_0$ by the lemma 1 in §3 (above). Hence the general term of $\mathfrak{O}^*$ is $\xi = x + yn\omega_0$ for x and y arbitrary; $\xi \equiv x (\text{mod } n)$ for all ξ in $\mathfrak{O}^*$, and, conversely, all such ξ have the form $x + yn\omega_0$. Q.E.D.

The integral domain $\mathfrak{O}^*$ corresponding to n is written $\mathfrak{O}_n$. Thus $\mathfrak{O}_1 = \mathfrak{O}$.

EXERCISE 6. In the field $R(\sqrt{-1})$ show that the residue classes of integers of $\mathfrak{O}_1$, $x + y\sqrt{-1}$, taken modulo 3, form a finite field (see §6 above) of nine elements. Show that the residue classes modulo 5 do not form a field. *Hint.* $5 = 2^2 + 1^2$.

EXERCISE 7. Write down the five residue classes of integers $x + y\sqrt{-1}$ of $\mathfrak{O}_1$ in $R(\sqrt{-1})$ modulo $2 + \sqrt{-1}$. Show that they form a field by showing a residue class containing each of the integers 0, 1, 2, 3, 4.

9. Basis of $\mathfrak{O}_n$

The integers in $\mathfrak{O}_n$ were seen to have the form $x + yn\omega_0$. This observation leads to several cases, depending on the residues of D_0 and of $D_0 n^2 = D$. We define

$$(1a) \qquad \omega = \frac{1 + \sqrt{D}}{2} = \frac{1 + n\sqrt{D_0}}{2}, \text{ if } D_0 \equiv D \equiv 1 \pmod 4, \qquad (n \text{ odd}),$$

(note $\omega = n\omega_0 - (n - 1)/2$);

$$(1b) \qquad \omega = \frac{\sqrt{D}}{2} = \frac{n\sqrt{D_0}}{2}, \text{ if } D \equiv 0,\ D_0 \equiv 1 \pmod 4, \qquad (n \text{ even}),$$

(note $\omega = n\omega_0 - n/2$); and

$$(2) \qquad \omega = \sqrt{D} = n\sqrt{D_0}, \quad \text{if } D_0 \not\equiv 1 \pmod 4, \qquad (\text{any } n).$$

In any case, $\mathfrak{O}_n = [1, \omega] = [1, n\omega_0]$. The details are left to the reader. Note, again, that $\mathfrak{O}_n$ in $R(\sqrt{m^2 D_0})$ is the same as $\mathfrak{O}_n$ in $R(\sqrt{D_0})$.

The designation of the letter $\mathfrak{O}$ for the integral domain has some historical importance going back to Gauss's work on quadratic forms. Gauss (1800) noted that for certain quadratic forms $Ax^2 + Bxy + Cy^2$ the discriminant need not be square-free, although A, B, and C are relatively prime. For example, $x^2 - 45y^2$ has $D = 4 \cdot 45$. The 4 was ignored for the reason that $4 \mid D$ necessarily by virtue of Gauss's requirement that B be even, but the factor of 3^2 in D caused Gauss to refer to the form as one of "order 3." Eventually, the forms corresponding to a value of D were called an "order" (*Ordnung*). Dedekind retained this word for what is here called an "integral domain."

The term "ring" is a contraction of "*Zahlring*" introduced by Hilbert (1892) to denote (in our present context) the ring generated by the rational integers and a quadratic integer η defined by

$$\eta^2 + B\eta + C = 0.$$

It would seem that the module $[1, \eta]$ is called a *Zahlring* because η^2 equals $-B\eta - C$ "circling directly back" to an element of $[1, \eta]$. This word has been maintained today. Incidentally, every Zahlring is an integral domain and the converse is true *for quadratic fields*.

EXERCISE 8. Show that the set of integers η in $\mathfrak{O}_1$ for which $\eta^p \equiv \eta \pmod{p}$, ($p$ prime) forms an integral domain directly from the definition.

EXERCISE 9. Specify this integral domain for different cases of $D_0 \pmod{4}$ [noting that $(D_0/p) \equiv D_0^{(p-1)/2} \pmod{p}$ according to Euler's lemma].

EXERCISE 10. Give an example of a ring contained in $\mathfrak{O}_1$ and not forming an integral domain. Can Theorem 1 (above) be generalized?

**10. Fields of Arbitrary Degree

The present course is devoted almost exclusively to quadratic fields, in which the basic ingredients of algebraic number theory are amply evident. Yet we should take a quick glance at fields generated by (say) the irreducible equation of arbitrary degree

$$a_0x^n + a_1x^{n-1} + \cdots + a_n = 0, \tag{1}$$

if only to see what lies beyond this course. This section, therefore, is wholly descriptive and the major results are unproved; they are of course, unnecessary for the later text.

For simplicity, start with the irreducible *cubic* equation

$$a_0x^3 + a_1x^2 + a_2x + a_3 = 0, \tag{2}$$

where a_0, a_1, a_2, a_3 are integers. This equation has three roots, which we call $\theta_1, \theta_2, \theta_3$ There are several possible fields we could consider; for example, $R(\theta_1)$, the field formed by performing rational operations of θ_1, with rational coefficients.

The field $R(\theta_1)$ consists entirely of elements of the form

$$\xi_1 = r_1 + r_2\theta_1 + r_3\theta_1^2, \tag{3}$$

where r_1, r_2 and r_3 are rational. This is not obvious (see Exercise 14). Furthermore, the field $R(\theta_1)$ may or may not contain θ_2 (see Exercises 15, 16, and 20). This is an alternative we tend to overlook in the quadratic case in which both roots of a quadratic must, of course, generate the same field, since they "share" the use of $\sqrt{D}$. At any rate, if $R(\theta_1)$ does not contain θ_2, a new field which can be called $R(\theta_1, \theta_2)$ is formed by rational operations on both θ_1 and θ_2. Then $R(\theta_1, \theta_2)$ is *larger* than $R(\theta_1)$ (in the sense that $R(\theta_1, \theta_2)$ has all elements of $R(\theta_1)$ and more elements in addition). Otherwise $R(\theta_1, \theta_2)$ is merely $R(\theta_1)$.

Generally, we can speak of θ and ϕ (instead of θ_1 and θ_2) as any two algebraic numbers with no specific relation between them, e.g., θ might satisfy (1), whereas ϕ satisfies an equation of degree m. Then there exists a number ψ of degree no greater than nm such that $R(\psi) = R(\phi, \theta)$ (see Exercise 17). If the degree of ψ is actually nm, we say one of two things: either $R(\psi)$ is a field of *degree m relative* to $R(\theta)$ or $R(\psi)$ is a field of *degree n relative* to $R(\phi)$. The fact that there are two such characterizations is extremely important[1] later on.

Here we mention another point, also easily taken for granted in the quadratic case, where any surd $(a + b\sqrt{D})/c$ generates the same field as $\sqrt{D}$, as long as $b \neq 0$. We must think, in general, of a field as an aggregate of elements, independently of generators, since in the cubic case there may be no special number like $\sqrt{D}$ which would seem to be the "logical" generator. In fact, in the preceding paragraph it may sometimes be more convenient to think of $R(\psi)$ as $R(\phi, \theta)$ or to think of two simultaneous generators instead of one, altogether.

Returning to the cubic field $R(\theta_1)$, every element ξ_1 satisfies a cubic equation of type (2), whose coefficients may be quite difficult to calculate (see Exercise 12). If the equation has $a_0 = 1$ or is *monic*, and all other coefficients are integral, then ξ_1 is called an *algebraic integer*, as in the quadratic case. We should like to be able to think of our field as an abstract collection of numbers having many possible generators; yet if θ_1 were chosen correctly formula (3) *might* include all *algebraic integers* in

[1] See the Concluding Survey.

the field if we restrict r_1, r_2, r_3 to all *rational integers. This cannot always be done.* The best we can do is to say that for ξ_1, an integer, the fractions r_1, r_2, and r_3 are integers or at worst have a denominator that must divide some *constant* integer Q.

This result need not be wholly mysterious. For instance, we write (3) for all three conjugates assuming that ξ_1, ξ_2, ξ_3 and θ_1, θ_2, θ_3 are algebraic integers. We also note the determinant:

$$(4) \qquad \begin{cases} \xi_1 = r_1 + r_2\theta_1 + r_3\theta_1^2 \\ \xi_2 = r_1 + r_2\theta_2 + r_2\theta_2^2 \\ \xi_3 = r_1 + r_2\theta_3 + r_3\theta_3^2 \end{cases} \qquad \Delta = \begin{vmatrix} 1 & \theta_1 & \theta_1^2 \\ 1 & \theta_2 & \theta_2^2 \\ 1 & \theta_3 & \theta_3^2 \end{vmatrix}.$$

If we eliminate r_2 and r_3, for instance, we find an expansion

$$(5) \quad r_1 = [\xi_1(\theta_2\theta_3^2 - \theta_3\theta_2^2) + \xi_2(\theta_3\theta_1^2 - \theta_1\theta_3^2) + \xi_3(\theta_1\theta_2^2 - \theta_2\theta_1^2)]/\Delta.$$

Now Δ can be expanded incidentally as

$$(6) \qquad \Delta = (\theta_1 - \theta_2)(\theta_2 - \theta_3)(\theta_3 - \theta_1).$$

There are similar expressions of r_2 and r_3, always with denominator Δ. We then use several results that are not proved here:

(a) The so-called *discriminant* of θ_1, $\Delta^2 = D$, is a rational integer (see Exercise 21).

(b) The algebraic integers form an integral domain.

(c) A rational fraction cannot be an algebraic integer unless it is a rational integer.

Thus each r_i can now be written in the form $\gamma_i/\Delta = \zeta_i/D$, where ζ_i ($= \Delta\gamma_i$) and γ_i are algebraic integers. Hence $Dr_i = \zeta_i$ is an ordinary (rational) integer z_i and $r_i = z_i/D$, which has the desired form, since $Q = D$, for example, serves as denominator. The numerators z_i are, of course, not arbitrary, any more than in the quadratic case [see (3) in §7 (above)].

It might suffice to say that the integral domain has a basis of n algebraic integers (which can be selected generally with much more difficulty than in the quadratic case). We have no occasion to do this here for $n > 2$.

The following exercises might clarify some of the difficulties to which we allude.

EXERCISE 11. Show that all powers of θ_1 are type (3) by induction. Assume $\theta_1^n = r_1^{(n)} + r_2^{(n)}\theta_1 + r_3^{(n)}\theta_1^2$ and multiply both sides by θ_1, using (2).

EXERCISE 12. Show that ξ_1 satisfies a cubic equation by showing $1, \xi_1, \xi_1^2, \xi_1^3$ to be numbers of type (3) and by subsequent elimination of the powers of θ_1.

EXERCISE 13. Show that if $\xi_i = r_1 + r_2\theta_i + r_3\theta_i^2$ expresses the three conjugates ($i = 1, 2, 3$), for ξ_1 of type (3), then show that $\xi_2\xi_3$ has an expansion of type (3). *Hint.*

$$\xi_2\xi_3 = (r_1 + r_2\theta_2 + r_3\theta_2^2)(r_1 + r_2\theta_3 + r_3\theta_3^2),$$

and note that from the root properties of (2)

$$\theta_2 + \theta_3 = -a_1/a_0 - \theta_1; \qquad \theta_2\theta_3 = -a_3/a_0\theta_1;$$

$$\theta_2^2 + \theta_3^2 = (\theta_2 + \theta_3)^2 - 2\theta_2\theta_3; \qquad 1/\theta_1 = -(a_0\theta_1^2 + a_1\theta_1 + a_2)/a_3;$$

$$\theta_2^2\theta_3 + \theta_3^2\theta_2 = (\theta_2^2 + \theta_3^2)(\theta_2 + \theta_3) - (\theta_2 + \theta_3)^3 + 3\theta_2\theta_3(\theta_2 + \theta_3).$$

EXERCISE 14. Show that all elements in $R(\theta_1)$ are type (3). (Note carefully if you are dividing by zero at any time!)

EXERCISE 15. Consider the *pure cubic* equation

$$x^3 = ab^2, \qquad (a \geq 1, b \geq 1, ab > 1),$$

where $(a, b) = 1$ and a and b have no square divisors. Show that $R(ab^2)^{1/3}$ does not contain $R(\rho(ab^2)^{1/3})$ where $\rho = (-1 + \sqrt{-3})/2$, an imaginary cube root of unity.

EXERCISE 16. Show $R(\rho(ab^2)^{1/3})$ does not contain $\rho^2(ab^2)^{1/3}$.

EXERCISE 17. Show that if $\psi = \rho + (ab^2)^{1/3}$ then $R(\psi)$ contains ρ and $(ab^2)^{1/3}$. Also write the rational equation defining ψ.

Hint. Solve for ρ by combining $(\psi - \rho)^3 = ab^2$ with $\rho^2 = -\rho - 1$ in the expression for ρ. (Do not "rationalize" the denominator.) The conjugates of ψ are, incidentally, $\rho + (ab^2)^{1/3}$, $\rho + \rho(ab^2)^{1/3}$, $\rho + \rho^2(ab^2)^{1/3}$, $\rho^2 + (ab^2)^{1/3}$, $\rho^2 + \rho(ab^2)^{1/3}$, $\rho^2 + \rho^2(ab^2)^{1/3}$.

EXERCISE 18. Show that the field generated by $\sqrt{a} + \sqrt{b} = \xi$ contains $\sqrt{ab}$ and $\sqrt{a}$ and $\sqrt{b}$. Show that ξ satisfies an equation of fourth degree.

EXERCISE 19. Show that the (cyclotomic) equation

$$(\lambda^7 - 1)/(\lambda - 1) = \lambda^6 + \lambda^5 + \lambda^4 + \lambda^3 + \lambda^2 + \lambda + 1 = 0$$

has as its six roots $\lambda^k = \exp 2\pi ik/7$ ($= \cos 2\pi k/7 + i \sin 2\pi k/7$), $1 \leq k \leq 6$.

EXERCISE 20. Show that $\mu_k = \lambda^k + 1/\lambda^k = 2\cos 2\pi k/7$ satisfies the equation

$$\mu^3 + \mu^2 - 2\mu - 1 = 0.$$

From this show that the three roots satisfy

$$\mu_2 = \mu_1^2 - 2, \qquad \mu_3 = \mu_2^2 - 2, \qquad \mu_1 = \mu_3^2 - 2.$$

(Thus $R(\mu_1) = R(\mu_2) = R(\mu_3)$.)

EXERCISE 21. Multiply the determinant Δ by its transpose (rows and columns interchanged) and verify that for rational S_i

$$D = \Delta^2 = \begin{vmatrix} S_0 & S_1 & S_2 \\ S_1 & S_2 & S_3 \\ S_2 & S_3 & S_4 \end{vmatrix}, \qquad S_i = \sum_{t=1}^{3} \theta_t^{\,i}.$$

EXERCISE 22 (Dedekind). Show that the number

$$\xi_1 = [1 + b(a^2b)^{1/3} + a(ab^2)^{1/3}]/3$$

is an algebraic integer in $R(ab^2)^{1/3}$ if $a^2 \equiv b^2 \pmod 9$. *Hint*. Show that the numbers

$$\xi_2 = [1 + b(a^2b)^{1/3}\rho + a(ab^2)^{1/3}\rho^2]/3$$
$$\xi_3 = [1 + b(a^2b)^{1/3}\rho^2 + a(a^2b)^{1/3}\rho]/3$$

are conjugates or that $(x - \xi_1)(x - \xi_2)(x - \xi_3)$ has rational integral coefficients.

chapter IV

Basis theorems

1. Introduction of n Dimensions

The main result of Chapter III, after the introduction of new terms, was a very simple one, namely the expressibility of a certain module $\mathfrak{O}_1$ by means of a basis $[1, \omega_0]$ and similarly for $\mathfrak{O}_n$, the integral domains associated with $\mathfrak{O}_1$ (in §7 and §9 of Chapter III).

Two questions are natural:

Is the situation as simple for the basis for an arbitrary module? Is there an easy relationship connecting different bases that can be used for an arbitrary module?

The answers are "generally" affirmative and lead to an interesting theory. For simplicity it is actually equally convenient to act in somewhat greater generality (taking more than two dimensions). The degree of generality achieved will also be useful in Chapter V when we make further applications.

2. Dirichlet's Boxing-in Principle

The general techniques for constructing a basis, however intimately connected with algebraic number theory, were not fully appreciated until very late (about 1896) when Minkowski, in his famous work *Geometry of Numbers*, showed in detail that considerable significance can be attached to the seemingly simple procedure of visualizing a module coordinatewise. Although the usefulness of Minkowski's techniques is not appreciated fully when restricted to the quadratic case, these techniques have a starkness and an appeal to fundamentals, which command recognition in their own right, as they bring out the importance of geometry as a tool of number theory.

A well-known earlier example of geometrical intuition is the following:

DIRICHLET'S BOXING-IN PRINCIPLE (1834)

If we have $g + 1$ *objects distributed among g boxes so that each box may have any number of these objects (or none at all), then at least one box will contain two objects.*

The principle is obvious. To apply it, let us consider the following version: if we have more than M^n points in an n-dimensional unit cube, where M is a positive integer, then two points exist each of whose n projections (coordinates) differ respectively by no more than $1/M$.

(Of course, a one-dimensional "cube" is a line segment, a two- dimensional "cube" is a square, etc.) For proof we simply divide each side into M parts yielding a total of M^n cubes. Then, if more than M^n points are present, two must lie in one cube.

3. Lattices

For an arbitrary module $\mathfrak{M}$ a *basis* was defined (Chapter III, §3) as a finite set of elements of $\mathfrak{M}$ (or vectors) $\mathbf{u}_1, \mathbf{u}_2, \cdots, \mathbf{u}_n$ for which the combinations denoted by

$$\mathbf{u} = x_1\mathbf{u}_1 + x_2\mathbf{u}_2 + \cdots + x_n\mathbf{u}_n, \tag{1}$$

for integral n-tuples $(x_1, x_2, \cdots, x_n)$ account for all elements of $\mathfrak{M}$. We also express (1) by saying $\mathbf{u}$ lies in the space *spanned* by $\mathbf{u}_1, \mathbf{u}_2, \cdots, \mathbf{u}_n$ (implying *integral* coefficients x_i).

An arbitrary module need not have a basis. For example, the set of all rational numbers (positive, negative, and zero) has no basis. [For, if $\mathbf{u}_1, \cdots, \mathbf{u}_n$ corresponded to fractions $p_1/q_1, \cdots, p_n/q_n$, then we could not obtain all fractions in (1), since the denominators are limited, as the reader can easily see.] Thus the elements of a module cannot be expected to be "too close" if the module has a basis. The matter of "not being too close" is expressed by means of two[1] terms: *finite dimensionality* and *discreteness.* This requires a series of definitions.

We first introduce *linear independence*: a finite set of vectors in $\mathfrak{M}$ $\mathbf{v}_1, \mathbf{v}_2, \cdots, \mathbf{v}_n$ is linearly *dependent* over the integers if rational integers $a_1, a_2, \cdots, a_n$, not all zero, exist for which

$$a_1\mathbf{v}_1 + a_2\mathbf{v}_2 + \cdots + a_n\mathbf{v}_n = \mathbf{0}; \tag{2}$$

[1] The "discreteness" implies separation in the usual sense, whereas "finite dimensionality" implies "noncrowding at ∞."

otherwise they are called *linearly independent*. The *dimensionality* of a vector space is then defined as the maximum number of linearly independent vectors.

A *norm* is a function of the vector $\mathbf{v}$ denoted by $\|\mathbf{v}\|$ with the properties (reminiscent of distance from the origin):

$$(2a) \qquad \|a\mathbf{v}\| = |a| \cdot \|\mathbf{v}\| \text{ for } a \text{ rational}^1, \qquad \text{(linearity)},$$

$$(2b) \qquad \|\mathbf{v}_1 + \mathbf{v}_2\| \le \|\mathbf{v}_1\| + \|\mathbf{v}_2\|, \qquad \text{(triangular inequality)},$$

$$(2c) \qquad 0 \le \|\mathbf{v}\| \text{ with equality only for } \mathbf{v} = \mathbf{0}, \qquad \text{(definiteness)}.$$

A *discrete* module is one in which a norm exists such that

$$(2d) \qquad \|\mathbf{v}\| \ge k \quad \text{when} \quad \mathbf{v} \ne \mathbf{0}$$

for k, a fixed, positive constant. There may be several norms satisfying this property (but all norms need not do so).

A *lattice*[2] is finally defined as a discrete, finite dimensional module. We shall use gothic symbols $\mathfrak{L}$, $\mathfrak{M}$, $\mathfrak{N}$, $\mathfrak{O}$ to denote lattices as well as modules.

We can easily check that the module of all integers ξ in $\mathfrak{O}$ has both properties and is therefore a lattice. We see that the dimensionality is two, since for every $\xi = x + y\omega_0$ the three quantities ξ, 1, ω_0 are linearly dependent (i.e., $1 \cdot \xi - x \cdot 1 - y \cdot \omega_0 = 0$), whereas 1 and ω_0 are linearly independent by the irrationality of $\sqrt{D}$. The discreteness follows from (many) choices of $\|\xi\|$, including

$$(3) \qquad \|\xi\| = \left[\frac{|\xi|^2 + |\xi'|^2}{2}\right]^{1/2}.$$

Here the properties ($2a$, b, c) are not wholly trivial. Property ($2a$) is easy; property ($2b$) is left to §4 below. To show property ($2d$), note that since $(|\xi| - |\xi'|)^2 \ge 0$, on expanding, we find

$$|\xi|^2 + |\xi'|^2 \ge 2\,|\xi\xi'| = 2\,|N(\xi)|.$$

Thus, unless $\xi = 0$, from the fact that $|N(\xi)| \ge 1$ it follows that for each ξ, $\|\xi\| \ge 1$. Furthermore, $\|\xi\| = \|\xi'\|$ and $\|a\| = |a|$, with rational a.

[1] As a matter of convenience, we define fractional combinations of vectors by saying $\mathbf{w}_1 = p/q\mathbf{w}_2$ means $q\mathbf{w}_1 = p\mathbf{w}_2$. The fact that this use of fractions is consistent is similar to the fact that the use of fractions in ordinary or modular arithmetic is consistent (see Chapter I, §1). Thus ($2a$) gives an extension of the norm symbol when a is rational.

[2] The geometrical idea of *lattice* was used by Gauss (1800) and is called *Gitter* (German), *treillis* (French), *reshetok* (Russian). An independent concept in algebra was introduced by Dedekind (1894) under the name *Dualgruppe*, more recently *Verband* (German). The English word *lattice* was unfortunately also used by algebraists for the other concept (adding to the confusion because Dedekind had been motivated by module theory !). In Russian the word *struktur* is widely used but not in French or English.

A *submodule* of a lattice is clearly a lattice, since the properties all carry over. It is called a *sublattice*. Thus any quadratic module is a lattice of dimension 2 or less and a sublattice of $\mathfrak{O}$.

4. Graphic Representation

To illustrate the ideas, note that the lattice $\mathfrak{O}_1$ of integers in a quadratic field can be represented in the real and imaginary cases by a suitable choice of coordinates. We can represent ξ by (ξ, ξ') in a real field (Figure 4.1) and by $(\text{Re } \xi, \text{Im } \xi)$ in a complex field (Figure 4.2).

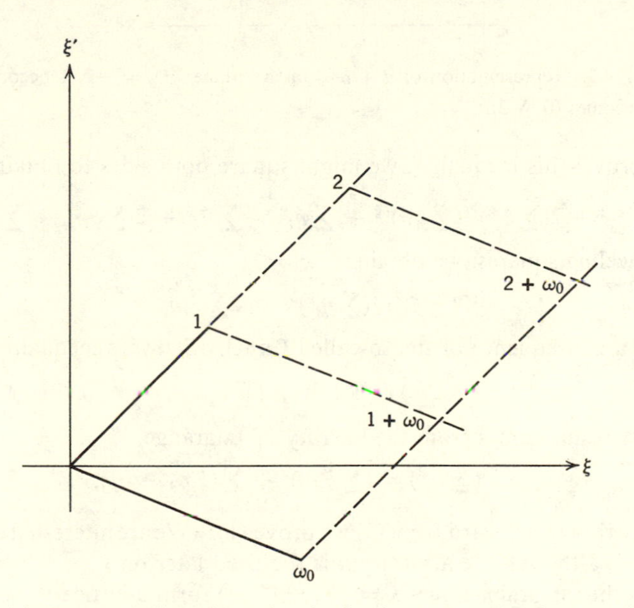

FIGURE 4.1. Representation of $a + b\omega_0$ in the plane: $D_0 = 5$, 1 becomes $(1, 1)$, ω_0 becomes $[(1 + \sqrt{5})/2, (1 - \sqrt{5})/2]$.

We can easily verify in both the real and the imaginary quadratic cases that $\|\xi\| = [(|\xi|^2 + |\xi'|^2)/2]^{1/2}$ is the distance from the point representing ξ to the origin divided by a constant factor to ensure that $\xi = \xi' = 1$ will be at unit distance. Thus property (2*b*) of §3 for the norm (3) of §3 is a consequence of this general result (expressed in n dimensions):

The triangular inequality. If $(\xi_1, \cdots, \xi_n)$, $(\eta_1, \cdots, \eta_n)$ are real n-tuples,

$$(1) \qquad \left(\sum_{i=1}^{n} \xi_i^2\right)^{1/2} + \left(\sum_{i=1}^{n} \eta_i^2\right)^{1/2} \geq \left[\sum_{i=1}^{n} (\xi_i + \eta_i)^2\right]^{1/2}.$$

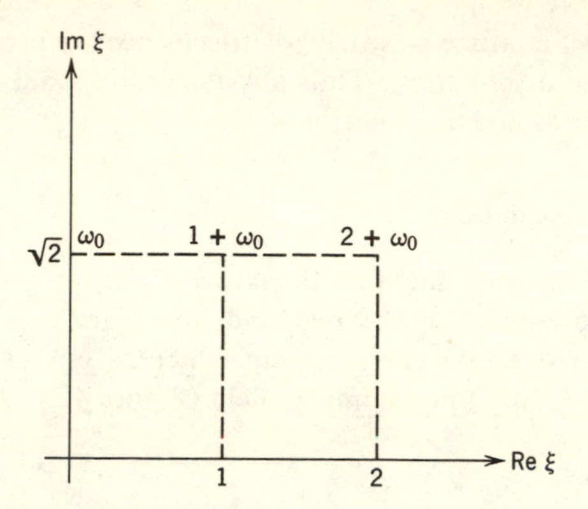

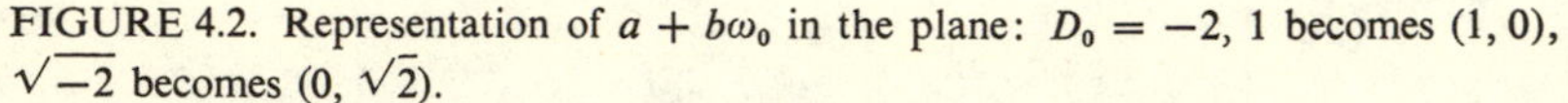

FIGURE 4.2. Representation of $a + b\omega_0$ in the plane: $D_0 = -2$, 1 becomes $(1, 0)$, $\sqrt{-2}$ becomes $(0, \sqrt{2})$.

To prove[1] this inequality, we might square both sides to obtain

$$\sum \xi_i^2 + 2(\sum \xi_i^2)^{1/2}(\sum \eta_i^2)^{1/2} + \sum \eta_i^2 \geq \sum \xi_i^2 + 2\sum \xi_i\eta_i + \sum \eta_i^2.$$

On canceling squares, we obtain

$$2(\sum \xi_i^2)^{1/2}(\sum \eta_i^2)^{1/2} \geq 2\sum \xi_i\eta_i.$$

This is a consequence of the so-called Cauchy-Schwarz inequality,

$$(\sum \xi_i^2)(\sum \eta_i^2) \geq (\sum \xi_i\eta_i)^2. \tag{2}$$

The last result comes from the identity of Lagrange,

$$(\sum \xi_i^2)(\sum \eta_i^2) = (\sum \xi_i\eta_i)^2 + \sum_{i>j} (\xi_i\eta_j - \xi_j\eta_i)^2. \tag{3}$$

Thus working backward from (3) we prove (1). (We are interested currently in $n = 2$, although the arbitrary n is required later on.) Q.E.D.

Thus the integral n-tuples $\mathbf{x} = (x_1, \cdots, x_n)$ form a lattice $\mathfrak{L}_n$ with norm $\|\mathbf{x}\| = (x_1^2 + \cdots + x_n^2)^{1/2}$. Note for $\mathbf{x} \neq 0$, $\|\mathbf{x}\| \geq 1$.

EXERCISE 1. Sketch the lattices for $\mathfrak{O}_1$ in the cases $D_0 = -5$ and $D_0 = 2$.

EXERCISE 2. Write out identity (3) for $n = 2$ and $n = 3$ and state when the equality can prevail in (2) and (1).

5. Theorem on Existence of Basis

THEOREM 1. Every lattice has a minimal basis.

We begin with some incidental remarks. First of all, the converse of the main theorem is a simple matter.

[1] All summations henceforth are from 1 to n on the indices.

THEOREM 2. A module with a basis (minimal or not) constitutes a lattice.

Proof. We consider the element $\mathbf{u}$ of $\mathfrak{M}$ represented by the basis as $\mathbf{u} = x_1\mathbf{u}_1 + \cdots + x_n\mathbf{u}_n$ as in (1) of §3 (above). In general, many representations are possible, but we define

$$\|\mathbf{u}\| = (x_1^2 + \cdots + x_n^2)^{1/2} \tag{1}$$

for the representation which yields the smallest possible value. To use the triangular inequality, note that if another $\mathbf{v}$ is represented by $\mathbf{v} = y_1\mathbf{u}_1 + \cdots + y_n\mathbf{u}_n$ for purposes of norm the vector $\mathbf{u} + \mathbf{v}$ would have the representation $(x_1 + y_1)\mathbf{u}_1 + \cdots + (x_n + y_n)\mathbf{u}_n$, among others (possibly). Thus

$$\|\mathbf{u} + \mathbf{v}\| \leq [\sum(x_i + y_i)^2]^{1/2} \leq (\sum x_i^2)^{1/2} + (\sum y_i^2)^{1/2} = \|\mathbf{u}\| + \|\mathbf{v}\|,$$

and the rest of the norm properties are verified quickly. The finite dimensionality is a form of Cramer's rule for systems of linear equations. Any $n + 1$ vectors of $\mathfrak{M}$ must have a linear relation with *integral* coefficients not all zero. (See Exercise 3 below.) Q.E.D.

As a major consequence of Theorem 1, using the minimal basis, we can replace the element of $\mathfrak{M}$, $\mathbf{u} = x_1\mathbf{u}_1 + \cdots + x_n\mathbf{u}_n$, by the vector of lattice $\mathfrak{L}_n$, $\mathbf{x} = (x_1, x_2, \cdots, x_n)$, whose components x_i are rational integers. The addition and subtraction of two vectors in $\mathfrak{M}$ by uniqueness, becomes an operation on the vectors in $\mathfrak{L}_n$. Thus, it follows that every lattice is equivalent[1] to a lattice $\mathfrak{L}_n$ of integral n-tuples.

To get right to the main result, Theorem 1, the central difficulty of the proof is that a lattice can be of dimension n, yet might not clearly have a basis of n elements. For example, the quadratic module

$$\mathfrak{M} = [35\sqrt{3}, 8 - 28\sqrt{3}, 6 - 21\sqrt{3}] \tag{2}$$

is generated by a set of 3 elements. It actually has a nonobvious, two-element basis, as we shall see as an exercise:

$$\mathfrak{M} = [10, 8 + 7\sqrt{3}]. \tag{3}$$

With this in mind, we proceed with the proof: let our lattice $\mathfrak{L}$ of dimension n have n linearly independent vectors $\mathbf{w}_1, \mathbf{w}_2, \cdots, \mathbf{w}_n$ (which still are not necessarily a basis). Any other vector $\mathbf{w}$ of the lattice $\mathfrak{L}$ need not be a linear *integral* combination of these $\mathbf{w}_i$, yet $\mathbf{w}$ satisfies a relation by virtue of linear dependence,

$$g\mathbf{w} = g_1\mathbf{w}_1 + g_2\mathbf{w}_2 + \cdots + g_n\mathbf{w}_n, \tag{4a}$$

[1] Yet in algebraic number theory the seemingly difficult norm (3) of §3 (above) can still be more useful than the simpler norm (1) for later purposes. (See Chapter VIII.)

where $g, g_1, g_2, \cdots, g_n$ are integers (which vary with $\mathbf{w}$). By changing signs, if necessary we can make $g > 0$. Such a representation is also *unique* to within constant factors, for, if

$$g'\mathbf{w} = g_1'\mathbf{w}_1 + \cdots + g_n'\mathbf{w}_n, \tag{4b}$$

we could obtain two representations of $gg'\mathbf{w}$ in integral coefficients from (4*a*) and (4*b*) by using a multiplier of g' and g on (4*a*) and (4*b*). By linear independence, the $\mathbf{w}_i$ must have the same coefficients each way, and $g'g_i = gg_i'$ or $g_i/g = g_i'/g'$.

We can show now that as w *ranges over* $\mathfrak{L}$, g *has only a finite number of (integral) values.* We are assuming, of course, that common divisors of $g, g_1, \cdots, g_n$ have been canceled out of (4*a*).

To determine this result, let us assume g takes on an infinite number of different values. We can first of all restrict values of g_i such that $0 \leq g_i < g$. Otherwise, we can use the division algorithm to write in each case $g_i = q_i g + g_i'$, where $0 \leq g_i' < g$, Then. if we write

$$\mathbf{w}^* = \mathbf{w} - \sum_{i=1}^{n} q_i \mathbf{w}_i,$$

we find for our new vector $\mathbf{w}^*$

$$g\mathbf{w}^* = \sum_{i=1}^{n} g_i'\mathbf{w}_i, \qquad 0 \leq g_i' < g. \tag{5}$$

The fractional coefficients for $\mathbf{w}^*$, with g the least common denominator,

$$(g_1'/g, g_2'/g, \cdots, g_n'/g)$$

are points of a unit n-dimensional cube. Let m be some integer. If there are more than m^n such points, by the Dirichlet boxing-in principle, these are distinct points $\mathbf{w}^*$, $\mathbf{w}\dagger$ of a cube, each of whose fractional coordinates is closer than $1/m$. Thus, writing the difference, we see

$$\mathbf{w}^* - \mathbf{w}\dagger = r_1\mathbf{w}_1 + r_2\mathbf{w}_2 + \cdots + r_n\mathbf{w}_n,\ |r_i| < 1/m, \tag{6a}$$

and by the triangular inequality, extended to n summands,

$$\|\mathbf{w}^* - \mathbf{w}\dagger\| = \left\|\sum_{i=1}^{n} r_i\mathbf{w}_i\right\| \leq \sum_{i=1}^{n} \|r_i\mathbf{w}_i\|,$$

$$\|\mathbf{w}^* - \mathbf{w}\dagger\| \leq \sum_{i=1}^{n} |r_i|\ \|w_i\| \leq \sum_{i=1}^{n} \|\mathbf{w}_i\|/m. \tag{6b}$$

But since m can be arbitrarily large, we shall have a vector $\mathbf{w}^* - \mathbf{w}\dagger = \mathbf{w}_0 \neq \mathbf{0}$, whose norm can be made arbitrarily small, causing a contradiction.

Q.E.D.

Hence we can write an arbitrary $\mathbf{w}$ in $\mathfrak{L}$ in the form

$$g\mathbf{w} = \sum_{i=1}^{n} g_i \mathbf{w}_i, \tag{7}$$

and, using the value $G = \max g$ with $G! = H$, we can write for any w in $\mathfrak{L}$

$$H\mathbf{w} = \sum_{i=1}^{n} G_i \mathbf{w}_i \qquad (H \text{ fixed}), \qquad G_i = g_i H/g, \tag{8}$$

although none of the fractions G_i/H need be reduced.

As a first step, in our lattice $\mathfrak{L}$ of dimension n, for any set of n linearly independent vectors $(\mathbf{w}_1, \cdots, \mathbf{w}_n)$, *an arbitrary vector of $\mathfrak{L}$, namely* $\mathbf{w}$, *is expressible by* (4) *with a fixed value of g, namely H.*

We next define a set of sublattices $\mathfrak{L}^{(1)}, \mathfrak{L}^{(2)}, \cdots, \mathfrak{L}^{(n)}$, as follows: $\mathfrak{L}^{(1)}$ is the set of vectors $\mathbf{v}$ of $\mathfrak{L}$ which are linearly dependent on $\mathbf{w}_1$; let $\mathfrak{L}^{(2)}$ be the set of vectors $\mathbf{v}$ of $\mathfrak{L}$ which are linearly dependent on $\mathbf{w}_1$ and $\mathbf{w}_2$. Clearly, $\mathfrak{L}^{(2)}$ contains $\mathfrak{L}^{(1)}$. More generally, let $\mathfrak{L}^{(k)}$ be the set of vectors $\mathbf{v}$ of $\mathfrak{L}$ which are linearly dependent on $\mathbf{w}_1, \cdots, \mathbf{w}_k$, or those $\mathbf{v}$ satisfying

$$H\mathbf{v} = x_1\mathbf{w}_1 + x_2\mathbf{w}_2 + \cdots + x_k\mathbf{w}_k, \qquad (x_i \text{ integers}). \tag{9a}$$

Of course, x_k (in fact *all* x_i) could vanish. Ultimately, $\mathfrak{L}^{(n)} = \mathfrak{L}$.

We now define $\mathbf{v}_k$ as some vector $\mathbf{v}$ of form (9*a*) in $\mathfrak{L}^{(k)}$ for which x_k takes on the minimum positive value (say) g_k for any k.

We assert that $[\mathbf{v}_1, \mathbf{v}_2, \cdots, \mathbf{v}_n]$ *constitutes a minimal basis of* $\mathfrak{L}$. This involves showing that any $\mathbf{v}$ belonging to $\mathfrak{L}$ can be represented uniquely as

$$\mathbf{v} = y_1\mathbf{v}_1 + y_2\mathbf{v}_2 + \cdots + y_n\mathbf{v}_n, \tag{10}$$

where the y_i are integers. We leave uniqueness as an exercise and prove the representability as follows:

LEMMA 1. For all $\mathbf{v}$ in $\mathfrak{L}^{(k)}$ represented by (9*a*), the values of x_k are all multiples of g_k, the minimum positive value of x_k.

Proof. To see this, we note that the values of x_k determined by (9*a*) are a module by the module property of $\mathfrak{L}^{(k)}$. The x_k are not all zero; e.g., with $\mathbf{v} = \mathbf{w}_k$, (9*a*) becomes

$$H\mathbf{w}_k = 0 + 0 + \cdots + H\mathbf{w}_k, \tag{9b}$$

and we see $x_k = H$. Thus by Lemma 1 in Chapter III, §3, the integers x_k are multiples of their minimum positive value g_k (which incidentally divides H). Q.E.D.

We prove the main result by induction. First, let $k = 1$. We then consider all $\mathbf{v}$ for which $H\mathbf{v}$ is of the form $x_1\mathbf{w}_1$. Here the x_i are the multiples of a minimal g_i, whence $H\mathbf{v}_1 = g_1\mathbf{w}_1$. Then for the variable $\mathbf{v}$ of $\mathfrak{L}^{(1)}$,

$\mathbf{v} = (x_1/g_1)\mathbf{v}_1$, in accordance with (10), (with $n = 1$). We next assume the representation to be valid if $\mathbf{v}$ lies in $\mathfrak{L}^{(k-1)}$ and we extend it to $\mathbf{v}$ in $\mathfrak{L}^{(k)}$ given by (9*a*). Here, too, the x_k are the multiples of some g_k for which $H\mathbf{v}_k = \cdots + g_k\mathbf{w}_k$. Thus, if $\mathbf{v}$ satisfies (9*a*), $\mathbf{v} - (x_k/g_k)\mathbf{v}_k$ lies in $\mathfrak{L}^{(k-1)}$, and step by step an expansion of $\mathbf{v}$ of type (10) is obtained (with $n = k$). Q.E.D.

As an illustration, we consider the module, $\mathfrak{M}$, of all pairs (x, y) of integers of the same parity, both odd or both even, i.e., $x \equiv y \pmod 2$. It is easily seen that $\mathbf{w}_1 = (2, 0)$ and $\mathbf{w}_2 = (0, 2)$ are linearly independent but no integral combination will yield $(1, 1)$, which is also in $\mathfrak{M}$. We note, however, $H = 2$, i.e., for any (x, y) in $\mathfrak{M}$,

$$2(x, y) = x(2, 0) + y(0, 2) = x\mathbf{w}_1 + y\mathbf{w}_2.$$

The minimal basis consists first of $(2, 0) = \mathbf{v}_1$, which is the shortest vector parallel to $\mathbf{w}_1$. To find $\mathbf{v}_2$, we ask for the smallest $|y| \neq 0$ for which $2(x, y)$ lies in $\mathfrak{M}$. This is given by $(x, y) = (1, 1) = \mathbf{v}_2$. Hence $\mathfrak{M} = [\mathbf{v}_1, \mathbf{v}_2]$. We could also take for $\mathbf{v}_2$ any $(x, 1)$ at all (for odd x). A systematic construction method (for $n = 2$) is deferred to §9 (below).

EXERCISE 3. Complete Theorem 2 (for $n = 2$) by showing that for any three vectors of $\mathfrak{M}$

$$\begin{cases} \mathbf{u} = a_1\mathbf{u}_1 + a_2\mathbf{u}_2 \\ \mathbf{v} = b_1\mathbf{u}_1 + b_2\mathbf{u}_2 \qquad a_i, b_i, c_i = \text{integers} \\ \mathbf{w} = c_1\mathbf{u}_1 + c_2\mathbf{u}_2 \end{cases}$$

there exists a linear relation $A\mathbf{u} + B\mathbf{v} + C\mathbf{w} = \mathbf{0}$ with *integral* coefficients not all zero. (Allow for some or all of the a_i, b_i, c_i to be zero.)

EXERCISE 4. Verify by inspection that every element shown in (2) is generated by some integral combination of elements in (3) *and conversely*. (*Hint*. $30 = 5(6 - 21\sqrt{3}) + 3(35\sqrt{3})$; $40 = 5(8 - 28\sqrt{3}) + 4(35\sqrt{3})$, etc.)

EXERCISE 5*a*. Show that the inequality (6*a*) can be proved without using fractional coefficients: use $G\mathbf{w}^\dagger = \sum_{i=1}^{n} G_i'\mathbf{w}_i$, $0 \le G_i' < G$; and write out $gG\mathbf{w}^* - gG\mathbf{w}^\dagger$.

EXERCISE 5*b*. Show that the representation (10) is unique or that if $\mathbf{v} = \mathbf{0}$ all $y_i = 0$. [*Hint*. For some representation of $\mathbf{v} = \mathbf{0}$ by (10) let k be the largest integer for which $y_k \neq 0$ and work back to the representation of $H\mathbf{v} = \mathbf{0}$ by (9*a*).]

EXERCISE 6*a*. Find the minimal basis of the module generated by the (redundant) integral vectors in each of the following two examples by inspection:

(*i*) $\mathbf{w}_1 = (1, 0, 0)$, $\mathbf{w}_2 = (0, 3, 0)$, $\mathbf{w}_3 = (0, 1, 1)$, $\mathbf{w}_4 = (0, 0, 3)$,

(*ii*) $\mathbf{w}_1 = (1, 0, 0, 0)$, $\mathbf{w}_2 = (0, 2, 1, 0)$, $\mathbf{w}_3 = (0, 0, 0, 4)$, $\mathbf{w}_4 = (0, 0, 1, 2)$, $\mathbf{w}_5 = (0, 1, 0, 1)$, $\mathbf{w}_6 = (0, 0, 2, 0)$, $\mathbf{w}_7 = (0, 4, 0, 0)$.

Here a minimal basis can be selected from among the generating elements listed, which is not always the case. Proceeding more generally with (*ii*), note

$\mathbf{w}_1, \mathbf{w}_2, \mathbf{w}_3, \mathbf{w}_4$ are four linearly independent vectors; hence we may write an arbitrary $\mathbf{w}$ in the module as

$$\mathbf{w} = x_1\mathbf{w}_1 + \cdots + x_7\mathbf{w}_7, \qquad (x_i \text{ integral}),$$
$$= y_1\mathbf{w}_1 + \cdots + y_4\mathbf{w}_4, \qquad (y_i \text{ not necessarily integral}).$$

On expanding we find relations between x_i and y_i reducing to

$$y_1 = x_1$$
$$y_2 = x_2 + x_5/2 + 2x_7$$
$$y_3 = x_3 + x_5/2 - x_6 + x_7$$
$$y_4 = x_4 - x_5/2 + 2x_6 - 2x_7$$

Then $\mathfrak{L}^{(1)}$ is defined by $y_2 = y_3 = y_4 = 0$, and a minimal $y_1 = 1$ occurs when $x_1 = 1$, other $x_i = 0$, hence $\mathbf{v}_1 = \mathbf{w}_1$; $\mathfrak{L}^{(2)}$ is defined by $y_3 = y_4 = 0$ (whence x_5 is even), and a minimal $y_2 = 1$ occurs when $x_2 = 1$, other $x_i = 0$, hence $\mathbf{v}_2 = \mathbf{w}_2$; $\mathfrak{L}^{(3)}$ is defined by $y_4 = 0$ (whence again x_5 is even), and a minimal $y_3 = 1$ occurs when $x_3 = 1$, other $x_i = 0$, hence $\mathbf{v}_3 = \mathbf{w}_3$; $\mathfrak{L}^{(4)}$ is defined with no restriction on y_i, and a minimal $y_4 = \frac{1}{2}(= -y_3 = -y_2)$ occurs when $x_5 = -1$, other $x_i = 0$, hence $\mathbf{v}_4 = (-\mathbf{w}_2 - \mathbf{w}_3 + \mathbf{w}_4)/2$. The $\mathbf{v}_1, \cdots, \mathbf{v}_4$ are another basis, expressed in terms of the independent set $\mathbf{w}_1, \cdots, \mathbf{w}_4$.

EXERCISE 6*b*. Show, by the above method, that a basis for the module generated by $\mathbf{w}_1 = (6, 8)$, $\mathbf{w}_2 = (8, 6)$, $\mathbf{w}_3 = (4, 4)$ is $\mathbf{w}_1$, $(\mathbf{w}_1 + \mathbf{w}_2)/7$.

EXERCISE 6*c*. Do likewise when $\mathbf{w}_1 = (a, b)$, $\mathbf{w}_2 = (b, a)$, $\mathbf{w}_3 = (c, c)$.

6. Other Interpretations of the Basis Construction

We preserve the usual notation with $\mathfrak{L}$, a given lattice of dimension n. It is convenient at times to use fractional notation; thus (8) of §5 (above) can be written as

$$\mathbf{w} = \sum_{i=1}^{n} (g_i/H)\mathbf{w}_i. \tag{1}$$

There are several equivalent forms of our basis construction.

THEOREM 3. A set of vectors $\mathbf{u}_1, \mathbf{u}_2, \cdots, \mathbf{u}_n$ constitutes a minimal basis of a given lattice $\mathfrak{L}$ of dimension n if and only if the relation

$$\mathbf{w} = \sum_{i=1}^{n} (x_i/G)\mathbf{u}_i \tag{2}$$

for $\mathbf{w}$ in $\mathfrak{L}$, implies each one of the fractions x_k/G is reducible to an integer.

Proof. Assume relation (2) to hold for some $\mathbf{w}$ with an x_k/G irreducible and $G > 1$. If $\mathbf{u}_1, \cdots, \mathbf{u}_n$ were a minimal basis, then we could write

$$\mathbf{w} = \sum_{i=1}^{n} y_i\mathbf{u}_i, \qquad (y_i \text{ integral}), \tag{3}$$

with some $y_k \neq x_k/G$. We have two representations, from (2) and (3), for $G\mathbf{w}$, contradicting the minimal basis property.

Conversely assume $\mathbf{u}_1, \cdots, \mathbf{u}_n$ is not a minimal basis by failure of linear independence. Then, if the set of $\mathbf{u}_i$ is linearly dependent for some integers a_i not all zero and

$$\mathbf{0} = \sum a_i \mathbf{u}_i, \tag{4}$$

then trivially the vector $\mathbf{w} = \mathbf{0}$ can be written

$$\mathbf{w} = \sum_{i=1}^{n} (a_i/G)\mathbf{u}_i \tag{5}$$

for any integer G at all (not zero)!

If the set $\mathbf{u}_i$ is linearly independent, but is not a minimal basis, by the method of proof of Theorem 1, for some fixed integer G, all $\mathbf{w}$ of $\mathfrak{L}$ can be written in the form (2). But if all x_i/G are integers, the set of $\mathbf{u}_i$ constitute a basis which must be minimal by uniqueness of coefficients (linear independence). Q.E.D.

We finally achieve a geometric construction if we return to the terminology of §5 where $\mathbf{w}_1, \cdots, \mathbf{w}_n$ represents a set of n independent vectors and $\mathfrak{L}^{(k)}$ represents the sublattice of vectors of $\mathfrak{L}$ linearly dependent on $\mathbf{w}_1, \cdots, \mathbf{w}_k$. An arbitrary vector of $\mathfrak{L}^{(k)}$ accordingly is

$$\mathbf{v} = \sum_{i=1}^{k} (x_i/H)\mathbf{w}_i \tag{6}$$

with rational (not necessarily integral) components x_i/H along $\mathbf{w}_i$.

THEOREM 4. A minimal basis can be constructed by induction as follows: a minimal basis of $\mathfrak{L}^{(1)}$ is a vector $\mathbf{v}_1$ linearly dependent on $\mathbf{w}_1$ and with minimum positive component along $\mathbf{w}_1$. Generally, a minimal basis of $\mathfrak{L}^{(k)}$ consists of the minimal basis of $\mathfrak{L}^{(k-1)}$ together with a vector $\mathbf{v}_k$ of minimum positive component along $\mathbf{w}_k$. Finally $\mathfrak{L}^{(n)} = \mathfrak{L}$.

Conversely, if the minimal basis of $\mathfrak{L}^{(k)}$ is formed by adjoining a vector $\mathbf{v}_k$ to the basis of $\mathfrak{L}^{(k-1)}$, then $\mathbf{v}_k$ must have minimal component along $\mathbf{w}_k$.

The proof of the converse is all that is really required. It follows from the fact that if the component of $\mathbf{v}^*$ along $\mathbf{w}_k$ is nonminimal, then the component is a multiple, s, of the minimum by the Lemma 1, §5. Then $\mathbf{v}^* = s\mathbf{v}_k$ belongs to $\mathfrak{L}^{(k-1)}$ and

$$\mathbf{v}^* = s\mathbf{v}_k + (a_1\mathbf{v}_1 + \cdots + a_{k-1}\mathbf{v}_{k-1}), \qquad |s| > 1. \tag{7}$$

Then $\mathbf{v}_k$ is expressible in terms of $\mathbf{v}_1, \cdots, \mathbf{v}_{k-1}, \mathbf{v}^*$, etc., in terms of a relationship of type (2) with an irreducible denominator s. Q.E.D.

GEOMETRICAL CONSTRUCTION

A minimal basis of $\mathfrak{L}^{(1)}$ consists of a vector $[\mathbf{v}_1]$ collinear with $\mathbf{w}_1$ and containing no point of $\mathfrak{L}$ except the two extremities of $\mathbf{v}_1$. If a minimal basis of $\mathfrak{L}^{(2)}$ is desired in the form $[\mathbf{v}_1, \mathbf{v}_2]$, then the parallelogram determined by $\mathbf{v}_1$ and $\mathbf{v}_2$ must contain no point of $\mathfrak{L}$ internally or on its boundary except the four vertices of the parallelogram. If a minimal basis of $\mathfrak{L}^{(3)}$ is desired in the form $[\mathbf{v}_1, \mathbf{v}_2, \mathbf{v}_3]$, then the parallelepiped determined by $\mathbf{v}_1$, $\mathbf{v}_2$, and $\mathbf{v}_3$ must contain no point of $\mathfrak{L}$ internally or on its boundary except the eight vertices of the parallelepiped, etc.

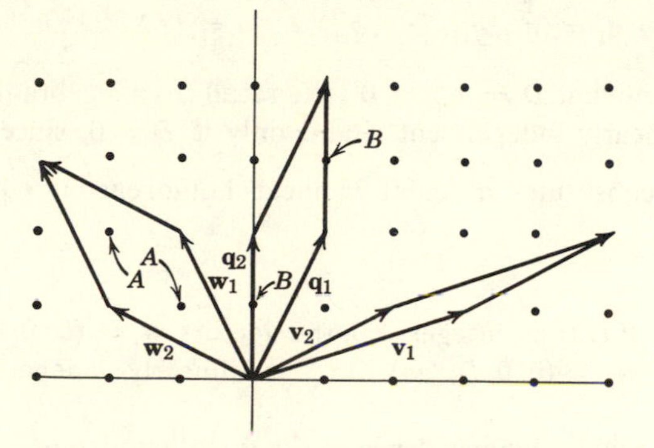

FIGURE 4.3. $\mathbf{v}_1$ and $\mathbf{v}_2$ are a basis of $\mathscr{L}_2$; $\mathbf{w}_1$ and $\mathbf{w}_2$ are not a basis of $\mathscr{L}_2$ (see A); $\mathbf{q}_1$ and $\mathbf{q}_2$ are not a basis of $\mathscr{L}_2$ (see B).

Note that the line is of minimum length, whereas the parallelogram and parallelepiped are of minimum "height," etc., for the minimal basis (see Figure 4.3).

7. Lattices of Rational Integers, Canonical Basis

We consider in more detail lattices $\mathfrak{M}$, which are sublattices of a lattice $\mathfrak{L}_n$, where $\mathfrak{L}_n$ is defined as the set of points $\mathbf{x}$ whose cartesian coordinates are arbitrary integers, with "distance" norm (see §4), Thus

$$\mathfrak{L}_n : \mathbf{x} = (x_1, x_2, \cdots, x_n), \qquad -\infty < x_i < +\infty.$$

Any submodule of $\mathfrak{L}_n$ "inherits" the lattice property because it inherits a norm as well as a finite dimensionality. (Clearly, any vectors of $\mathfrak{M}$ that are linearly dependent in $\mathfrak{L}_n$ are linearly dependent in $\mathfrak{M}$ by the same equation of type (4) in §5 and vice versa). We can regard $\mathfrak{L}_n$ as a lattice with special minimal basis

$$(1a)\qquad \begin{cases} \mathbf{u}_1 = (1, 0, 0, \cdots, 0), \\ \mathbf{u}_2 = (0, 1, 0, \cdots, 0), \\ \cdots\cdots\cdots\cdots\cdots\cdots \\ \mathbf{u}_n = (0, 0, 0, \cdots, 1), \end{cases}$$

so that

$$(1b)\qquad \mathbf{x} = \mathbf{u}_1 x_1 + \mathbf{u}_2 x_2 + \cdots + \mathbf{u}_n x_n.$$

We assume the sublattice $\mathfrak{M}$ is of dimension n, i.e., that it has n independent vectors $\mathbf{q}_1, \mathbf{q}_2, \cdots, \mathbf{q}_n$. We define the components

$$(2)\qquad \mathbf{q}_i = (q_{i1}, q_{i2}, \cdots, q_{in}), \qquad i = 1, 2, \cdots, n,$$

and the determinant $D = |q_{ij}| \neq 0$. We recall from algebra the result that the $\mathbf{q}_i$ are linearly independent if and only if $D \neq 0$, since the relation $\sum_{i=1}^{n} \mathbf{q}_i x_i = 0$ constitutes a set of n linear homogeneous equations in n unknowns x_i.

LEMMA 2. If D is an integer $\neq 0$, the vectors $\mathbf{w}_1 = (D, 0, 0, \cdots)$, $\mathbf{w}_2 = (0, D, 0, \cdots)$, $\mathbf{w}_3 = (0, 0, D, \cdots)$, $\cdots$, are linearly independent and lie in $\mathfrak{M}$.

Proof. The linear independence of the $\mathbf{w}_i$ follows from

$$(3)\qquad \sum_{i=1}^{n} a_i \mathbf{w}_i = (a_1 D, a_2 D, a_3 D, \cdots).$$

Trivially, $\sum_{i=1}^{n} a_i \mathbf{w}_i = \mathbf{0}$ implies all $a_i = 0$. The fact that the $\mathbf{w}_i$ lie in $\mathfrak{M}$ follows from the result that for each k we can solve the system

$$(4a)\qquad \sum_{i=1}^{n} b_i \mathbf{q}_i = \mathbf{w}_k$$

for integral b_i. To do this, we write the j-component of system (4a)

$$(4b)\qquad \sum_{i=1}^{n} b_i q_{ij} = D\delta_{kj},$$

where $\delta_{kj} = 1$ if $k = j$ and 0 if $k \neq j$. If Q_{ij} is the cofactor of q_{ij}, it is clear that $b_i = Q_{ik}$ satisfies the system (4a) or (4b) from elementary determinant theory, e.g.,

$$(4c)\qquad \sum_{j=1}^{n} Q_{ik} q_{ij} = D\delta_{kj}. \qquad \text{Q.E.D.}$$

LEMMA 3. A minimal basis for $\mathfrak{M}$ is provided by the so-called canonical basis:

$$(5)\qquad \begin{cases} \mathbf{v}_1 = (v_{11}, 0, 0, \cdots, 0) \\ \mathbf{v}_2 = (v_{21}, v_{22}, 0, \cdots, 0) \\ \mathbf{v}_3 = (v_{31}, v_{32}, v_{33}, \cdots, 0) \\ \cdots\cdots\cdots\cdots\cdots\cdots\cdots\cdots \\ \mathbf{v}_n = (v_{n1}, v_{n2}, v_{n3}, \cdots, v_{nn}). \end{cases}$$

where $\mathbf{v}_1$ is defined as the vector in the space spanned by $\mathbf{u}_1$ which lies in $\mathfrak{M}$ and has the smallest possible value of v_{11}; likewise $\mathbf{v}_2$ is defined as the vector in the space spanned by $\mathbf{u}_1$ and $\mathbf{u}_2$ which lies in $\mathfrak{M}$ and has the smallest positive value of v_{22}; $\mathbf{v}_3$ is defined as the vector in the space spanned by $\mathbf{u}_1$, $\mathbf{u}_2$, $\mathbf{u}_3$, which lies in $\mathfrak{M}$ and has smallest positive value of v_{33}, etc.

Furthermore, we can choose

$$(6)\qquad \begin{cases} 0 \le v_{21} < v_{11}, \\ 0 \le v_{31} < v_{11}, \qquad 0 \le v_{32} < v_{22}, \\ 0 \le v_{ji} < v_{ii}, \qquad (j > i). \end{cases}$$

These conditions determine a minimal basis uniquely.

Proof. The first part is obvious under the interpretation of the construction in terms of the $\mathbf{w}_i$ given in Lemma 2. (Making $v_{ii} > 0$ is a trivial matter, since any vector $\mathbf{v}$ can be replaced by $-\mathbf{v}$ for the purpose of defining $\mathbf{v}_i$.) The inequalities (6) can be proved by the divisor-quotient method. For instance, take $n = 3$. The inequality $0 \le v_{32} < v_{22}$ can be ensured by starting with a v_{32} not necessarily satisfying this inequality and writing $v_{32}/v_{22} = q + r/v_{22}$ where r, the remainder, satisfies $0 \le r < v_{22}$. Hence $\mathbf{v}_3 - q\mathbf{v}_2$ lies in $\mathfrak{M}$ has the same (third) component v_{33} as $\mathbf{v}_3$ but has r instead of v_{32}, satisfying the inequality. Likewise, some combination $\mathbf{v}_3 - q\mathbf{v}_1$ (or $\mathbf{v}_2 - q\mathbf{v}_1$) can take care of v_{31} (or v_{21}).

The uniqueness of this minimal basis is proved by the fact that all v_{ii} are minimal, hence unique (see the geometric construction of §6). Thus, if $\mathfrak{M}$ had two canonical bases with unequal $\mathbf{v}_i$ the difference $\mathbf{v}$ of the $\mathbf{v}_i$ would have the ith component and succeeding ones 0. Hence the vector $\mathbf{v}$ is eligible for consideration as a $\mathbf{v}_j$ for some $j < i$. But the jth component of $\mathbf{v}$ is the difference of two values of v_{ji} and is therefore less than the corresponding v_{ii}, contradicting the minimum property of v_{ii}. Thus all v_{ji} are also equal, making $\mathbf{v}_i$ the same. Q.E.D.

We introduce the integral matrix (a_{ij}), which is called *unimodular* if its determinant is ± 1 and *strictly unimodular* if the determinant is $+1$. We then consider a more general basis relation.

LEMMA 4. Two lattices of minimal bases $[\mathbf{v}_1, \mathbf{v}_2, \cdots, \mathbf{v}_n]$ and $[\mathbf{v}_1', \mathbf{v}_2', \cdots, \mathbf{v}_n']$ in $\mathfrak{L}_n$ are equivalent if and only if the bases are connected by a *unimodular* relationship:

$$\mathbf{v}_i = \sum_{j=1}^{n} a_{ij}\mathbf{v}_j', \qquad 1 \leq i \leq n, \tag{7}$$

$$\mathbf{v}_i' = \sum_{j=1}^{n} b_{ij}\mathbf{v}_j, \qquad 1 \leq i \leq n, \tag{8}$$

with integral coefficients.

Proof. Assume the lattices to be equivalent. The basis elements $\mathbf{v}_i$ are individually in $[\mathbf{v}_1', \mathbf{v}_2', \cdots, \mathbf{v}_n']$, and (7) are valid, likewise (8). Hence (a_{ij}), (b_{ij}), the matrices of two transformations, combine to produce the identical transformation of $[\mathbf{v}_1, \cdots, \mathbf{v}_n]$ onto itself. Thus

$$|a_{ij}| \cdot |b_{ij}| = +1,$$

and $|a_{ij}| = |b_{ij}| = \pm 1$, since all coefficients and determinants are integers.

Conversely, the system (7) can be reversed. If q_{ij} is the cofactor of a_{ij}, the substitution easily yields

$$\sum_{i=1}^{n} \mathbf{v}_i q_{ik} = \mathbf{v}_k' \, |a_{ij}| = \pm \mathbf{v}_k'$$

to form (8) for the unimodular case. Thus each set of n vectors spans the others from the unimodular property. Q.E.D.

LEMMA 5. Two different minimal bases of $\mathfrak{M}$, a sublattice of $\mathfrak{L}_n$, have the same determinant except possibly for sign, i.e., if $\mathbf{v}_i = [v_{i1}, \cdots, v_{in}]$, $\mathbf{v}_i' = [v_{i1}', \cdots, v_{in}']$ for $1 \leq i \leq n$, then $|v_{ij}| = \pm\, |v_{ij}'|$.

Proof. By the product theorem for determinants, if

$$v_{ik} = \sum_{i=1}^{n} a_{ij} v_{ik}',$$

then

$$|v_{ik}| = |a_{ij}| \, |v_{ik}'|. \qquad \text{Q.E.D.}$$

We shall call this common absolute value D. Two canonical bases with the same D are not necessarily equivalent, since all we have done is to specify the product $v_{11}v_{22} \cdots v_{nn}$. (In fact, in so doing, we completely ignore the nondiagonal integers v_{ij}.)

8. Sublattices and Index Concept

The preceding theory of submodules (or sublattices) of $\mathfrak{L}_n$ can be carried over to submodules (or sublattices) of any given lattice $\mathfrak{L}$. All we need do is write the minimal basis of the lattice as n vectors

$$\mathfrak{L} = [\mathbf{u}_1, \mathbf{u}_2, \cdots, \mathbf{u}_n]$$

and identify the element $\mathbf{x} = x_1\mathbf{u}_1 + \cdots + x_n\mathbf{u}_n$ with $\mathbf{x} = [x_1, x_2, \cdots, x_n]$ the element of $\mathfrak{L}_n$. Thus operations on elements of $\mathbf{x}$ in $\mathfrak{M}$ correspond in a one-to-one fashion with operations on elements of $\mathbf{x}$ in $\mathfrak{L}_n$. (We continue our restriction to sublattices of dimension n.)

From this it follows that if the modules $\mathfrak{L}$, $\mathfrak{M}$, $\mathfrak{N}$ have the property that $\mathfrak{L}$ includes $\mathfrak{M}$ and $\mathfrak{M}$ includes $\mathfrak{N}$ then a set of minimal bases can be found

$$(1) \qquad \begin{cases} \mathfrak{L} = [\mathbf{u}_1, \cdots, \mathbf{u}_n], \\ \mathfrak{M} = [\mathbf{v}_1, \cdots, \mathbf{v}_n], \\ \mathfrak{N} = [\mathbf{w}_1, \cdots, \mathbf{w}_n], \end{cases}$$

such that

$$(2) \qquad \begin{cases} \mathbf{v}_i = \sum_{j=1}^{n} a_{ij}\mathbf{u}_j, \\ \mathbf{w}_i = \sum_{j=1}^{n} b_{ij}\mathbf{v}_j, \end{cases}$$

or, if (c_{ij}) represents the product of the matrices (b_{ij}) and (a_{ij}), then

$$(3) \qquad \mathbf{w}_i = \sum_{j=1}^{n} c_{ij}\mathbf{u}_j.$$

Thus we have the following lemma:

LEMMA 6. **If $\mathfrak{L}$ includes $\mathfrak{M}$ and $\mathfrak{M}$ includes $\mathfrak{N}$, then if the minimal basis of $\mathfrak{L}$ in $\mathfrak{M}$ has determinant A and the basis of $\mathfrak{M}$ in $\mathfrak{N}$ has determinant B, the basis of $\mathfrak{L}$ in $\mathfrak{N}$ has determinant AB.**

We next consider any two modules in $\mathfrak{L}_n$ namely $\mathfrak{M}$ and $\mathfrak{L}$, where $\mathfrak{M}$ is contained in $\mathfrak{L}$. Then, if vectors $\mathbf{x}$ and $\mathbf{y}$ belong to $\mathfrak{L}$, we say

$$(4) \qquad \mathbf{x} \equiv \mathbf{y} \pmod{\mathfrak{M}}$$

if $\mathbf{x} - \mathbf{y}$ belongs to $\mathfrak{M}$. By the module property, this definition permits addition and subtraction, hence multiplication by a rational integer (but not necessarily by an algebraic integer). This definition is also consistent with the ordinary congruence modulo $\mathfrak{M}$ if $\mathfrak{L}$ is the (one-dimensional) module of rational integers. Congruent vectors constitute an equivalence class in the same sense as congruent integers.

The number of different classes will be seen to be finite. It is called the *index* and is written $j = [\mathfrak{L}/\mathfrak{M}]$. Thus the classes form a finite module of j elements which is denoted by the quotient symbol $\mathfrak{L}/\mathfrak{M}$.

LEMMA 7. **If $\mathfrak{L}$ contains $\mathfrak{M}$, the index j of $\mathfrak{L}/\mathfrak{M}$ is the absolute value of determinant of the basis expression for $\mathfrak{M}$ in terms of the basis of $\mathfrak{L}$.**

Proof. It suffices to take $\mathfrak{L} = \mathfrak{L}_n$ by the opening remarks of §7 (above). We suppose $\mathfrak{M}$ to have a canonical basis as described in Lemma 2, §7. We show every vector of $\mathfrak{L}_n$ is congruent modulo $\mathfrak{M}$ to one of the following $|D| = v_{11}v_{22} \cdots v_{nn}$ *incongruent* vectors:

$$\mathbf{t} = (t_1, t_2, \cdots, t_n), \quad 0 \leq t_1 < v_{ii}. \tag{5}$$

We begin by taking an arbitrary vector in $\mathfrak{L}_n$, namely, $\mathbf{w} = (w_1, \cdots, w_n)$. We then subtract a sufficiently large multiple of $\mathbf{v}_n$ (see Lemma 3 of §7), according to the familiar division algorithm: letting $w_n = v_{nn}q + t_n$, we see that $\mathbf{w} - q\mathbf{v}_n$ has the nth component t_n satisfying (5). Then working with just $(n - 1)$ first components of $\mathbf{w} - q\mathbf{v}_n$, we produce the $(n - 1)$th component t_{n-1} in an expression of the form $\mathbf{w} = q\mathbf{v}_n - q'\mathbf{v}_{n-1}$. Thus, by induction, $\mathbf{w}$ is congruent modulo $\mathfrak{M}$ to a vector of type (5). No two vectors of type (5) are congruent modulo $\mathfrak{M}$ (see Exercise 8 below). Q.E.D.

Lemma 8 is a corollary.

LEMMA 8. If $j = 1$ in Lemma 7, $\mathfrak{M} = \mathfrak{L}$.

The index, by definition, is independent of basis. It constitutes an *invariant* concept to replace the determinant (which is seemingly dependent on the particular basis).

EXERCISE 7. Prove that the index j has the property that, for any element $\mathbf{v}$ of $\mathfrak{L}$, $j\mathbf{v}$ belongs to $\mathfrak{M}$. Show this two ways, first by using the determinant property and second by using the subgroup definition (i.e., that $\mathfrak{M}$ is an additive subgroup of $\mathfrak{L}$).

Thus if $\mathfrak{M}$ is a module in $\mathfrak{D}_n$, by taking $\mathbf{v} = 1$, show that any modulo contains the integer j, equal to its index.

EXERCISE 8. Show that no two vectors of type (5) are congruent modulo $\mathfrak{M}$ (by showing the difference between any two such vectors is either $\mathbf{0}$ or inexpressible by the canonical basis of Lemma 3, §7).

9. Application to Modules of Quadratic Integers

A module $\mathfrak{M}$ of quadratic integers in $\mathfrak{D}$, an integral domain, now presents three alternatives:

(a) The module may consist wholly of zero.

(b) The module may be of dimension one.

In these cases $\mathfrak{M}$ consists of multiples of a quadratic or rational integer (possibly 0) and is relatively uninteresting.

(c) In the nontrivial case the canonical basis, by Lemma 3, is

$$\mathfrak{M} = [a, b + c\omega], \qquad a > b \geq 0, \qquad c > 0. \tag{1}$$

Here a is the smallest absolute value of a rational integer in $\mathfrak{M}$ (not 0), and $b + c\omega$ is a term of $\mathfrak{M}$ with the smallest positive coefficient of ω.

In practice, a module might be written with an excessive number of elements, and our object is to choose clever combinations of these elements to form a *minimal basis* of the original module, as illustrated in (2) and (3) of §5 (above). Thus we may be confronted with a module having the (nonminimal) basis shown:

$$\mathfrak{M} = [\xi_1, \xi_2, \cdots, \xi_t],$$

We can augment the set by any

$$\xi_{t+1} = \sum_{i=1}^{t} a_i \xi_i \tag{2}$$

for integral a_i, and we can remove any ξ_j which happens to be noticed as a linear combination of others. These operations are sufficient for the construction of a basis. Indeed, the desired minimal basis vectors, belonging to the module, have the form (2); moreover, once *they* are put in, the original vectors may be dropped, since they are necessarily linear combinations of the minimal basis vectors!

We want to obtain the basis in canonical form, so we set

$$\mathfrak{M}_t = [a_1 + b_1\omega, a_2 + b_2\omega, \cdots, a_t + b_t\omega] \tag{3}$$

and proceed by induction. Thus

$$\mathfrak{M}_2 = [a_1 + b_1\omega, a_2 + b_2\omega]. \tag{4a}$$

The most general element of $\mathfrak{M}_2$ is

$$x_1(a_1 + b_1\omega) + x_2(a_2 + b_2\omega) = (x_1a_1 + x_2a_2) + (x_1b_1 + x_2b_2)\omega;$$

hence the most general rational integer of $\mathfrak{M}_2$ is a multiple of $x_1'a_1 + x_2'a_2 = a$, where x_1' and x_2' are the smallest integers not both zero which permit $x_1'b_1 + x_2'b_2$ to equal zero, or $x_1'/x_2' = -b_2/b_1$ reduced to lowest terms. Futhermore, the term with the smallest nonzero coefficient of ω has form $b + c\omega$ where $c = \gcd(b_1, b_2)$, since c divides all $x_1b_1 + x_2b_2$ and can be expressed as such a combination. Finally

$$\mathfrak{M}_2 = [a, b + c\omega]. \tag{4b}$$

Now if we consider $\mathfrak{M}_3 = [a_1 + b_1\omega, a_2 + b_2\omega, a_3 + b_3\omega]$, then $\mathfrak{M}_3 = [a, b + c\omega, a_3 + b_3\omega]$, and we can reduce $[b + c\omega, a_3 + b_3\omega]$ by the same procedure to the form $[a', b' + c'\omega]$; thus

$$\mathfrak{M}_3 = [a, a', b' + c'\omega]. \tag{4c}$$

Then we note $[a, a'] = [\gcd(a, a')] = [a'']$. This, in fact, is the Euclidean algorithm. Hence, $\mathfrak{M}_3 = [a'', b' + c'\omega]$. Likewise $\mathfrak{M}_t$ can be reduced to canonical form step by step.

EXERCISE 9. Reduce to canonical form, the following bases:

$$\mathfrak{M}_2 = [35\sqrt{3}, 8 - 28\sqrt{3}],$$
$$\mathfrak{M}_3 = [35\sqrt{3}, 8 - 28\sqrt{3}, 6 - 21\sqrt{3}],$$
$$\mathfrak{M}_4 = [35\sqrt{3}, 8 - 28\sqrt{3}, 6 - 21\sqrt{3}, 85].$$

EXERCISE 10. Show that for every linear transformation

$$\begin{cases} x' = ax + by \\ y' = cx + dy \end{cases} \qquad ad - bc = n(\neq 0)$$

a unique set of integers A, B, C, P, Q, R, S exists such that the transformation can be expressed as a result of the transformation

$$\begin{cases} x' = Ax'' + By'' \\ y' = \qquad\quad Cy'' \end{cases} \qquad \begin{matrix} AC = |n| \\ 0 \leq B < A \end{matrix}$$

together with the transformation

$$\begin{cases} x'' = Px + Qy \\ y'' = Rx + Sy \end{cases} \qquad PS - QR = \pm 1.$$

Hint. Make use of a canonical basis $(A, 0)$, (B, C) for the module in $\mathfrak{L}_2$ generated by the vectors $\mathbf{w}_1 = (a, c)$, $\mathbf{w}_2 = (b, d)$.

10. Discriminant of a Quadratic Field

Let a module $\mathfrak{M}$ in $\mathfrak{O}_1$ for $R(\sqrt{D})$ be expressed with a two-element basis

$$\mathfrak{M} = [\xi_1, \xi_2], \tag{1}$$

where ξ_1 and ξ_2 belong to $\mathfrak{O}_1$. We define

$$\Delta = \Delta(\mathfrak{M}) = \begin{vmatrix} \xi_1 & \xi_2 \\ \xi_1' & \xi_2' \end{vmatrix} = \xi_1\xi_2' - \xi_1'\xi_2 \tag{2}$$

as the *different* of the module. We next see that the different $\Delta \neq 0$ if and only if ξ_1 and ξ_2 are linearly independent (equivalently if and only if the dimension of $\mathfrak{M}$ is two). For we see that $\Delta = \xi_2\xi_2'[(\xi_1/\xi_2) - (\xi_1/\xi_2)']$; hence $\Delta = 0$ if and only if (ξ_1/ξ_2) is equal to its conjugate and consequently to a rational fraction (say) r/s. This violates linear independence (with integral coefficients) for ξ_1 and ξ_2 (e.g., $r\xi_2 - s\xi_1 = 0$).

The module of all integers of an integral domain $\mathfrak{O}_n$ (in the terminology of Chapter III, §9) is

$$\mathfrak{O}_n = [1, n\omega_0]. \tag{3a}$$

It has a different

$$\text{(3b)}\qquad \Delta = \Delta(\mathfrak{O}_n) = n(\omega_0 - \omega_0') = \begin{cases} n\sqrt{D_0}, & \text{if } D_0 \equiv 1 \pmod 4, \\ 2n\sqrt{D_0}, & \text{if } D_0 \not\equiv 1 \pmod 4. \end{cases}$$

A more important quantity is

$$\text{(4)}\qquad d(\mathfrak{M}) = d = \Delta^2,$$

the *discriminant* of the module $\mathfrak{M}$, which we see is always an integer:

$$\text{(5)}\qquad d(\mathfrak{O}_n) = \begin{cases} n^2 D_0 & \text{if } D_0 \equiv 1 \pmod 4, \\ 4n^2 D_0 & \text{if } D_0 \not\equiv 1 \pmod 4. \end{cases}$$

In general, if $\mathfrak{M}$ is a module of integers in $\mathfrak{O}_n$,

$$\text{(6)}\qquad \Delta(\mathfrak{M}) = j\Delta(\mathfrak{O}_n), \quad \text{and} \quad d(\mathfrak{M}) = j^2 d(\mathfrak{O}_n),$$

where $j = [\mathfrak{O}_n/\mathfrak{M}]$, the index of $\mathfrak{M}$ in $\mathfrak{O}_n$. This follows from the determinant multiplication property. Very often, for convenience, we refer to the discriminant d of $\mathfrak{O}_1$ as the "discriminant of the field $R(\sqrt{D_0})$" or as the "fundamental discriminant" (of this field).

Thus, starting with $d(\mathfrak{O}_1) = d$, we note either

$$D_0 = d \equiv 1 \pmod 4 \quad \text{or} \quad D_0 = d/4 \not\equiv 1 \pmod 4,$$

and $D_0 (\neq 1)$ is square-free. We can now unify past notation by writing

$$\text{(7)}\qquad \mathfrak{O}_n = [1, \omega_n], \qquad n \geq 1,$$

where

$$\text{(8)}\qquad \omega_n = n\left(\frac{d - \sqrt{d}}{2}\right),$$

and

$$\Delta(\mathfrak{O}_n) = n\sqrt{d} \quad \text{whereas} \quad d(\mathfrak{O}_n) = n^2 d.$$

EXERCISE 11. By a direct comparison with Chapter III, §9, verify (7).

EXERCISE 12. Verify (6) directly by calculating $\Delta(\mathfrak{M})$ for $\mathfrak{M} = [a, b + c\omega_n]$.

**11. Fields of Higher Degree

The problem of constructing a minimal basis of the algebraic integers in fields of degree n is one that in general is subject to the "existence" type of argument and amply illustrates the power of the abstract methods of this chapter. Suppose we start with (say) the field $R(\theta)$ where θ is of degree n; then, as we noted in Chapter III, §10, the quantities

$$\text{(1)}\qquad 1, \theta, \theta^2, \cdots, \theta^{n-1}$$

are linearly independent, using integral coefficients, and the general integer of $R(\theta)$ must be of the type

$$\xi = r_0 + r_1\theta + \cdots + r_{n-1}\theta^{n-1} \tag{2}$$

where r_i are rational numbers of denominator $\leq Q$ by §5 (above). Then a perfectly good norm is

$$\|\xi\| = [r_0^2 + \cdots r_{n-1}]^{1/2}, \tag{3}$$

and if $\xi \neq 0$ some $r_k \neq 0$ and, necessarily,

$$\|\xi\| \geq |r_k| \geq 1/Q(> 0). \tag{4}$$

satisfying the discreteness condition. From this, the *existence* of the minimal basis follows.

We observe that this proof is not based on an explicit construction; in fact, there is almost no theory to help us! In a general way we know, from Theorem 3 of §6 (above), that a set of integers $\omega_0, \omega_1, \cdots, \omega_{n-1}$ of $R(\theta)$ will *fail* to be a minimal basis if and only if another integer ω of $R(\theta)$ exists for which

$$G\omega = x_0\omega_0 + \cdots + x_{n-1}\omega_{n-1}, \tag{5}$$

where x_i, $G > 1$, are integers and G cannot be "divided out." This statement is of great practical value in establishing the basis of higher fields, where illustrations are deferred to the bibliography. It might suffice to note in the quadratic case the fact that ω is an integer in the familiar example $2\omega = 1 + \sqrt{-3}$. This indicates again that $[1, \sqrt{-3}]$ does not form a minimal basis of the algebraic integers of $R(\sqrt{-3})$.

**chapter V

Further applications of basis theorems

STRUCTURE OF FINITE ABELIAN GROUPS

1. Lattice of Group Relations

In earlier work we restricted ourselves, as a matter of convenience, to abelian groups which are representable as the product of cyclic groups. This representability was easily achieved in Chapter I, §5, for the reduced-residue class group.

According to the famous theorem of Kronecker (1877), cited in Chapter I, §7, *every finite abelian group is decomposable into the product of cyclic groups*. We shall give proof by using a type of lattice basis.

Let the elements of a commutative group of order h be written as

$$\mathbf{a}_1(=\mathbf{e}), \mathbf{a}_2, \mathbf{a}_3, \cdots, \mathbf{a}_h.$$

Then we consider all h-tuples of integers (positive, negative, or zero) in $\mathfrak{L}_h$

(1) $$(x_1, x_2, \cdots, x_h),$$

and we fix our attention on $\mathfrak{L}_*$, the set of all h-tuples for which a group relation is determined (using the multiplicative operation)

(2) $$\mathbf{a}_1^{x_1}\mathbf{a}_2^{x_2}\cdots\mathbf{a}_h^{x_h}=\mathbf{e}.$$

First of all, we note that $\mathfrak{L}_*$ is a lattice. Let us suppose that

$$\mathbf{a}_1^{y_1}\mathbf{a}_2^{y_2}\cdots\mathbf{a}_h^{y_h} = \mathbf{e}.$$

Then

$$\mathbf{a}_1^{x_1\pm y_1}\mathbf{a}_2^{x_2\pm y_2}\cdots\mathbf{a}_h^{x_h\pm y_h} = \mathbf{e}.$$

Thus the sum or difference of two h-tuples in $\mathfrak{L}_*$ is also in $\mathfrak{L}_*$.

Now $\mathfrak{L}_*$ will obviously contain certain simple elements. For example, if g_i is the order of $\mathbf{a}_i$, $\mathbf{a}_i^{g_i} = \mathbf{e}$, and $\mathfrak{L}_*$ contains the h vectors

$$(3)\qquad \begin{cases}(0, \cdots, 0, g_i, 0, \cdots, 0),\\ (g_i \text{ at the } i\text{th position, } 1 \le i \le h).\end{cases}$$

Thus, $g_1 = 1$ (for $\mathbf{a}_1 = \mathbf{e}$), yielding $(1, 0, \cdots, 0)$. Since we have h linearly independent vectors (3), we can conclude $\mathfrak{L}_*$ has a finite index in $\mathfrak{L}_h$. (It will later turn out that this index is h.)

Starting from the h vectors (3), we have a set that is very easy to augment to a finite set of vectors completely determining $\mathfrak{L}_*$, at least in theory, for the components of any vector of $\mathfrak{L}_*$ are each determined modulo g_i. We can then examine every single $(x_1, x_2, \cdots, x_h)$ of $\mathfrak{L}_h$ where $0 \le x_i < g_i$ by the group properties to see if (2) holds.

EXERCISE 1. For the cyclic group $\mathbf{Z}(3)$ the elements are $\mathbf{a}_1 = \mathbf{e} = \mathbf{a}_2^3$, $\mathbf{a}_2^2 = \mathbf{a}_3$, $\mathbf{a}_3^2 = \mathbf{a}_2$. Then show that the elements of $\mathfrak{L}_*$ are generated by the set of vectors

$$(1, 0, 0),\qquad (0, 3, 0),\qquad (0, 0, 3),\qquad (0, 1, 1).$$

EXERCISE 2. For the "four-group" $\mathbf{Z}(2) \times \mathbf{Z}(2)$ the elements are $(\mathbf{a}_1 =)\mathbf{e}$, $\mathbf{a}_2$, $\mathbf{a}_3$, $(\mathbf{a}_4 =)\mathbf{a}_2\mathbf{a}_3$, where $\mathbf{a}_2^2 = \mathbf{a}_3^2 = \mathbf{e}$. Then show that the corresponding relations are, similarly, given by

$$(1, 0, 0, 0),\qquad (0, 2, 0, 0),\qquad (0, 0, 2, 0),\qquad (0, 0, 0, 2),\qquad (0, 1, 1, 1).$$

EXERCISE 3. Do likewise for the group $\mathbf{Z}(4)$.

EXERCISE 4. Do likewise for the group $\mathbf{Z}(5)$.

2. Need for Diagonal Basis

Thus we can then find a basis of $\mathfrak{L}_*$ in triangular (canonical) form:

$$(1)\qquad \begin{cases} v_1 = (a_{11}, 0, 0, \cdots, 0),\\ v_2 = (a_{21}, a_{22}, 0, \cdots, 0),\\ v_3 = (a_{31}, a_{32}, a_{33}, \cdots, 0), \quad 0 \leqq a_{ij} \leqq a_{jj}\\ \cdots\cdots\cdots\cdots\cdots\cdots\\ v_h = (a_{h1}, a_{h2}, a_{h3}, \cdots, a_{hh}),\\ a_{11}a_{22}\cdots a_{hh} = H.\end{cases}$$

This means that all the interrelations between elements come as combinations (powers, products, and quotients) of

$$(2)\qquad \begin{cases} \mathbf{a}_1^{a_{11}} & = \mathbf{e}, \\ \mathbf{a}_1^{a_{21}}\mathbf{a}_2^{a_{22}} & = \mathbf{e}, \\ \cdots\cdots\cdots & \\ \mathbf{a}_1^{a_{h1}}\mathbf{a}_2^{a_{h2}}\mathbf{a}_3^{a_{h3}} \cdots \mathbf{a}_h^{a_{hh}} & = \mathbf{e}. \end{cases}$$

If we had achieved the "diagonalized" stage in which $a_{ij} = 0$, for $j \neq i$, then we could say all interrelations come from cyclic groups of order a_{jj} generated by $\mathbf{a}_j^{a_{jj}} = \mathbf{e}$. This is exactly the same as saying that the group decomposes into a product of cyclic groups (even though we find many $a_{jj} = 1$). But for this purpose, the canonical basis is not good enough.

We must use a type of "double" reduction. We must reduce the basis of both $\mathfrak{L}_*$ and $\mathfrak{L}_h$ simultaneously in order to obtain a *diagonal* array in relations (1) and (2) instead of the triangular array.

3. Elementary Divisor Theory

We consider, in a general context, a lattice $\mathfrak{M}$ which is a sublattice of $\mathfrak{L}_n$ of finite index. Then a basis of $\mathfrak{M}$, namely, $[\mathbf{v}_1, \cdots, \mathbf{v}_n]$ can be written in terms of the basis $[\mathbf{u}_1, \cdots, \mathbf{u}_n]$ of $\mathfrak{L}_n$ by making use of a matrix $(a_{ij}) = \mathbf{A}$. Thus

$$(1)\qquad \mathfrak{M} \begin{cases} \mathbf{v}_1 = a_{11}\mathbf{u}_1 + a_{12}\mathbf{u}_2 + \cdots + a_{1n}\mathbf{u}_n, \\ \mathbf{v}_2 = a_{21}\mathbf{u}_1 + a_{22}\mathbf{u}_2 + \cdots + a_{2n}\mathbf{u}_n, \\ \cdots\cdots\cdots \\ \mathbf{v}_s = \cdots + a_{sk}\mathbf{u}_k + \cdots + a_{sm}\mathbf{u}_m + \cdots + a_{sn}\mathbf{u}_n, \\ \cdots\cdots\cdots \\ \mathbf{v}_t = \cdots + a_{tk}\mathbf{u}_k + \cdots + a_{tm}\mathbf{u}_m + \cdots + a_{tn}\mathbf{u}_n, \\ \cdots\cdots\cdots \\ \mathbf{v}_n = a_{n1}\mathbf{u}_1 + a_{n2}\mathbf{u}_2 + \cdots + a_{nn}\mathbf{u}_n. \end{cases}$$

where

$$(2)\qquad \mathfrak{L}_n \begin{cases} \mathbf{u}_1 = (1, 0, \cdots, 0), \\ \mathbf{u}_2 = (0, 1, \cdots, 0), \\ \cdots\cdots \\ \mathbf{u}_n = (0, 0, \cdots, 1). \end{cases}$$

Now the matrix $\mathbf{A} = (a_{ij})$ is a representative of the relation between the basis $\mathbf{u}_i$ of $\mathfrak{L}_n$ and the basis $\mathbf{v}_i$ of $\mathfrak{M}$. Yet (a_{ij}) still depends on the two bases, which can be chosen with some degree of arbitrariness. For instance, we could replace the set of vectors $(\mathbf{u}_1, \cdots, \mathbf{u}_n)$ by a unimodular transform of the set; likewise for $(\mathbf{v}_1, \cdots, \mathbf{v}_n)$.

Let us consider special unimodular transformations of the bases of $\mathfrak{L}_n$ or $\mathfrak{M}$. Here q is an integer and s, t, k, m denote special indices ($s \neq t$, $k \neq m$):

$$
(3) \quad \begin{cases}
\text{(a)} \quad \mathbf{v}_s \pm q\mathbf{v}_t \to \mathbf{v}_s, & (\mathbf{v}_i \text{ unchanged for } i \neq s), \\
\text{(b)} \quad \mathbf{u}_k \pm q\mathbf{u}_m \to \mathbf{u}_k, & (\mathbf{u}_i \text{ unchanged for } i \neq k), \\
\text{(c)} \quad \mathbf{v}_s \to -\mathbf{v}_s, & (\mathbf{v}_i \text{ unchanged for } i \neq s), \\
\text{(d)} \quad \mathbf{u}_k \to -\mathbf{u}_k, & (\mathbf{u}_i \text{ unchanged for } i \neq k), \\
\text{(e)} \quad \mathbf{v}_s \leftrightarrow \mathbf{v}_t, & (\mathbf{v}_i \text{ unchanged for } i \neq s, t), \\
\text{(f)} \quad \mathbf{u}_k \leftrightarrow u_m, & (\mathbf{u}_i \text{ unchanged for } i \neq k, m).
\end{cases}
$$

Here the symbol $\to$ should be read "replaces" and $\mathbf{u}_s \leftrightarrow \mathbf{u}_i$ denotes an interchange. The transformations are generally reminiscent of admissible operations in evaluating a determinant, but here the important thing is that the operations are all *reversible* in integral coefficients. For example, the inverse of (a) is the operation $\mathbf{v}_s \mp q\mathbf{v}_t \to \mathbf{v}_s$ (opposite sign) and likewise for (b). All the other operations are their own inverses. Thus the elementary operations described are *unimodular* and in any combination constitute a unimodular transformation on $\mathbf{u}_i$ and $\mathbf{v}_i$. The value of the determinant $|a_{ij}|$ is preserved except possibly for sign.

Let us observe the effect of these six rules on the matrix A:

$$
(4) \quad \begin{cases}
\text{(a)} \ a_{sk} \pm qa_{tk} \to a_{sk}, & a_{ik} \text{ unchanged for } i \neq s; & k = 1, 2, \cdots n; \\
\text{(b)} \ a_{sm} \pm qa_{sk} \to a_{sm}, & a_{sj} \text{ unchanged for } j \neq m; & s = 1, 2, \cdots n; \\
\text{(c)} \ a_{sk} \to -a_{sk}, & a_{ik} \text{ unchanged for } i \neq s; & k = 1, 2, \cdots n; \\
\text{(d)} \ a_{sk} \to -a_{sk}, & a_{sj} \text{ unchanged for } j \neq k; & s = 1, 2, \cdots n; \\
\text{(e)} \ a_{sk} \leftrightarrow a_{tk}, & a_{ik} \text{ unchanged for } i \neq s, t; & k = 1, 2, \cdots n; \\
\text{(f)} \ a_{sk} \leftrightarrow a_{sm}, & a_{sj} \text{ unchanged for } j \neq k, m; & s = 1, 2, \cdots n.
\end{cases}
$$

We can easily recognize (a): the addition (subtraction) of q times the t-row to (from) the s-row ($t \neq s$); (b): the addition (subtraction) of q times the k-column to (from) the m-column ($k \neq m$); (c) and (d): the change in sign of a row or column; (e) and (f): the interchange of two rows and columns.

THEOREM ON ELEMENTARY DIVISORS

The exercise of rules (a) *through* (f) *can reduce the matrix A to purely diagonal form.*

Proof. We consider the smallest positive matrix element, other than zero, which can be created by the exercise of these rules. Call it z_1; then by the exercise of rules (e) and (f) we can bring z_1 into the upper left-hand corner to replace a_{11}. Now, all elements of the first row (or column) must be divisible by z_1. To see this, we shall show $z_1 \mid a_{12}$, for, by the exercise of rule (b) (with $s = k = 1$, $m = 2$), we can form a matrix in which $a_{12} - qa_{11} \rightarrow a_{12}$ for any q positive or negative, whereas the other elements of the first row and column are unchanged. Clearly, if q is the quotient and r the remainder in the division of a_{12} by $a_{11} (= z_1)$, then $a_{12} = qa_{11} + r$. Thus $r \rightarrow a_{12}$, but $0 \leq r < a_{11} (= z_1)$ and $r = 0$. This proof, incidentally, shows how to make the first row and first column equal to zero, except for the corner element z_1.

Thus the new matrix has the following form:

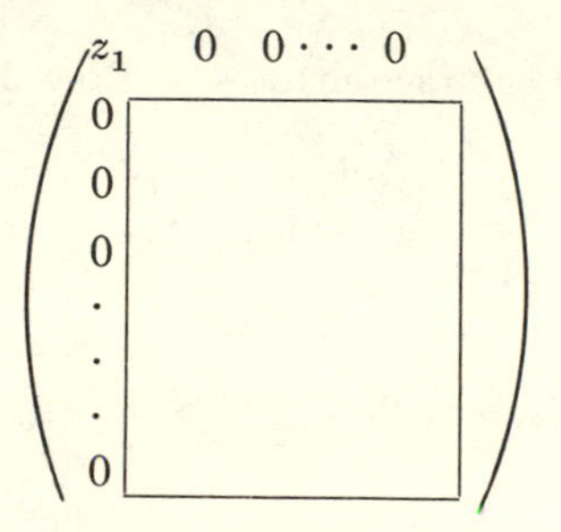

It transforms a *new* basis of $\mathfrak{L}_n = [\mathbf{u}_1', \cdots, \mathbf{u}_n']$ into a new basis of $\mathfrak{M} = [\mathbf{v}_1', \cdots, \mathbf{v}_n']$. We repeat the operations, using $[\mathbf{u}_2', \cdots, \mathbf{u}_n']$ and $[\mathbf{v}_2', \cdots, \mathbf{v}_n']$, leaving $\mathbf{u}_1'$, $\mathbf{v}_1'$ alone. We can then achieve a value z_2 in the upper left-hand corner of the $(n - 1)$ by $(n - 1)$ matrix with zeros in the second row and column. Ultimately, we obtain a diagonal matrix with values $z_1, z_2, \cdots, z_n$ along the diagonal. This tells us that a new basis of $\mathfrak{M}$ is expressible as "diagonal-term" multiples of a new basis of $\mathfrak{L}_n$.

Q.E.D.

Thus bases of $\mathfrak{L}_n$ and $\mathfrak{M}$ can always be chosen as

$$\text{(5)} \qquad \mathfrak{L}_n = [\mathbf{w}_1, \cdots, \mathbf{w}_n], \qquad \mathfrak{M} = [z_1\mathbf{w}_1, \cdots, z_1\mathbf{w}_n].$$

The more complete theory of elementary *divisors* tells us that $z_1 \mid z_2$, $z_2 \mid z_3$, etc., hence the term. It is nevertheless clear that the value of the determinant or index of $\mathfrak{M}$ in $\mathfrak{L}_n$ is $z_1 z_2 \cdots z_n$.

As a method of keeping track of the operations, note that we might write down a new basis for $\mathfrak{L}_n$ and $\mathfrak{M}$ in set (3) every time we use one of the rules in set (4).

ILLUSTRATIVE EXAMPLE

Problem. To reduce to elementary divisors, the transformation

$$(6)\qquad \begin{cases} \mathbf{v}_1 = 2\mathbf{u}_1 + 2\mathbf{u}_2 \\ \mathbf{v}_2 = \mathbf{u}_1 + 4\mathbf{u}_2 \end{cases} \qquad \begin{pmatrix} 2 & 2 \\ 1 & 4 \end{pmatrix}.$$

Solution. Clearly 1 is the value of z_1. The rest is straightforward. (Recall that the $\rightarrow$ means "replaces" not "becomes.") Use operation (e), or interchange of rows, on the original system:

$$\begin{cases} \mathbf{v}_1 = 2\mathbf{u}_1 + 2\mathbf{u}_2 \\ \mathbf{v}_2 = \mathbf{u}_1 + 4\mathbf{u}_2. \end{cases} \quad \text{Let} \begin{cases} \mathbf{v}_2 \rightarrow \mathbf{v}_1 \\ \mathbf{v}_1 \rightarrow \mathbf{v}_2, \end{cases} \quad \text{e.g.,} \begin{cases} \mathbf{v}_1 = \mathbf{v}_2' \\ \mathbf{v}_2 = \mathbf{v}_1'. \end{cases}$$

Use operation (b), or subtraction of 4 × Row 2 from Row 1, on the resulting system:

$$\begin{cases} \mathbf{v}_1' = \mathbf{u}_1 + 4\mathbf{u}_2 \\ \mathbf{v}_2' = 2\mathbf{u}_1 + 2\mathbf{u}_2. \end{cases} \quad \text{Let} \begin{cases} \mathbf{u}_1 - 4\mathbf{u}_2 \rightarrow \mathbf{u}_1 \\ \mathbf{u}_2 \rightarrow \mathbf{u}_2, \end{cases} \quad \text{e.g.,} \begin{cases} \mathbf{u}_1 = \mathbf{u}_1' - 4\mathbf{u}_2' \\ \mathbf{u}_2 = \mathbf{u}_2'. \end{cases}$$

Use operation (a), or subtraction of 2 × Row 1 from Row 2, on the resulting system:

$$\begin{cases} \mathbf{v}_1' = \mathbf{u}_1' \\ \mathbf{v}_2' = 2\mathbf{u}_1' - 6\mathbf{u}_2'. \end{cases} \quad \text{Let} \begin{cases} \mathbf{v}_1' \rightarrow \mathbf{v}_1' \\ \mathbf{v}_2' - 2\mathbf{v}_1' \rightarrow \mathbf{v}_2', \end{cases} \quad \text{e.g.,} \begin{cases} \mathbf{v}_1' = \mathbf{v}_1'' \\ \mathbf{v}_2' - 2\mathbf{v}_1' = \mathbf{v}_2''. \end{cases}$$

Use operation (d), or column sign-change, on the resulting system:

$$\begin{cases} \mathbf{v}_1'' = \mathbf{u}_1' \\ \mathbf{v}_2'' = -6\mathbf{u}_2'. \end{cases} \quad \text{Let} \begin{cases} \mathbf{u}_1' \rightarrow \mathbf{u}_1' \\ \mathbf{u}_2' \rightarrow \mathbf{u}_2', \end{cases} \quad \text{e.g.,} \begin{cases} \mathbf{u}_1' = \mathbf{u}_1'' \\ \mathbf{u}_2' = -\mathbf{u}_2''. \end{cases}$$

Finally

$$(7)\qquad \begin{cases} \mathbf{v}_1'' = \mathbf{u}_1'' \\ \mathbf{v}_2'' = 6\mathbf{u}_2'' \end{cases} \qquad \begin{pmatrix} 1 & 0 \\ 0 & 6 \end{pmatrix},$$

in terms of the transformations in the right-hand column above:

$$(8a)\qquad \begin{cases} \mathbf{u}_1'' = \mathbf{u}_1 + 4\mathbf{u}_2, \\ \mathbf{u}_2'' = \quad\; - \mathbf{u}_2, \end{cases}$$

$$(8b)\qquad \begin{cases} \mathbf{v}_1'' = \quad\;\; \mathbf{v}_2, \\ \mathbf{v}_2'' = \mathbf{v}_1 - 2\mathbf{v}_2. \end{cases}$$

Thus, for our illustrative example, (5) becomes

(9) $\mathfrak{L}_2 = [\mathbf{u}_1, \mathbf{u}_2] = [\mathbf{u}_1'', \mathbf{u}_2''], \qquad \mathfrak{M} = [\mathbf{v}_1, \mathbf{v}_2] = [\mathbf{v}_1'', \mathbf{v}_2''] = [\mathbf{u}_1'', 6\mathbf{u}_2''].$

Matrices can be used more formally[1] but in the few practical problems here they are unwieldy.

EXERCISE 5. Prove the divisor property $z_1 \mid z_2$, $z_2 \mid z_3$, etc., by noticing from (5) that z_1 must divide the remaining matrix of order $(n - 1)$, etc.

EXERCISE 6. Verify that the following matrices provide a canonical basis for the lattices $\mathfrak{L}_*$ of Exercises 1 to 4 above. (See Exercise 6*a*, Chapter IV.)

$$\text{(i)} \begin{pmatrix} 1 & 0 & 0 \\ 0 & 3 & 0 \\ 0 & 1 & 1 \end{pmatrix} \quad \text{(ii)} \begin{pmatrix} 1 & 0 & 0 & 0 \\ 0 & 2 & 0 & 0 \\ 0 & 0 & 2 & 0 \\ 0 & 1 & 1 & 1 \end{pmatrix} \quad \text{(iii)} \begin{pmatrix} 1 & 0 & 0 & 0 \\ 0 & 4 & 0 & 0 \\ 0 & 2 & 1 & 0 \\ 0 & 1 & 0 & 1 \end{pmatrix} \quad \text{(iv)} \begin{pmatrix} 1 & 0 & 0 & 0 & 0 \\ 0 & 5 & 0 & 0 & 0 \\ 0 & 3 & 1 & 0 & 0 \\ 0 & 2 & 0 & 1 & 0 \\ 0 & 1 & 0 & 0 & 1 \end{pmatrix}$$

EXERCISE 7. Perform the reduction for the foregoing matrices, obtaining the new bases as in the illustrative problem.

4. Basis Theorem for Abelian Groups

KRONECKER'S THEOREM

Every finite abelian group can be decomposed into the product of cyclic groups.

Proof. Let

$$(1) \qquad \mathbf{a}_1(= \mathbf{e}), \mathbf{a}_2, \cdots, \mathbf{a}_h$$

denote the elements of the group $\mathbf{G}$. Let us construct a *canonical* basis for $\mathfrak{L}_*$, the lattice of group relations determined by equating to $\mathbf{e}$ the elements

$$(2) \qquad \mathbf{a}_1', \mathbf{a}_2', \cdots, \mathbf{a}_h',$$

with the triangular form manifest as

$$(3) \qquad \mathbf{a}_i' = \mathbf{a}_1^{x_{i1}}\mathbf{a}_2^{x_{i2}} \cdots \mathbf{a}_i^{x_{ii}}, \qquad (1 \le i \le h).$$

Now we apply elementary divisor theory to matrix (x_{ij}). This means that we can find bases of $\mathfrak{L}^*$ and $\mathfrak{L}_h$ (written for group elements, rather than exponents):

$$(4) \qquad \mathbf{A}_i' = (\mathbf{a}_1')^{r_{i1}}(\mathbf{a}_2')^{r_{i2}} \cdots (\mathbf{a}_h')^{r_{ih}}, \qquad 1 \le i \le h, \quad |r_{ij}| = \pm 1,$$

$$(5) \qquad \mathbf{A}_i = \mathbf{a}_1^{s_{i1}}\mathbf{a}_2^{s_{i2}} \cdots \mathbf{a}_h^{s_{ih}}, \qquad 1 \le i \le h, \quad |s_{ij}| = \pm 1.$$

[1] The formal reduction theory is due to C. Hermite (1851) and H. J. S. Smith (1861) (see treatises on algebra in the bibliography).

in terms of which the system (3) becomes $\mathbf{A}_i' = \mathbf{A}_i^{z_i}$ by (5). Hence

(6) $$\mathbf{A}_1^{z_1} = \mathbf{A}_2^{z_2} = \cdots = \mathbf{A}_h^{z_h} = \mathbf{e}$$

for $z_1, \cdots, z_h$ the set of elementary divisors of (x_{ij}). Now system (6) represents a set of generators of all relations among the elements of the abelian group. By the basis property, $|s_{ij}| = \pm 1$, so that the $\mathbf{a}_i$ are all expressible in terms of $\mathbf{A}_i$, e.g.,

(7) $$\mathbf{a}_i = \mathbf{A}_1^{t_{i1}}\mathbf{A}_2^{t_{i2}} \cdots \mathbf{A}_h^{t_{ih}}, \qquad 1 \leq i \leq h.$$

The group $\mathbf{G}$ is generated by h cyclic groups (6) in the generating elements $\mathbf{A}_i$. Q.E.D.

Actually, many of the z_i are 1, as an easy illustration will show. For instance, in the case of the four-group $\mathbf{Z}(2) \times \mathbf{Z}(2)$, a basis of $\mathfrak{L}_*$ is reduced

$$\text{from} \quad \begin{pmatrix} 1 & 0 & 0 & 0 \\ 0 & 2 & 0 & 0 \\ 0 & 0 & 2 & 0 \\ 0 & 1 & 1 & 1 \end{pmatrix} \quad \text{to} \quad \begin{pmatrix} 1 & 0 & 0 & 0 \\ 0 & 1 & 0 & 0 \\ 0 & 0 & 2 & 0 \\ 0 & 0 & 0 & 2 \end{pmatrix}.$$

The left-hand matrix indicates that the group $\mathbf{Z}(2) \times \mathbf{Z}(2)$ is generated by the relations

(8) $$\begin{cases} \mathbf{a}_1 = \mathbf{e}, \\ \mathbf{a}_2^2 = \mathbf{e}, \\ \mathbf{a}_3^2 = \mathbf{e}, \\ \mathbf{a}_2\mathbf{a}_3\mathbf{a}_4 = \mathbf{e}. \end{cases}$$

The right-hand matrix indicates the generators:

$$(9a) \begin{cases} \mathbf{A}_1 = \mathbf{e}, \\ \mathbf{A}_2 = \mathbf{e}, \\ \mathbf{A}_3^2 = \mathbf{e}, \\ \mathbf{A}_4^2 = \mathbf{e}, \end{cases} \qquad (9b) \begin{cases} \mathbf{A}_1 = \mathbf{a}_1, \\ \mathbf{A}_2 = \mathbf{a}_2\mathbf{a}_3\mathbf{a}_4, \\ \mathbf{A}_3 = \mathbf{a}_2, \\ \mathbf{A}_4 = \mathbf{a}_3. \end{cases} \qquad (9c) \begin{cases} \mathbf{a}_1 = \mathbf{A}_1, \\ \mathbf{a}_2 = \mathbf{A}_3, \\ \mathbf{a}_3 = \mathbf{A}_4, \\ \mathbf{a}_4 = \mathbf{A}_2\mathbf{A}_3\mathbf{A}_4. \end{cases}$$

EXERCISE 8. Write out the results (9*a*), (9*b*) in the other groups of Exercises 1 to 4 and 6.

EXERCISE 9. Prove $z_1 z_2 \cdots z_h = h$ from (6).

5. Simplification of Result

A fairly elaborate but elementary equivalence theory of cyclic decomposition enables us to find several different forms of the basis. For instance, in Chapter I, §7, we saw $\mathbf{Z}(6) = \mathbf{Z}(2) \times \mathbf{Z}(3)$, but an abelian group of order 4 might be equivalent to $\mathbf{Z}(4)$ or $\mathbf{Z}(2) \times \mathbf{Z}(2)$ (not both).

By the discussion in that section, once the cyclic structure is achieved we can assure ourselves that each cyclic group will be of order p^k for p prime by breaking up each cyclic group into factors whose order is p^k. Then we could achieve the following structure for an arbitrary abelian group $\mathbf{G}$ of order h:

$$\mathbf{G} = \mathbf{Z}(p_1^{k_1}) \times \mathbf{Z}(p_2^{k_2}) \times \cdots, \tag{1}$$

where p_i are primes, not necessarily distinct, and $k_i > 0$; $h = \Pi p_i^{k_i}$. A further theory, too lengthy to repeat here, shows in effect that the decomposition (1) is *unique* except for order of factors.

The uniqueness is not needed directly for applications in this book, but it enables us to know the values of the elementary divisors z_i ahead of time from the divisibility properties in Exercise 5, §3 (above), for a given group structure. For example, let

$$\mathbf{G} = \mathbf{Z}(p^a) \times \mathbf{Z}(p^b) \times \mathbf{Z}(p^c) \times \mathbf{Z}(q^d) \tag{2}$$

where p and q are different primes and $0 < a \leq b \leq c$, $0 < d$. This leads to a matrix of (arbitrarily large) order $h = p^{a+b+c}q^d$ with h elementary divisors. They are, however, known as follows:

$$1 = z_1 = \cdots = z_{h-3}, \qquad p^a = z_{h-2}, \qquad p^b = z_{h-1}, \qquad p^c q^d = z_h, \tag{3}$$

since only then will §4 lead to a structure (where $z_i \mid z_{i+1}$)

$$\mathbf{G} = \mathbf{Z}(p^a) \times \mathbf{Z}(p^b) \times \mathbf{Z}(p^c q^d) \tag{4}$$

consistent with the definition (2) of $\mathbf{G}$.

In our applications the groups which arise will be even sufficiently simple that the cyclic decomposition may be deduced "by inspection."

GEOMETRIC REMARKS ON QUADRATIC FORMS

6. Successive Minima

In our previous definition of a lattice $\mathfrak{M}$ we introduced at least one norm $\|\mathbf{v}\|$ defined for the vectors of $\mathfrak{M}$ to serve as evidence of discreteness.

We shall carry these ideas a bit further. We shall talk only of the lattice $\mathfrak{L}_n$ of all integral n-tuples

$$\mathbf{v} = (x_1, \cdots, x_n), \tag{1}$$

and we shall specify $\|\mathbf{v}\|$ as a very important type of function, which we now define.

Call a quadratic form in integral coefficients

$$(2) \quad Q(x_1, x_2, \cdots, x_n) = a_{11}x_1^2 + a_{12}x_1x_2 + a_{22}x_2^2 + \cdots + a_{ii}x_i^2 + a_{ij}x_ix_j + \cdots + a_{nn}x_n^2, \qquad (i < j),$$

positive definite if for all integral values except $\mathbf{v} = \mathbf{0}$

$$(3) \qquad Q(x_1, \cdots, x_n) > 0.$$

The definition of positive definiteness is seen to be equivalent to one we would obtain if we considered real values of x_i instead of just integral ones, and there are well-known techniques of algebra for determining positive definiteness. A sufficient condition is that a real transformation matrix exists, (α_{ij}), such that

$$(4) \qquad x_i = \sum_{j=1}^{n} \alpha_{ij}\xi_j, \qquad |\alpha_{ij}| \neq 0$$

and

$$(5) \qquad Q(x_1, \cdots, x_n) = \sum_{i=1}^{n} \xi_i^2.$$

(This condition is also necessary, but we refer the student to the bibliography for further references in algebra.)

We take

$$(6) \qquad \|\mathbf{v}\| = Q(x_1, \cdots, x_n)^{1/2}.$$

Then we can show, among other things, that $\|\mathbf{v}\|$ has the properties of the norm of Chapter IV, §3, and also that for any T the inequality

$$(7) \qquad Q(x_1, \cdots, x_n) \leq T$$

holds only for a finite number of $\mathbf{v}$. The latter statement follows from the fact that by (5) $|\xi_i| \leq T^{1/2}$ and by (4) the coordinates x_i are bounded once T is assigned:

$$(8) \qquad |x_i| \leq \sum_{j=1}^{n} |\alpha_{ij}|\,|\xi_i| \leq T^{1/2}\left(\sum_{j=1}^{n} |\alpha_{ij}|\right).$$

We now define successive minima as follows: consider all $\mathbf{v}$ in $\mathfrak{L}_n$. For these vectors $\mathbf{v}$, the norm $\|\mathbf{v}\|$ has a positive minimum m_1, which it achieves for at most a finite number of $\mathbf{v}$ by the property just established. Trivially, if $\|\mathbf{v}_1\| = m_1$ so does $\|-\mathbf{v}_1\| = m_1$. We ask, less trivially, how many linearly *independent* $\mathbf{v}$ have the same minimum m_1. Let there be k of them, $\mathbf{v}_1, \mathbf{v}_2, \cdots, \mathbf{v}_k$. For instance, if $Q = x_1^2 + x_2^2 + 3x_3^2$, then $m_1 = 1$ (say) at $\mathbf{v} = (1, 0, 0)$ and $\mathbf{v}_2 = (0, 1, 0)$.

Let $k < n$; then consider the set of vectors $\mathbf{v}$ linearly independent of $\mathbf{v}_1, \cdots, \mathbf{v}_k$, and let m_{k+1} be the minimum of $\|\mathbf{v}\|$ for this set. The minimum may be established at $\mathbf{v}_{k+1}, \cdots, \mathbf{v}_l$ (additional linearly independent vectors).

Then if $l < n$ we continue to ask for vectors $\mathbf{v}$ linearly independent of $\mathbf{v}_1, \cdots, \mathbf{v}_l$, for which $\|\mathbf{v}\|$ takes its next minimum m_{l+1}, etc.

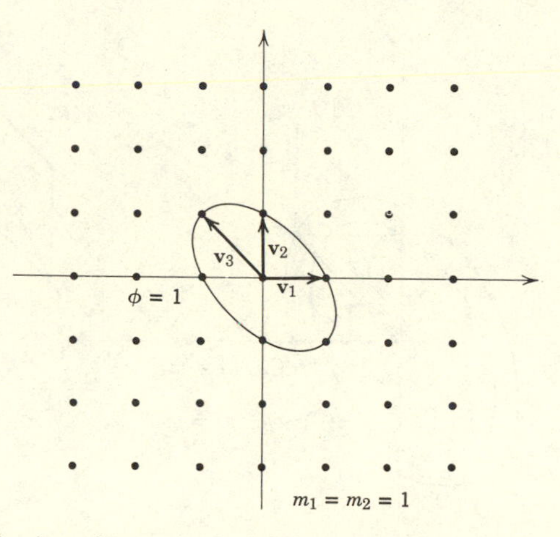

FIGURE 5.1. $\phi(x, y) = x^2 + xy + y^2$. Minimal sets for $\phi(x, y)$ are $[\mathbf{v}_1, \mathbf{v}_2]$, $[\mathbf{v}_1, \mathbf{v}_3]$, $[\mathbf{v}_2, \mathbf{v}_3]$.

We write for simplicity $m_t = \|\mathbf{v}_t\|$ for all t; thus always

$$(9) \qquad m_1 = m_2 = \cdots = m_k \leq m_{k+1} = \cdots = m_l \leq \cdots \leq m_n.$$

Figure 5.1, in two dimensions, shows the several minima for

$$(10) \qquad \phi(x, y) = x^2 + xy + y^2 = (x + y/2)^2 + (3^{1/2}y/2)^2,$$

where $m_1 = m_2$ and Figure 5.2 shows a case in which $m_1 < m_2$.

The natural question that arises is this: *does the set* $\mathbf{v}_1, \mathbf{v}_2, \cdots, \mathbf{v}_n$ *have to be a basis of* $\mathfrak{L}_n$? *The answer is affirmative when* $n = 1, 2, 3$. This serves as a great convenience in a manner we shall soon describe. *The answer is negative when* $n \geq 4$, as we shall see below. This startling result somehow tells us that four-dimensional space is more "pliable" than lower dimensional space, and it was probably the first indication that out of the theory of lattice points would grow a special branch of number theory called the "geometry of numbers."[1]

[1] This name was introduced by Minkowski (1893). There is a confusing cognate, "geometric number theory," which Landau used (1929) to denote the process of counting *large* numbers of lattice points in a region, as we do in Chapter X.

EXERCISE 10. Show $\|\mathbf{v}\|$ satisfies the definitions of norm in Chapter IV, §3.

EXERCISE 11. Show the following form is positive definite:

$$Q(x, y, z) = x^2 + y^2 + z^2 + xy + yz + xz.$$

Hint. $$Q(x, y, z) = \left[x + \left(\frac{y+z}{2}\right)\right]^2 + \frac{3y^2 + 3z^2 + 2yz}{4}.$$

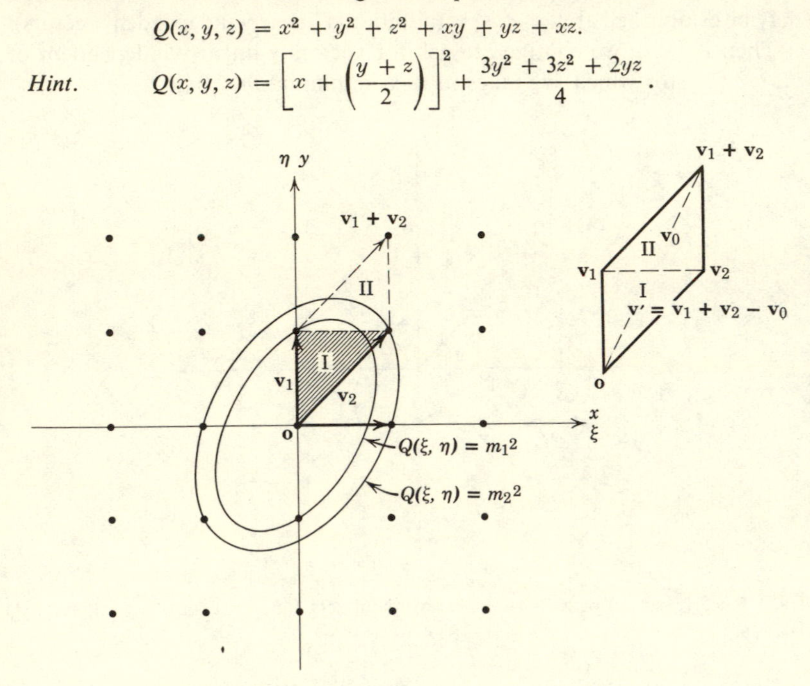

FIGURE 5.2. For proof of theorem in §7.

EXERCISE 12. Do likewise for

$$Q(x, y, z, t) = x^2 + y^2 + z^2 + t^2 + xy + yz + xt + xz + yt + zt$$
$$= \left(x + \frac{y+z+t}{2}\right)^2 + \text{etc.}$$

(In each case the number of variables is decreased one at a time, each time a square is completed, and removed).

7. Binary Forms

Let us consider how the earlier remarks apply to

$$Q(x, y) = Ax^2 + Bxy + Cy^2, \tag{1}$$

a positive definite quadratic form satisfying the conditions in Chapter III, §1;

$$A > 0, \qquad D = B^2 - 4AC < 0. \tag{2}$$

We find the successive minima m_1 and m_2 belonging to $[Q(x, y)]^{1/2}$ and corresponding to

$$\mathbf{v}_1 = (x_1, y_1) \quad \text{and} \quad \mathbf{v}_2 = (x_2, y_2).$$

These points lie on ellipses $Q(\xi, \eta) = m_1^2$ and m_2^2. (See Figure 5.2.) We wish to prove the earlier assertion that $\mathbf{v}_1, \mathbf{v}_2$ is a *basis* of $\mathfrak{L}_2$.

If we regard $\mathbf{v}_1$ and $\mathbf{v}_2$ as the $\mathbf{w}_1$ and $\mathbf{w}_2$ of the geometrical construction in Chapter IV, §5, we see that the basis property of $[\mathbf{v}_1, \mathbf{v}_2]$ amounts to the fact that the parallelogram bounded by the corners $\mathbf{0}, \mathbf{v}_1, \mathbf{v}_2, \mathbf{v}_1 + \mathbf{v}_2$ has no other lattice points in its interior or boundary than the four vertices. Otherwise, the parallelogram would contain a lattice point of $\mathfrak{L}_2$ other than the vertices. It can be seen next in Figure 5.2 that triangle I (shaded) would contain a lattice point $\mathbf{v}'$ other than its vertices, (for, if such a lattice point $\mathbf{v}_0$ were in triangle II then $\mathbf{v}' = \mathbf{v}_1 + \mathbf{v}_2 - \mathbf{v}_0$ would be in triangle I). By the convexity of the ellipse (see Exercise 13 below), triangle I lies interior to the locus $Q(\xi, \eta) = m_2^2$ and thus $Q(\mathbf{v}')^{1/2} < m_2^2$, which contradicts the definition of m_2 (as $\mathbf{v}'$ is clearly not collinear with $\mathbf{v}_1$). Q.E.D.

In Chapter XII we treat in greater detail the problem of expressing the form $Q(x, y)$ under a different basis. For the time being, it might suffice to notice that we are in effect writing

$$(x, y) = x(1, 0) + y(0, 1), \tag{3}$$

or we are using $\mathbf{u}_1 = (1, 0)$ and $\mathbf{u}_2 = (0, 1)$ as a basis of $\mathfrak{L}_2$.

Suppose we wanted to use $\mathbf{v}_1, \mathbf{v}_2$ as a basis of $\mathfrak{L}_2$:

$$\begin{cases} \mathbf{v}_1 = (x_1, y_1) = x_1\mathbf{u}_1 + y_1\mathbf{u}_2, \\ \mathbf{v}_2 = (x_2, y_2) = x_2\mathbf{u}_1 + y_2\mathbf{u}_2. \end{cases} \tag{4}$$

Then by Lemma 4 in Chapter IV, §7, $\mathbf{v}_1, \mathbf{v}_2$ constitutes a basis exactly when

$$x_1y_2 - x_2y_1 = \pm 1. \tag{5}$$

In terms of a new basis, the variables become

$$X\mathbf{v}_1 + Y\mathbf{v}_2 = x\mathbf{u}_1 + y\mathbf{u}_2; \tag{6}$$

or as (X, Y) varies over all $\mathfrak{L}_2$, the integral couple (x, y), given by

$$\begin{cases} x = Xx_1 + Yx_2, \\ y = Xy_1 + Yy_2. \end{cases} \tag{7}$$

also varies over all $\mathfrak{L}_2$. Note that we can solve back for integral (X, Y) by virtue of the determinant condition (5). Now, using (7), we see

$$Q(x, y) = A(Xx_1 + Yx_2)^2 + B(Xx_1 + Yx_2)(Xy_1 + Yy_2) + C(Xy_1 + Yy_2)^2, \tag{8}$$

$$Q(x, y) = A_0X^2 + B_0XY + C_0Y^2. \tag{9}$$

Here the substitution $(X, Y) = (1, 0)$ and $(X, Y) = (0, 1)$ in (7) and (9) yields

$$A_0 = Q(x_1, y_1) = m_1^2, \tag{10a}$$

$$C_0 = Q(x_2, y_2) = m_2^2. \tag{10b}$$

Thus, since the successive minima correspond to $\mathbf{v}_1$ and $\mathbf{v}_2$, which form a basis, we can write a quadratic form in new variables in such a way that the minima m_1 and m_2 are manifest[1] as the coefficients A_0 and C_0. This is essentially Lagrange's (1773) type of reduction procedure.

EXERCISE 13. Show that the interior of triangle I is given by the points $\lambda\mathbf{v}_1 + \mu\mathbf{v}_2$, where these inequalities hold: $0 < \lambda$, $0 < \mu$, $\lambda + \mu < 1$. Hence, using the distance property for the norm, show $Q(\mathbf{v}') < m_2^2$. (We can restrict ourselves to rational values of λ and μ.)

EXERCISE 14. Show that if $\mathbf{v}_1$, $\mathbf{v}_2$, and $\mathbf{v}_3$ are vectors of $\mathfrak{L}_2$ for which $[\mathbf{v}_1, \mathbf{v}_2]$, $[\mathbf{v}_2, \mathbf{v}_3]$, $[\mathbf{v}_3, \mathbf{v}_1]$ are bases of $\mathfrak{L}_2$ then for some choice of signs $\pm\mathbf{v}_1 \pm \mathbf{v}_2 \pm \mathbf{v}_3 = 0$.

EXERCISE 15. Show that if $Q(\mathbf{v}_1) = Q(\mathbf{v}_2) = Q(\mathbf{v}_3) = m_1$ for three vectors independent in pairs, $Q(\mathbf{v}) = x^2 \pm xy + y^2$, except for a constant factor.

EXERCISE 16. If $Q(x, y) = Ax^2 + Bxy + Cy^2$, $(D < 0)$, and $0 < A \leq C$, show $m_1^2 \leq A$ and $m_2^2 \leq C$. How can the values of m_1 and m_2 be determined by graphing $Q(x, y) = C$?

EXERCISE 17. Reduce by Lagrange's procedure the forms

$$5x^2 - 16xy + 13y^2, \qquad 3x^2 + 5xy + 3y^2.$$

8. Korkine and Zolatareff's Example

The Lagrange type of reduction theory was extended to positive definite quadratic forms in *three* variables without any extraordinary occurrence, by Gauss and Seeber (1831). For example, the form $x^2 + y^2 + z^2 + xy + yz + xz$ has three equal minima $m_1 = m_2 = m_3 = 1$ for $(x, y, z) = (1, 0, 0)$, $(0, 1, 0)$, $(0, 1, 0)$; $(1, -1, 0)$, $(0, 1, -1)$, $(1, 0, -1)$ and their negatives. Any three of these, indeed, are dependent or form a basis of $\mathfrak{L}_3$.

[1] As we shall note in Chapter XII, §4, $B_0^2 - 4A_0C_0 = D$, so that B_0 is also determined, except for sign, by the successive minima.

In (1872), however, Korkine and Zolatareff showed that the positive definite form in *four* variables

$$Q(x, y, z, t) = x^2 + y^2 + z^2 + t^2 + xt + yt + zt \tag{1}$$

has minima $m_1 = m_2 = m_3 = m_4 = 1$ at the obvious points $\mathbf{v} = (1,0,0,0)$, $\mathbf{v}_2 = (0, 1, 0, 0)$ and $\mathbf{v}_3 = (0, 0, 1, 0)$, as well as at $\mathbf{v}_4 = (-1, -1, -1, 2)$, which do *not* form a basis of $\mathfrak{L}_4$ when taken together. This was the first indication that the *lattices* in n dimensional space would have interesting properties for each n that would depend on n with number-theoretic irregularity.

Minkowski continued the study of quadratic forms and ultimately generalized his results to the study of forms for which the "unit sphere" $|Q(x, y, \cdots)| \leq 1$ is a "convex" solid. The work was undertaken for purposes of studying quadratic forms and algebraic numbers but was later taken over by the British school more or less as a fascinating end in itself. Mordell, Davenport, and Mahler considered the use of nonconvex bodies (or norms that do not satisfy the distance inequality). The important point historically is the parting of the ways between algebraic number theory and quadratic forms (whose erstwhile synthesis is our main objective here).

EXERCISE 18. Verify that the form (1) is a positive definite.

Hint. $3Q(x, y, z, t) = F(x) + F(y) + F(z)$, where $F(x) = 3x^2 + 3xt + t^2$. For which quadruples does it achieve its minima?

PART 2

IDEAL THEORY IN QUADRATIC FIELDS

chapter VI

Unique factorization and units

1. The "Missing" Factors

From our introductory survey it is clear that the representation $Q(x, y) = m$ is closely related to the representation of factors of m and that these factorizations are reflected in those of algebraic numbers. It is therefore natural to ask when an integral domain $\mathfrak{O}_n$ for a quadratic field $R(\sqrt{D_0})$ will display unique factorization into unfactorable elements which could then be called "primes."

The answer is usually negative; the unfactorable elements do not suffice.

Yet unique factorization can be accomplished by the introduction of "ideal" elements. For instance, in Hilbert's example (see Chapter III, §5), we saw that the set of positive integers g, where $g \equiv 1 \pmod 4$, displays no unique factorization until one discovers additional "ideal" primes q, where $q \equiv -1 \pmod 4$.

The "ideal" elements in algebraic number theory were introduced by Kummer (1857), who found that he needed unique factorization in order to help prove certain cases of Fermat's last theorem. Kummer's ideal elements were actual numbers (like the primes congruent to -1 modulo 4), which belonged to a more inclusive field[1] than the one in which the factorization was attempted.

[1] Some details occur in the section on class-fields in the concluding survey.

Dedekind soon discovered (1871) that the objective of unique factorization could be achieved by constructing (in the same quadratic field) a special type of module for the "ideal" element.

We shall first see to what extent unique factorization fails in quadratic fields.

2. Indecomposable Integers, Units, and Primes

We begin by considering *divisibility in an integral domain* $\mathfrak{O}$ (as defined in Chapter III, §8): we say that an element ξ_1 of $\mathfrak{O}$ *divides* an element ξ of $\mathfrak{O}$ *in* $\mathfrak{O}$ if an element ξ_2 of $\mathfrak{O}$ exists such that

$$\xi = \xi_1\xi_2. \tag{1}$$

We next define a *unit in* $\mathfrak{O}$ as an element which divides 1 in $\mathfrak{O}$. (Note 1 is necessarily in $\mathfrak{O}$.) Then a unit trivially divides in $\mathfrak{O}$ any element of $\mathfrak{O}$. We finally designate as *indecomposable in* $\mathfrak{O}$ any nonunit element ξ for which a factorization (1) is possible in $\mathfrak{O}$ only when ξ_1 or ξ_2 is a unit in $\mathfrak{O}$. If ξ_2 is a unit in $\mathfrak{O}$ we say ξ and ξ_1 are *associates* in $\mathfrak{O}$.

We can now prove the following lemmas (as exercises):

LEMMA 1. An element of $\mathfrak{O}$ is a unit in $\mathfrak{O}$ if and only if its reciprocal lies in $\mathfrak{O}$.

LEMMA 2. The units of $\mathfrak{O}$ are closed under multiplication and division.

We define a *prime in* $\mathfrak{O}$ as a nonunit element π in $\mathfrak{O}$ with the property that if π divides the product of two elements α and β in $\mathfrak{O}$, then π divides α or β in $\mathfrak{O}$. For a prime this property would hold inductively for any number of factors in the product.

LEMMA 3. All primes in $\mathfrak{O}$ are indecomposable in $\mathfrak{O}$.

THEOREM 1. If an element of $\mathfrak{O}$ is expressible as the product of a finite number of primes in $\mathfrak{O}$, it is uniquely expressible as such by rearranging factors and identifying associates of primes.

Proof. If α is an element with two such decompositions,

$$\alpha = \pi_1\pi_2 \cdots \pi_s = \pi_1{}^*\pi_2{}^* \cdots \pi_t{}^*,$$

we can cancel π_1 into some $\pi_i{}^*$ by noticing that $\pi_1 \mid \alpha$. (The primes π_1 and $\pi_i{}^*$ are then associates in $\mathfrak{O}$.) The proof proceeds by induction as in elementary number theory, and, incidentally, $s = t$. Q.E.D.

If $\mathfrak{O}$ is the ring of rational integers, 1 and -1 are clearly the only units in $\mathfrak{D}$ and the definition of prime is seen to be precisely that of an indecomposable element, taken with positive sign by convention.

From now on we shall take $\mathfrak{D}$ as some integral domain $\mathfrak{D}_n$ belonging to the field of $R(\sqrt{D_0})$ (as defined in Chapter III, §9), and we shall usually be able to omit reference to $\mathfrak{D}$. Unfortunately, the indecomposables are not necessarily primes, as is evidenced from the nonunique factorization into indecomposables in §6 (below).

EXERCISE 1. Prove Lemmas 1, 2, and 3.

3. Existence of Units in a Quadratic Field

Before discussing indecomposable integers, we must obtain more information about units. We shall be aided by further specialized knowledge about $\mathfrak{D}_n$.

LEMMA 4. The units of $\mathfrak{D}_n$ are precisely those integers of $\mathfrak{D}_n$ whose norm is $+1$ or -1.

Proof. If η is a unit, then an integer ω exists for which $\eta\omega = 1$. Therefore $N(\eta)\, N(\omega) = 1$ and by the result for rational units $N(\eta) = \pm 1$ (and, incidentally, $\omega = \eta' N(\eta)$). The converse is likewise easy. Q.E.D.

Thus associates have the same norm in absolute value. The converse is *not* necessarily true, as we shall see.

Another way of stating Lemma 4 is that η (in $\mathfrak{D}_n$) satisfies an equation $\eta^2 - A\eta \pm 1 = 0$ (A is a rational integer) precisely when η is a unit in $\mathfrak{D}_n$. Still another way is to say that η is a unit in $\mathfrak{D}_n$ precisely when η and $1/\eta$ are algebraic integers (and units) in $\mathfrak{D}_n$. (Note that if η belongs to $\mathfrak{D}_n$ so does $\pm 1/\eta = A - \eta$.)

THEOREM 2. Any given nonzero integer in $\mathfrak{D}_n$ can be expressed as a product of a finite number of indecomposables of $\mathfrak{D}_n$ in at least one manner.

Proof. Any α of $\mathfrak{D}_n$ which is decomposable can be factored in $\mathfrak{D}_n$ as $\alpha = \alpha_1\alpha_2$ with α_1 and α_2 nonunits. Hence $|N(\alpha)| = |N(\alpha_1)|\, |N(\alpha_2)| >$ $|N(\alpha_1)|$ and $|N(\alpha_2)|$, yielding factors in $\mathfrak{D}_n$ of decreasing norm, to which the factorization process is extended. Clearly, decreasing norms make this process finite. Q.E.D.

IMAGINARY FIELDS

The study of units immediately distinguishes real and imaginary quadratic fields. The imaginary case is easier.

THEOREM 3. For an imaginary quadratic field the only units for $\mathfrak{O}_1$ are as follows:

$$\eta = \pm\sqrt{-1}, \pm 1 \text{ for } R(\sqrt{-1}) \qquad \text{(four units)},$$

$$\eta = \frac{\pm 1 \pm \sqrt{-3}}{2}, \pm 1 \text{ for } R(\sqrt{-3}) \qquad \text{(six units)},$$

$$\eta = \pm 1 \text{ for other cases} \qquad \text{(two units)}.$$

For any $\mathfrak{O}_n$ $(n > 1)$ the only units are $\eta = \pm 1$ (two units).

Proof. The most general integer of $\mathfrak{O}_1$ can be written for a square-free $D_0 < 0$ as $\omega = (a + b\sqrt{D_0})/2$ where a and b agree in parity if $D_0 \equiv 1 \pmod 4$ or are both even if $D_0 \not\equiv 1 \pmod 4$. Then we must solve

$$4N(\omega) = a^2 + b^2(-D_0) = \pm 4.$$

Since $D_0 < 0$, a solution with $b \neq 0$ is possible only for $D_0 = -3, -1$; these cases are enumerated.

In the expression for $\mathfrak{O}_n$ for $n > 1$ the value $|D_0|$ is multiplied by n^2 (see Chapter III, §9); hence the conclusion. Q.E.D.

REAL FIELDS

In a real quadratic field, however, units other than ± 1 always exist. In fact, this is true in any quadratic integral domain $\mathfrak{O}_n$, as we shall see.

LEMMA 5. If ξ_1 and ξ_2 are two real nonzero quantities and if the ratio ξ_1/ξ_2 is irrational, then for any positive integer T we can find integers A and B (not both zero) for which

$$\text{(1)} \qquad \begin{cases} |A\xi_1 + B\xi_2| \leq (|\xi_1| + |\xi_2|)/T, \\ |A| \leq T, \ |B| \leq T. \end{cases}$$

Proof. This is one of Dirichlet's earliest applications of his boxing-in principle. We begin by assuming $\xi_1 > 0$, $\xi_2 > 0$, and we define the form

$$\text{(2)} \qquad f(a, b) = a\xi_1 + b\xi_2, \qquad 0 \leq a \leq T, \qquad 0 \leq b \leq T.$$

We note

$$\text{(3)} \qquad f(a, b) \neq f(a', b') \quad \text{if} \quad (a, b) \neq (a', b'),$$

since otherwise $(a - a')\xi_1 + (b - b')\xi_2 = 0$, contradicting the irrationality of ξ_1/ξ_2. Now $f(a, b)$ takes on $(T + 1)^2$ different values as a and b vary from 0 to T. These values lie in the interval between 0 and

$(|\xi_1| + |\xi_2|)T$. Next, we divide that interval into T^2 segments $I_1, I_2, \cdots, I_{T^2}$ with

$$I_j: (j-1)\lambda \leq x \leq j\lambda, \qquad \lambda = T(|\xi_1| + |\xi_2|)/T^2. \tag{4}$$

We find that since $(T+1)^2 > T^2$ there must be two values of $f(a, b)$ in the same interval (say) I_s, or $|f(a, b) - f(a', b')| \leq \lambda$. Thus, if $A = a - a'$, $B = b - b'$, we find integers A, B that are not both 0 for which

$$|f(A, B)| \leq \lambda, \qquad |A| \leq T, \ |B| \leq T. \tag{5}$$

If ξ_1 and ξ_2 are not positive, a minor modification of signs is made. Q.E.D.

COROLLARY. If D is a positive integer, not a perfect square, then a fixed integer m exists for which the equation

$$A^2 - B^2D = m \tag{6}$$

has infinitely many solutions in integers (A, B).

Proof. By the preceding lemma, we can find integers A and B (not both zero), for which

$$\begin{cases} |A - B\sqrt{D}| \leq (1 + \sqrt{D})/T \\ |A| \leq T, \qquad |B| \leq T. \end{cases} \tag{7}$$

for any positive integer T. Furthermore,

$$|A + B\sqrt{D}| \leq |A| + |B\sqrt{D}| \leq T(1 + \sqrt{D}) \tag{8}$$

and, multiplying inequalities (7) and (8),

$$|A^2 - B^2D| \leq (1 + \sqrt{D})^2. \tag{9}$$

Now, if only a finite number of pairs of integers (A, B) occur as $T \to \infty$, it could not be true that $|A - B\sqrt{D}|$ can be made arbitrarily small without equaling zero. Therefore, there must be infinitely many different pairs of integers (A, B) occurring as $T \to \infty$ for which $A^2 - B^2D$ is bounded, and there must be at least one m for which (6) has infinitely many solutions. Q.E.D.

LEMMA 6. Under the condition that D is not a perfect square, there must exist at least one pair of integers (a, b) for which

$$a^2 - b^2D = 1, \qquad (a \neq \pm 1). \tag{10}$$

Proof. It is easy to insist that $A > 0$ in the preceding corollary for the infinitude of (A, B); otherwise there would also be only a finite number of solutions for which $A < 0$, and, of course, there is only a finite number for $A = 0$.

We take $(m^2 + 1)$ of the solutions, $(A > 0)$, and we separate them into m^2 classes according to the residues of (A, B) modulo m. There must then be at least two in one class, by Dirichlet's boxing-in principle,

$$A_1 \equiv A_2, \qquad B_1 \equiv B_2 \pmod{m}. \tag{11}$$

We set up $\eta_1 = A_1 + B_1\sqrt{D}$; $\eta_2 = A_2 + B_2\sqrt{D}$ and find

$$N(\eta_1) = N(\eta_2) = m, \qquad \eta_1 \equiv \eta_2 \pmod{m}. \tag{12}$$

Of course, $N(\eta_1/\eta_2) = 1$, but we must still show that $\eta_1/\eta_2 = \xi$ is an *integer*. Since $\eta_2\eta_2' = m$, we write

$$\xi = 1 + (\eta_1 - \eta_2)/\eta_2 = 1 + (\eta_1 - \eta_2)\eta_2'/m \tag{13}$$

and $\eta_1 - \eta_2$ is actually divisible by m. Explicitly,

$$\xi = 1 + \left(\frac{A_1 - A_2}{m} + \sqrt{D}\,\frac{B_1 - B_2}{m}\right)(A_2 - B_2\sqrt{D}) = a + b\sqrt{D} \tag{14}$$

on expanding. We easily see that $\eta_1 \neq \eta_2$, since $(A_1, B_1) \neq (A_2, B_2)$ and $\eta_1 \neq -\eta_2$, for A_1 and A_2 are both positive. Thus (10) holds. Q.E.D.

4. Fundamental Units

We now consider the set of all units of any real quadratic $\mathfrak{O}_n$. The set contains at least one nontrivial unit ($\neq \pm 1$) by Lemma 6, (above), since, with $D = n^2 D_0$, $\mathfrak{O}_n$ clearly contains all integers $a + b\sqrt{D}$ (see Chapter III, §9). Consider the set of all units of $\mathfrak{O}_n$ (of norm $+1$ and -1). This set of units symbolized by $\{\rho\}$ is a multiplicative group. We next consider the set of values $\mathfrak{U} = \{\log|\rho|\}$ which becomes an *additive* group. For example, the inverse is $-\log|\rho| = \log|\rho'|$, since $|N(\rho')| = |N(\rho)| = 1$. Likewise, since $N(\rho_1)N(\rho_2) = N(\rho_1\rho_2)$, then $\log|\rho_1| + \log|\rho_2| = \log|\rho_1\rho_2|$. We shall see that the set of values $\mathfrak{U}$ constitutes a lattice. Then we use its basis.

LEMMA 7. For any algebraic integer ω, of $\mathfrak{O}_n$ (not necessarily a unit), both values $\log|\omega|$, $\log|\omega'|$ cannot be arbitrarily close to zero unless they both equal zero.

Proof. Referring to Figure 4.1, we see $|\omega| - 1$ and $|\omega'| - 1$ cannot both be made arbitrarily small without being zero; for the points $(1, 1)$, $(1, -1)$, $(-1, 1)$, $(-1, -1)$ must all be separated by at least a finite constant distance from the point (ω, ω'), according to the lattice property of $\mathfrak{O}_n$. Q.E.D.

THEOREM 4. The set $\mathfrak{U} = \{\log |\rho|\}$ for ρ a unit of $\mathfrak{O}_n$ constitutes a one-dimensional lattice.

Proof. Since $\log |\rho'| = -\log |\rho|$, we conclude from Lemma 7 that for some constant $k > 0$

$$(1) \qquad \big| \log |\rho| \big| \geq k \quad \text{if} \quad \rho \neq \pm 1.$$

Furthermore, the quantities $\log |\rho|$ are all linearly dependent (with integral coefficients) on any nonvanishing element value (say) $\log |\rho_0|$. For otherwise some $\log |\rho_1|$ would exist for which $\xi = \log |\rho_1| / \log |\rho_0|$ is *irrational*, and, by Lemma 5, we could make the absolute value of the function $f(a, b) = (a \log |\rho_1| + b \log |\rho_0|)$ arbitrarily small in absolute value, indeed $< k$. Then the unit $\rho = \rho_1^a \rho_0^b (\neq \pm 1)$ would violate inequality (1). Thus it must be possible for $f(a, b)$ to vanish for integral a and b not both zero.

We therefore have the discreteness conditions for a lattice. (See Chapter IV, §3.) Q.E.D.

The (minimal) basis of $\mathfrak{U}$ is a one-dimensional vector written as $\log \eta_n$. We put this result in "antilog" form.

COROLLARY. There exists a special unit η_n in any $\mathfrak{O}_n$ such that all units ρ in $\mathfrak{O}_n$ are given by

$$\rho = \pm \eta_n{}^m, \qquad m = 0, \pm 1, \pm 2, \pm 3, \cdots.$$

This unit η_n is called the *fundamental unit* if (for standardization) $\eta_n > 1$ also. According to this definition; η_n' might be positive or negative, depending on $N(\eta_n)$. Note this η_n is precisely the unit which minimizes $\log \rho$ for ρ an arbitrary unit > 0.

As an illustration in $R(\sqrt{5})$, we can prove $\eta_1 = (1 + \sqrt{5})/2$, $\eta_1{}^2 = (3 + \sqrt{5})/2$, $\eta_1{}^3 = 2 + \sqrt{5}$. Note that $\eta_1{}^3$ is the fundamental unit of $\mathfrak{O}_2$, since, by the inclusion of $\mathfrak{O}_n$ in $\mathfrak{O}_1$, the units of $\mathfrak{O}_n$ are all to be found among the units of $\mathfrak{O}_1$. Note also that $N(\eta_1) = -1$. $N(\eta_1{}^2) = +1$. The set of integers where $N(\omega) = 1$ is $\omega = \pm \eta_1{}^{2n}$, in $\mathfrak{O}_1$ whenever $N(\eta_1) = -1$.

The answer to the question whether $N(\eta_n) = +1$ or $N(\eta_n) = -1$ for the fundamental unit in an arbitrary $\mathfrak{O}_n$ is not known completely. One can prove with ease that in $\mathfrak{O}_1$ (hence in $\mathfrak{O}_n$) $N(\eta_1) = +1$ if D_0 has any prime divisor $q \equiv -1 \pmod 4$; For, if $\eta_1 = (T + V\sqrt{D_0})/2$, then the equation $T^2 - D_0 U^2 = -4$ leads to

$$T^2 \equiv -4 \pmod q$$

as $q \mid D_0$, causing a contradiction. Furthermore, $N(\eta_1) = -1$, if $D = D_0$

is a *prime* $\equiv 1 \pmod 4$, but this is a much deeper result, which we prove in Chapter XI, §2. If D_0 has no prime divisor $\equiv -1 \pmod 4$ the value of $N(\eta)$ is generally unknown, and becomes a matter of vital concern in Chapter XI, §2 (below).

The table of fundamental units in the appendix is evidence of the irregularity of the *size* of the fundamental unit, which is governed by even weaker rules than the *sign* of the norm.

EXERCISE 2. Give a separate proof of Lemma 6 based on the idea that ω must satisfy an equation with discrete coefficients (integral values of $\omega + \omega'$ and $\omega\omega'$).

EXERCISE 3. Show $\eta = (1 + \sqrt{5})/2$ is the smallest unit $\eta_1\ (>1)$ for all real quadratic fields whatever. *Hint*. Consider all equations $x^2 - Ax - 1 = 0$. Solve for $\eta > 1$ and consider $d\eta/dA$. Likewise take $x^2 - Ax + 1 = 0$ and between these two equations choose the smallest η. Find the next six smallest units $\eta_1\ (>1)$ for quadratic fields (of whatever discriminant).

EXERCISE 4. Construct the first 8 powers of $\eta_1 = (1 + \sqrt{5})/2$ and tell which are fundamental units of some $\mathfrak{O}_n$. Note that η_u and η_v can be equal even if $u \neq v$.

5. Construction of a Fundamental Unit

As a practical matter, one could not use the Dirichlet boxing-in principle to construct a unit or even to construct numbers of equal norm. One normally would use the method of continued fractions, but this method has the disadvantage of being incapable of generalization to fields of higher degree. We shall find an "irregular" but pragmatic method in Chapter IX, using factorizations. In the meantime, we shall show how to verify units once they are found (say) in Table III (appendix).

LEMMA 8. Of all the units ρ in $\mathfrak{O}_n$, the fundamental unit minimizes $|\rho + \rho'|$.

Proof. Let ρ satisfy the equation, with A written for $\rho + \rho'$,

$$\rho^2 - A\rho + N(\rho) = 0, \qquad N(\rho) = \pm 1.$$

Then we can take $A > 0$ by a choice of sign on ρ. We find that, whether $N(\rho)$ is $+1$ or -1, the root ρ that satisfies $\rho > 1$ is

$$\rho = [A + \sqrt{A^2 - 4N(\rho)}]/2. \tag{1}$$

Thus ρ obviously increases monotonically with A once $N(\rho)$ is chosen as $+1$ or -1. We just have to show that the minimum value of $\rho + \rho'$ for which $N(\rho) = +1$ exceeds the minimum value of $\rho + \rho'$ for which

$N(\rho) = -1$. Clearly, if $N(\eta) = -1$, we need note only $\eta' = -1/\eta$, whereas $(\eta')^2 = 1/\eta^2$. Thus

$$\eta^2 + (\eta')^2 = \eta^2 + 1/\eta^2 > \eta^2 > \eta > \eta - 1/\eta = \eta + \eta'. \tag{2}$$

Hence the η of norm -1 has the smaller value of $\eta + \eta'$. Q.E.D.

In practice, then, if some unit ρ_0 is known to satisfy

$$\rho^2 - A\rho \pm 1 = 0, \qquad \text{for } A = A_0, \tag{3}$$

we need only ask which of the equations for $0 \leq A \leq |A_0|$ has the smallest A and also produces a unit of $\mathfrak{O}_n$. The work can be lightened considerably, since A is restricted by

$$A^2 \mp 4 \equiv 0 \pmod{D_0}. \tag{4}$$

THEOREM 5. The fundamental unit of $\mathfrak{O}_n$ can be found by considering the smallest integer T for which

$$T^2 - DU^2 = \pm 4 \qquad T > 0,\ U > 0, \tag{5}$$

where $D = n^2D_0$, and U is taken even when n is even and $D_0 \not\equiv 1 \pmod 4$. Then the fundamental unit of $\mathfrak{O}_n$ is $(T + U\sqrt{D})/2$ and the most general unit is $\pm[(T + U\sqrt{D})/2]^m$. Here $D = n^2D_0$ in the usual notation.

Proof. Note $\rho = (T + U\sqrt{D})/2$ satisfies the equation $\rho^2 - T\rho \pm 1 = 0$, of type (3) (above). (See Chapter III, §9.) Q.E.D.

To construct a fundamental unit of $\mathfrak{O}_n$ from a fundamental unit of $\mathfrak{O}_1$, we require a special result, which will be approached in a mode of greater generality.

LEMMA 9. If η_1 is a unit of $\mathfrak{O}_1$, then for some exponent t, $0 < t \leq n^2$, $\eta_1^{\,t}$ belongs to $\mathfrak{O}_n$.

Proof. Note that the most general integer of $\mathfrak{O}_1$ is $a + b\omega_0$, where a and b are integers; hence we consider n^2 residue classes based on the residue of a and b modulo n. If we list $\eta_1^{\,s}$, where $s = 0, 1, \cdots, n^2$, we have $n^2 + 1$ units of which two must belong to one class. Thus $\eta_1^{\,s_1} \equiv \eta_1^{\,s_2} \pmod n$, for $0 \leq s_1 < s_2 \leq n^2$. Now we let $s_2 - s_1 = t$, and we find for some integer λ in $\mathfrak{O}_1$

$$\eta_1^{\,s_2} = \eta_1^{\,s_1} + \lambda n; \tag{6}$$

hence, transposing $\eta_1^{\,s_1}$ and multiplying by the *integer* $\eta_1^{-s_1}$, we see

$$\eta_1^{\,t} = 1 + n\lambda\eta_1^{-s_2} \equiv \text{rational modulo } n. \tag{7}$$

Thus $\eta_1^{\,t}$ belongs to $\mathfrak{O}_n$, by the basic definition of $\mathfrak{O}_n$ in Chapter III, §8. Q.E.D.

EXERCISE 5*a*. Find the fundamental unit of $\mathfrak{O}_1$ for $R(\sqrt{13})$, starting with $\rho = 18 - 5\sqrt{13}$ and considering equations of type (4). Is this ρ the fundamental unit of $\mathfrak{O}_5$?

EXERCISE 5*b*. Find the fundamental unit of $\mathfrak{O}_1$, $\mathfrak{O}_2$, $\mathfrak{O}_3$, $\mathfrak{O}_4$, and $\mathfrak{O}_5$ for $R(\sqrt{2})$ starting with $\eta_1 = 1 + \sqrt{2}$.

EXERCISE 6. Prove that every unit λ of $\mathfrak{O}_n$ can be written as $\lambda = r + n\mu$, where μ belongs to $\mathfrak{O}_1$ and $r^2 \equiv \pm 1 (\text{mod } n)$.

EXERCISE 7. Show that a unit of $R(\sqrt{D_0})$, (D_0 square-free), can have the form $(a + b\sqrt{D_0})/2$ for a and b odd only when $D_0 \equiv 5 \pmod 8$. Show that the congruence is not sufficient by consulting the units in Table III (appendix).

6. Failure of Unique Factorization into Indecomposable Integers

In order to appreciate the complexities of the theory of quadratic fields, we must accept the fact that unique factorization is not generally valid.

A typical case is presented by the field $R(\sqrt{-5})$. Here we observe as an illustration

$$21 = 3 \cdot 7 = (1 + 2\sqrt{-5})(1 - 2\sqrt{-5}). \tag{1}$$

The factors shown are all indecomposable. Otherwise, if we write

$$\xi = \xi_1\xi_2 = 3, 7, \quad \text{or} \quad (1 \pm 2\sqrt{-5}),$$

taking norms, $N(\xi) = N(\xi_1)\,N(\xi_2) = 9, 49,$ or 21. With the most general $\xi_i = a + b\sqrt{-5}$, we find that we must solve

$$a^2 + 5b^2 = 3 \text{ or } 7. \tag{2}$$

Thus ξ is not decomposable into two factors (each with $N(\xi_i) > 1$), since (2) is an impossible equation (by trial and error).

Now, $1 + 2\sqrt{-5}$ divides the product of two indecomposables, 3 and 7, but does not divide either one, since $N(1 + 2\sqrt{-5}) = 21$, which does not divide $N(3)(= 9)$ or $N(7)(= 49)$. *There is no unique factorization into indecomposable algebraic integers in* $R(\sqrt{-5})$.

Our current state of knowledge of modules holds some hope that they can provide the answer. We say in the introductory survey that in contrast to the impossible (2)

$$2a^2 + 2ab + 3b^2 = \tfrac{1}{2}N[2a + (1 + \sqrt{-5})b] \tag{3}$$

does represent 3 and 7 for an obvious choice $(a, b) = (0, 1)$ and $(1, 1)$. A multiplicative theory of modules will accordingly be developed.

Lest the reader jump at conclusions, we should note that the factorizations involved in "indecomposability" are not always trivial. For instance, it will be seen later on that $R(\sqrt{6})$ has unique factorization. Note by way of verification that no difficulty arises here with regard to the statement

$$6 = (\sqrt{6})^2 = 3 \cdot 2. \tag{4}$$

Actually, we can show $\sqrt{6}$, 3, and 2 are all consistently decomposable. For instance, $6 = -(3 + \sqrt{6})(3 - \sqrt{6})(2 + \sqrt{6})(2 - \sqrt{6})$, whereas

$$\begin{cases} (3 + \sqrt{6})(3 - \sqrt{6}) = 3, \ N(3 \pm \sqrt{6}) = 3, \\ (2 + \sqrt{6})(2 - \sqrt{6}) = -2, \ N(2 \pm \sqrt{6}) = -2, \\ (3 - \sqrt{6})(2 + \sqrt{6}) = -(3 + \sqrt{6})(2 - \sqrt{6}) = \sqrt{6}. \end{cases} \tag{5}$$

It is then an easy result to see that 6 now has the same *four* factors in (4) either way!

The matter is still not settled because the statement "$6 = (\sqrt{6})^2$" in (4) should not be acceptable unless we can show 2 and 3 to be associates of perfect squares in $R(\sqrt{6})$, ($\sqrt{2}$ and $\sqrt{3}$ are excluded from the field).

We note

$$\begin{cases} (3 + \sqrt{6})/(3 - \sqrt{6}) = 5 + 2\sqrt{6} = \rho_1, \\ -(2 + \sqrt{6})/(2 - \sqrt{6}) = 5 + 2\sqrt{6} = \rho_1, \end{cases} \tag{6}$$

where ρ_1 is a *unit*. Thus the factors of 2 and 3 are *associates*. Hence, finally, making use of the unit in (6), we write (5) as

$$\begin{cases} 3 = (3 - \sqrt{6})^2 \rho_1, \\ 2 = (2 - \sqrt{6})^2 \rho_1, \\ \sqrt{6} = -(3 - \sqrt{6})(2 - \sqrt{6})\rho_1, \end{cases} \tag{7}$$

which is a wholly satisfactory explanation of how statement (4) really leads to a unique factorization.

With these remarks in mind, we turn our attention to some cases in which unique factorization succeeds and is easily demonstrated.

EXERCISE 8. Show that, if p and q are primes, then, in $R(\sqrt{-pq})$, $-pq = \sqrt{-pq}\,\sqrt{-pq} = -(p)(q)$ represents a factorization in two irreconcilable ways into indecomposable algebraic integers.

EXERCISE 9. Assume for odd primes p and q that $(q/p) = (p/q) = -1$ (in Legendre symbols). Then show that for *neither* sign is $px^2 - qy^2 = \pm 4$ solvable. From this show that $pq = \sqrt{pq}\,\sqrt{pq} = (p)(q)$ is a "nonunique" pair of factorizations into indecomposables in $R(\sqrt{pq})$.

EXERCISE 10. Show that in $R(\sqrt{14})$ the relation $14 = (\sqrt{14})^2 = 7 \cdot 2$ does *not* violate unique factorization by finding integers of norms equal in absolute value to 7 and 2 by trial and error and by showing that 7 and 2 are associates of squares.

EXERCISE 11. Show $10 = (\sqrt{10})^2 = 2 \cdot 5$ leads to nonunique factorization. *Hint*. Show the unsolvability of $x^2 - 10y^2 = \pm 2, \pm 5$.

*7. Euclidean Algorithm

The basis of unique factorization in rational number theory is the Euclidean algorithm, which we now reformulate for an integral domain $\mathfrak{O}_n$.

An integral domain $\mathfrak{O}_n$ is *Euclidean* if given any two elements α and β $(\neq 0)$ of $\mathfrak{O}_n$; an element γ of $\mathfrak{O}_n$ can be used as "quotient" so that the "remainder" on dividing α/β is of smaller norm than β. Symbolically,

$$|N(\alpha - \beta\gamma)| < |N(\beta)|. \tag{1}$$

THEOREM 6. If $\mathfrak{O}_n$ is Euclidean, then any element of $\mathfrak{O}_n$ can be expressed uniquely as a finite product of indecomposable elements, ignoring units and the order of the factors.

Proof. We first factor any element of $\mathfrak{O}_n$ into indecomposables (by Theorem 2, §3 above). We then show that the indecomposables are primes (by Lemma 11 below), so that Theorem 1 applies. Here the proof goes in stages that the reader can easily recognize by recalling elementary number theory.

LEMMA 10. If α and β are two elements of $\mathfrak{O}_n$, a *Euclidean integral domain*, then they have a greatest common divisor, gcd $(\alpha, \beta) = \gamma$ in the sense that any ρ which divides α and β divides γ, and conversely. We can write

$$\gamma = \alpha\xi + \beta\eta, \tag{2}$$

where ξ and η belong to $\mathfrak{O}_n$.

Proof. Consider the set of elements $\alpha\xi + \beta\eta = f(\xi, \eta)$. Let the element of smallest norm (in absolute value) be

$$\gamma = f(\xi_0, \eta_0) = \alpha\xi_0 + \beta\eta_0. \tag{3}$$

Then we assert that any $f(\xi, \eta)$ where ξ and η belong to $\mathfrak{O}_n$ is a multiple of γ. Otherwise, let $\gamma_1 = f(\xi_1, \eta_1)$ where γ_1 is not a multiple of γ. Clearly, then, by the Euclidean algorithm, for some ρ, $|N(\gamma_1 - \rho\gamma)| < |N(\gamma)|$ and thus $0 \neq \gamma_1 - \rho\gamma = f(\xi_1 - \rho\xi_0, \eta_1 - \rho\eta_0)$ has smaller (absolute) norm than γ, contradicting the definition of γ.

It therefore follows that γ satisfies the property of the theorem for

$\gamma \mid \alpha = f(1, 0)$, $\gamma \mid \beta = f(0, 1)$ whereas any ρ that divides α and β necessarily divides $\gamma = [(\alpha/\rho)\xi_0 + (\beta/\rho)\eta_0]\rho$. Q.E.D.

LEMMA 11. If π is an indecomposable element of $\mathfrak{O}_n$ and $\pi \mid \alpha\beta$, then $\pi \mid \alpha$ or $\pi \mid \beta$.

Proof. If $\pi \nmid \alpha$, then, since π is indecomposable, gcd $(\pi, \alpha) = 1$. By Lemma 10, we have

$$(4) \qquad 1 = \xi_1\pi + \xi_2\alpha$$

for integers ξ_1, ξ_2 of $\mathfrak{O}_n$. Then $\beta = \xi_1\pi\beta + \xi_2\alpha\beta = \pi(\xi_1\beta + \xi_2(\alpha\beta/\pi))$ and $\pi \mid \beta$. Q.E.D.

*8. Occurrence of the Euclidean Algorithm

If we divide (1) of §7 (above) by $N(\beta)$, we see that the Euclidean algorithm for $\mathfrak{O}_n$ states the following: for any fraction α/β formed by elements of $\mathfrak{O}_n$ an integer γ of $\mathfrak{O}_n$ exists for which

$$(1) \qquad |N(\alpha/\beta - \gamma)| < 1.$$

We note that the denominator of α/β can always be rationalized so that $\alpha/\beta = \alpha\beta'/B$, where $B = N(\beta)$. Thus the most general fraction in $\mathfrak{O}_n = [1, \omega]$ is

$$(2) \qquad \frac{\alpha}{\beta} = \frac{A_1 + A_2\omega}{B},$$

whereas $\gamma = a + b\omega$. Thus, when A_1, A_2, $B(> 0)$ are given integers, we are trying to find integers (a, b) such that

$$(3) \qquad \left|N\left(\left[\frac{A_1}{B} - a\right] + \omega\left[\frac{A_2}{B} - b\right]\right)\right| < 1.$$

COMPLEX CASE ($D < 0$)

In order to cover all cases of $\mathfrak{O}_n$ (for $n \geq 1$), we can take for D any negative integer; then we can always take $\omega = \sqrt{D}$, while we can also take $\omega = (\sqrt{D} + 1)/2$ when $D \equiv 1 \pmod 4$. This includes all cases of $\mathfrak{O}_n$ in Chapter III, §9.

Either way, we take the plane with the complex integers as represented in Chapter IV and for each lattice point ξ we lay out the region consisting of points closer, in the ordinary Euclidean sense, to ξ than to any other lattice point. This region is called a *zone*. When $\omega = \sqrt{D}$, the zones are simply

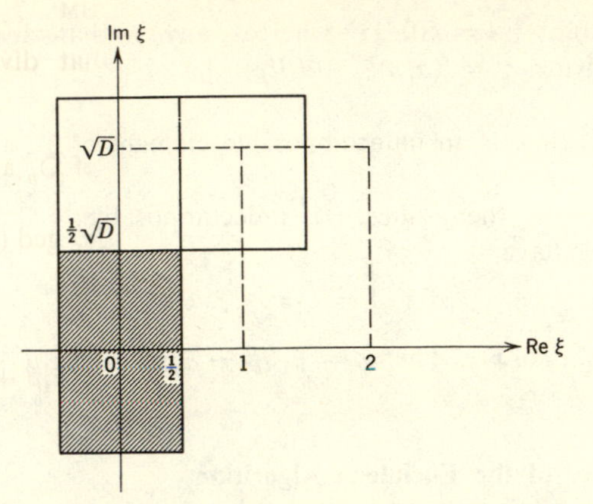

FIGURE 6.1

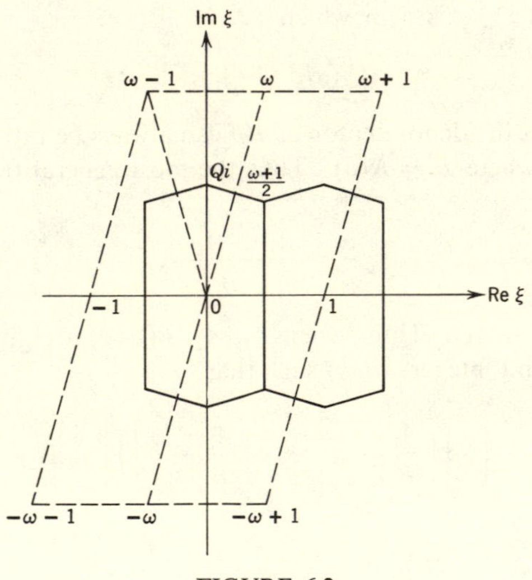

FIGURE 6.2

small rectangles, as shown in Figure 6.1. The farthest point from the origin in the zone around the origin has distance

$$(4) \qquad \left|\frac{1}{2} + \frac{i}{2}\sqrt{|D|}\right| = \left[\frac{1 + |D|}{4}\right]^{1/2}, \qquad (D < 0).$$

In the case in which $\omega = (1 + \sqrt{D})/2$ the zones are formed by taking the perpendicular bisectors in the parallelograms, as shown in Figure 6.2.

The farthest point from the origin in the zone about the origin is easily seen to be Qi the point of the imaginary axis equidistant from 0 and ω. Thus

$$|Qi| = |Qi - \omega| = \left| Qi - \frac{1}{2} - \frac{i\sqrt{|D|}}{2} \right|.$$

$$Q^2 = (Q - \sqrt{|D|}/2)^2 + \tfrac{1}{4}.$$

(5a) $$Q = (|D| + 1)/4\sqrt{|D|}, \qquad D \equiv 1 \pmod 4, \qquad D < 0.$$

(5b) $$-Qi = (1 - D)/4\sqrt{D}.$$

Now, since the norm of ξ in $R(\sqrt{D})$ is precisely the Euclidean distance $((\mathrm{Re}\ \xi)^2 + (\mathrm{Im}\ \xi)^2)^{1/2}$, it follows from inequality (1) that a necessary and sufficient condition for the existence of a Euclidean algorithm is that we be able to put the zone about the origin *wholly* inside a unit circle (so that not even the extremities of the zone lie on the circle). We note, for example, that the extreme point of the zone corresponds by (5b) to a fraction in the field. Thus it is required that the expression (4) or (5a) be less than 1. By a simple exercise in inequalities we now discover:

THEOREM 7. The only imaginary Euclidean integral domains are $\mathfrak{O}_1$, the ring of all integers in $R(\sqrt{D_0})$ for $D_0 = -1, -2, -3, -7, -11$.

THE REAL CASE $(D > 0)$

We cover the cases of $\mathfrak{O}_n$, as we did for the complex case, except D is no perfect square. But here, we have a much more difficult problem, for the zones are very complicated. For example, if $\omega = \sqrt{D}$, inequality (3) becomes

(6a) $$\left| \left(\frac{A_1}{B} - a\right)^2 - D\left(\frac{A_2}{B} - b\right)^2 \right| < 1,$$

which could be satisfied by an infinite number of integers (a, b) for a given pair of fractions A_1/B and A_2/B. We therefore have trouble in saying (6a) is *impossible*, although we can check that it is possible "often enough" to establish the Euclidean algorithm in many cases.

THEOREM 8. The sets of $\mathfrak{O}_1$ of all integers in $R(\sqrt{2})$, $R(\sqrt{3})$ are Euclidean.

Proof. Let A_1/B, A_2/B be given; then we need only choose a and b as the closest integers to these fractions:

(7) $$|A_1/B - a| \leq \tfrac{1}{2}, \qquad |A_2/B - b| \leq \tfrac{1}{2}, \qquad \text{thus, with } D = 2, 3,$$

(8) $$-D/4 \leq (A_1/B - a)^2 - D(A_2/B - b)^2 \leq 1/4. \qquad \text{Q.E.D.}$$

In a few cases we can show the nonexistence of the Euclidean algorithm directly: for instance, for $\mathfrak{O}_2$ in $R(\sqrt{5})$; then the integers of $\mathfrak{O}_2$ are of the form $a + b\sqrt{5}$. If we let $\alpha/B = (1 + \sqrt{5})/2$, we see that for any γ in $\mathfrak{O}_2$, $|N(\alpha/B - \gamma)| \geq 1$ by virtue of the fact that α/B happens to be an integer in $\mathfrak{O}_1$, and the norm is therefore a *rational* integer. Hence the Euclidean algorithm is seen to be inoperative for that particular value of α/B.

It happens to be true (although we omit proof) that $\mathfrak{O}_1$ for $R(\sqrt{6})$ has the Euclidean algorithm, To illustrate one case, take $\alpha = 1 + \sqrt{6}$, $B = 2$. We wish to choose $\gamma = a + b\sqrt{6}$ so that in accordance with (1) and (6*a*)

$$(6b) \qquad |(\tfrac{1}{2} - a)^2 - 6(\tfrac{1}{2} - b)^2| < 1.$$

Now if we looked for (a, b) "close to" $(\frac{1}{2}, \frac{1}{2})$ we would hardly think of choosing $a = 2, b = 0$, which actually satisfies (6*b*). Yet, we could have even chosen $a + b\sqrt{6} = 14 + 6\sqrt{6}$, which looks much farther away in the lattice Figure 4.1 in the Euclidean sense but not in the sense of the norm (6*a*).

We next consider cases in which $\omega = (1 + \sqrt{D})/2$ and $D \equiv 1 \pmod 4$, hence $D \geq 5$; and we write $\alpha/B = A_1/B + \omega A_2/B$. We search for the integer $\gamma = a + b\omega$ satisfying (1). Now

$$(9) \quad N\left(\frac{\alpha}{B} - \gamma\right) = \left(\frac{A_1}{B} - a\right)^2 + \left(\frac{A_1}{B} - a\right)\left(\frac{A_2}{B} - b\right) - \frac{D-1}{4}\left(\frac{A_2}{B} - b\right)^2.$$

Thus, if we take a and b as the closest integers to A_1/B and A_2/B, we find we are dealing with new variables

$$(10) \qquad P = A_1/B - a, \qquad Q = A_2/B - b,$$

which satisfy

$$(11) \qquad -\tfrac{1}{2} \leq P \leq \tfrac{1}{2}, \qquad -\tfrac{1}{2} \leq Q \leq \tfrac{1}{2}.$$

LEMMA 12. Consider the function

$$(12) \qquad f(P, Q) = P^2 + PQ - sQ^2,$$

where s is a real constant > 1 and P and Q are restricted by condition (11). Then

$$(13) \qquad \max|f(P, Q)| = |f(-\tfrac{1}{4}, +\tfrac{1}{2})| = (4s + 1)/16.$$

Proof. The maximum of $|f(P, Q)|$ is achieved on the boundary of the square defined by condition (11) because of the homogeniety. Thus, since $f(tP, tQ) = t^2 f(P, Q)$, the larger $|t|$ becomes, the larger $|f|$ becomes. For the boundary we use the symmetry involved in the condition $f(P, Q) = f(-P, -Q)$; hence we can consider $f(P, \frac{1}{2})$ and $f(\frac{1}{2}, Q)$, for $-\frac{1}{2} \le P \le \frac{1}{2}$ and $-\frac{1}{2} \le Q \le \frac{1}{2}$. We can differentiate and we find

$$\partial f(P, \tfrac{1}{2})/\partial P = 0 \quad \text{at} \quad (P, Q) = (-\tfrac{1}{4}, +\tfrac{1}{2}),$$

$$\partial f(\tfrac{1}{2}, Q)/\partial Q = 0 \quad \text{at} \quad (P, Q) = (\tfrac{1}{2}, 1/(4s)).$$

Then comparing with "end points", $|f(-\frac{1}{4}, \frac{1}{2})| = (4s + 1)/16, f(\frac{1}{2}, 1/(4s)) = (4s + 1)/(16s)$, $|f(-\frac{1}{2}, \frac{1}{2})| = s/4$, and $|f(\frac{1}{2}, \frac{1}{2})| = |(2 - s)/4|$, we see the maximum is as indicated in (13). Q.E.D.

THEOREM 9. The set of all integers $\mathfrak{O}_1$ in $R(\sqrt{5})$ or $R(\sqrt{13})$ are Euclidean.

Proof. We just verify that with $s = (D - 1)/4$ these values of D make $(4s + 1)/16 < 1$ and are $\equiv 1 \pmod 4$; thus Lemma 12 applies. Q.E.D.

The Euclidean algorithm is, however, *not excluded* when $D > 16$, for the "closest" (a, b) to $(A_1/B, A_2/B)$ need not be the ones given by (10) and (11) for Lemma 12.

It has been proved that the Euclidean algorithm is valid for a variety of cases including the integral domain of all integers $(\mathfrak{O}_1)$ of the fields of $R(\sqrt{D_0})$ for[1]

$$D_0 = 2, 3, 5, 6, 7, 11, 13, 17, 19, 21, 29, 33, 37, 41, 57, 73.$$

Recently Davenport (1946) proved that these are the only such fields.

There are unique factorization fields that are not Euclidean in the real and in the complex case. The first is $D_0 = 14$ and $D_0 = -19$. Further information can be found in Table 3 in the appendix.

EXERCISE 12. Verify that $\mathfrak{O}_1$ in $R(\sqrt{-5})$ is not Euclidean directly by taking a ratio α/B of two numbers of type $a + b\sqrt{-5}$ and showing no suitable γ exists for (1).

EXERCISE 13. Do the same for $\mathfrak{O}_2$ in $R(\sqrt{-3})$ (where the integers are $a + b\sqrt{-3}$).

EXERCISE 14. For which complex non-Euclidean integral domains $\mathfrak{O}_n$ will the crucial α/B [for which no γ satisfies equation (1)] necessarily be the extremities of a zone?

EXERCISE 15. Verify that $r^2 - 6s^2 = R^2 - 6S^2$ if $R + S\sqrt{6} = (r + s\sqrt{6})(5 + 2\sqrt{6})^n$. From this find an additional value of (a, b) for which (6*b*) holds that is further from (0, 0) than (14, 6) when $D = 6$.

[1] Attention is called to the fact that $D_0 = 97$ had been incorrectly listed in the literature for several years. (See bibliography.)

EXERCISE 16. Show that the equations $A^2 - 14B^2 = \pm 1, \pm 2, \pm 3$ cannot have solutions in which both A and B are odd by considering both sides modulo 8 and modulo 3. From this show that with $\alpha/B = (1 + \sqrt{14})/2$ no γ exists to satisfy (1) and thus $\mathfrak{O}_1$ for $R(\sqrt{14})$ is not Euclidean.

*9. Pell's Equation

The study of units goes back to antiquity and precedes all other results in algebraic number theory. Euclid, in one way or another, knew formulas like

$$1 = (\sqrt{2} - 1)(\sqrt{2} + 1)$$

by properties of circles. Given a circle of unit radius, at an exterior point of distance $\sqrt{2}$ from the center tangents of length 1 can be drawn. (One need only visualize a circumscribing square with diagonal.) The next step was taken by Archimedes (who worked with $\sqrt{3}$), but who, in essence, built up recursion formulas. For instance, let

$$u_n + v_n\sqrt{2} = (1 + \sqrt{2})^n, \tag{1}$$

then

$$u_n{}^2 - 2v_n{}^2 = (-1)^n, \tag{2}$$

as we see by taking norms. Actually, the binomial expansion was not present in those times but this type of formula was discovered:

$$(u_n + v_n\sqrt{2})(1 + \sqrt{2}) = (u_n + 2v_n) + \sqrt{2}(u_n + v_n) = u_{n+1} + v_{n+1}\sqrt{2},$$

$$\begin{cases} u_{n+1} = u_n + 2v_n, \\ v_{n+1} = u_n + v_n, \end{cases} \qquad (u_1 = v_1 = 1). \tag{3}$$

From the last two equations we learn that by substitution

$$u_{n+1}^2 - 2v_{n+1}^2 = -(u_n{}^2 - 2v_n{}^2)$$

directly, without even the intervention of radicals.

Historically, the equation

$$x^2 - my^2 = \pm 1 \tag{4}$$

attracted a great deal of interest. Eventually Euler named it after Pell (a seventeenth century mathematician). There is a straightforward computational procedure for determining all solutions by continued fractions, which we do not consider here. The important feature for us to recognize, assuming $m > 0$ but not a perfect square, is that all such solutions come from units in $\mathfrak{O}_1$ the integral domain associated with $R(\sqrt{m})$.

EXERCISE 17. Write out a general rule for determining exactly to which integral domains $\mathfrak{D}_n$ the unit $x + y\sqrt{m}$ belongs if x and y satisfy (4).

EXERCISE 18. Find the general solution of the six equations

$$x^2 - 20y^2 = 4, -4, \pm 4, 1, -1, \pm 1$$

in terms of powers of some fundamental unit. *Hint.* Start with $(1 + \sqrt{5})/2$.

EXERCISE 19. Show that if $\eta = (a + b\sqrt{D})/2$ is a unit of norm ± 1, η^3 has the form $a' + b'\sqrt{D}$ (with "no denominator"). *Hint.* Show $\eta^3 = \eta(a^2 \mp 1) \mp a$.

**10. Fields of Higher Degree

Although we devote most of our attention to quadratic fields, in the process we would do well to note briefly to what extent the material becomes unified under the study of fields of arbitrary degree. In fact, the theory of units illustrates this unification. Indeed, the persistent dichotomy between indefinite and definite quadratic forms seems less severe when referred to fields of arbitrary degree. We shall merely make some relevant statements (without proof).

A unit is, in general, an algebraic integer which divides 1.

THEOREM 10 (Dirichlet). Let a field $R(\theta)$ be generated over the rationals by means of an algebraic number θ which has r real and $2s$ complex conjugates (so that $r + 2s = n$, the degree of the equation for θ). Then in the integral domain $\mathfrak{D}$ corresponding to $R(\theta)$, for some definite root, the most general unit is given by assigning integral exponents t_i in

$$\omega = \rho^{t_0}\eta_1^{\,t_1} \cdots \eta_m^{\,t_m}, \qquad (m = r + s - 1). \tag{1}$$

Here ρ is an imaginary root of unity of finite degree. Thus $\rho^g = 1$, whereas $\eta_1, \cdots, \eta_m$ are a set of-so-called fundamental units in $R(\theta)$ (which cannot be replaced by fewer units). In effect t_0 is determined modulo g, but the other t_i take on all integral values and all the units are uniquely given by formula (1).

We observe the quadratic case:

If $D < 0$, $m = 0$ (there is no fundamental unit), but when $D = -1$, $\rho = i$ and $\rho^4 = 1$; when $D = -3$, $\rho = \frac{1}{2} + \sqrt{-\frac{3}{2}}$ and $\rho^6 = 1$; otherwise, $\rho = -1$ and $\rho^2 = 1$.

If $D > 0$, $m = 1$ (or there is always a fundamental unit), and $\rho = -1$, $(\rho^2 = 1)$.

Despite superficially promising appearances, it takes much more than this formula to unify indefinite and definite quadratic forms. Indeed, it is necessary to re-examine the foundations in a manner well advanced[1] beyond the present work.

[1] See the Concluding Survey.

To conclude this section, we note that the problem of finding units is in general extremely important and also extremely difficult. The key to Fermat's last theorem, indeed, is buried in the problem of finding "cyclotomic" units in $R(\rho)$ where $\rho = \exp 2\pi i/p$, a primitive pth root of unity for p prime ≥ 5. It is seen that the field $R(\rho)$ is of degree $p - 1$, since the equation

$$(2) \quad f(\rho) = (\rho^p - 1)/(\rho - 1) = \rho^{p-1} + \rho^{p-2} + \cdots + \rho + 1 = 0$$

happens to be irreducible. It is easily shown that the imaginary roots of unity are ρ^{t_0}, and we further observe that $r = 0$, $s = (p - 1)/2$ as all $(p - 1)$ conjugates of ρ are imaginary (namely $\exp 2\pi k i/p = \rho^k$, $k = 1, 2, \cdots, p - 1$). There are many real units that one can construct, such as $\xi_t = \rho^t + 1/\rho^t = 2 \cos 2\pi t/p$. Yet a complete set of fundamental units is generally unknown. As an example of irregular behavior, the set $\xi_1, \xi_2, \cdots \xi_{s-1}$ does serve as a fundamental system for $p = 5, 7, 11$, but not for larger p such as 17. (Compare Exercise 19, Chapter III, §10, and Exercise 22, below).

EXERCISE 20. Verify that each of the following are units for the appropriate field: $1 + 2^{1/3} + 2^{2/3}$, $4 + 3 \cdot 3^{1/3} + 2 \cdot 3^{2/3}$. Note that the conjugates are formed as in Chapter III, §10, Exercise 22.

EXERCISE 21. Show that in (2), the quantities ρ^k are all roots unless $p \mid k$. Show that $\rho^t + 1/\rho^t = (\rho^t + i)(\rho^t - i)/\rho^t$ is a unit by using

$$\prod_{k=1}^{p-1} (\rho^k - a) = f(a).$$

EXERCISE 22. Show that when $p = 17$ the roots $\xi_t = 2 \cos 2\pi t/17$, $(1 \leq t \leq 7)$ are not fundamental by showing $\xi_3 \xi_5 \xi_6 \xi_7 = -1$.

chapter VII

Unique factorization into ideals

1. Set Theoretic Notation

The failure of unique factorization led Dedekind (1871) to the introduction of "ideal" factors which consist of special types of modules rather than individual "idealized" numbers. Before going into detail, it is necessary to review the basic terminology of sets now commonly accepted, which was introduced primarily for this purpose.[1]

For convenience, we use module notation $\mathfrak{M}, \mathfrak{N}, \cdots$ to denote the sets (restricted to sets of algebraic integers with the notation $\alpha, \beta, \cdots$ for the elements).

We say one set of elements $\mathfrak{M}$ *contains* another set $\mathfrak{N}$, i.e.,

$$\mathfrak{M} \supseteq \mathfrak{N} \quad \text{or} \quad \mathfrak{N} \subseteq \mathfrak{M}$$

if every element of $\mathfrak{N}$ belongs to $\mathfrak{M}$. The converse relation may or may not hold. If $\mathfrak{M}$ includes $\mathfrak{N}$ but $\mathfrak{N}$ does not include $\mathfrak{M}$, we write $\mathfrak{M} \supset \mathfrak{N}$ or $\mathfrak{N} \subset \mathfrak{M}$, and we call this *strict* inclusion. If $\mathfrak{M} \supseteq \mathfrak{N}$ and $\mathfrak{N} \supseteq \mathfrak{M}$, we say $\mathfrak{M} = \mathfrak{N}$ or the sets are the same.

If an element α *belongs* to $\mathfrak{M}$, we write

$$\alpha \in \mathfrak{M} \quad \text{or} \quad \mathfrak{M} \ni \alpha.$$

Actually the $\in$ and $\ni$ behave very much like the $\subseteq$ and $\supseteq$ and only the fear

[1] The set-theoretic concept, indeed, proved so satisfactory that Dedekind later (1872) introduced a set-theoretic definition of irrationals, known as the "Dedekind cut."

of esoteric logical paradoxes causes us to denote "membership" by a different symbol than "inclusion."

The set consisting of a finite or infinite set of elements would be denoted by $\{\alpha, \beta\}, \{\alpha, \beta, \gamma, \cdots\}$. Thus $\{\alpha\} \subset \mathfrak{M}$ means precisely the same as $\alpha \in \mathfrak{M}$. For instance, $\mathfrak{M} \subseteq \mathfrak{N}$ will mean that $\alpha \in \mathfrak{M}$ implies $\alpha \in \mathfrak{N}$.

Actually, the set terminology has been subsequently enlarged to become a calculus of propositions with symbols for implication, conjunction (and), disjunction (or), etc. We shall refrain from overindulgence in symbolic language. We shall, however, use the negations $\not\subset$, $\not\supset$, $\notin$, $\not\ni$, as well as $\neq$.

The union of two sets $\mathfrak{M}$ and $\mathfrak{N}$ is the set consisting of all elements in $\mathfrak{M}$, $\mathfrak{N}$ or both. The union is denoted by $\mathfrak{M} \cup \mathfrak{N}$.

The intersection of two sets $\mathfrak{M}$ and $\mathfrak{N}$ is the set of elements common to both. The intersection is denoted by $\mathfrak{M} \cap \mathfrak{N}$.

We define the product $\alpha\mathfrak{M}$ as the aggregate $\{\alpha\xi\}$ where $\xi \in \mathfrak{M}$. Thus $\alpha(\beta\mathfrak{M}) = (\alpha\beta)\mathfrak{M}$, etc.

We define the conjugate of a set $\mathfrak{M}$ as the set of conjugates denoted by $\mathfrak{M}'$. Thus $(\alpha\mathfrak{M})' = \alpha'\mathfrak{M}'$, etc.

There is a vast literature on Boolean algebra dedicated to the manipulations of the symbols $\cup$, $\cap$, $\supseteq$, $\supset$, $\subseteq$, $\subset$, $=$, $\in$, $\ni$, and negations. We shall carry this out only for a special type of module where the operations are quite rewarding in their consequences.

EXERCISE 1. Show that the intersection of two modules is also a module as well as the product of a module by an algebraic integer using the set-theoretic notation whenever possible.

2. Definition of Ideals

We start with $\mathfrak{O}_n$, a quadratic integral domain. We define an *ideal* $\mathfrak{a}$ *in* $\mathfrak{O}_n$ (denoted by lower-case gothic letters) as a module in $\mathfrak{O}_n$ with the special property that if $\xi \in \mathfrak{O}_n$ then $\xi\alpha \in \mathfrak{a}$. In symbols, if

(1) $$\alpha, \beta \in \mathfrak{a}, \qquad \xi \in \mathfrak{O}_n,$$

then,

(2) $$\alpha \pm \beta \in \mathfrak{a} \text{ (property valid for modules)},$$

(3) $$\alpha\xi \in \mathfrak{a} \text{ (property distinguishing ideals)}.$$

There are two ways to look at ideals. One way is to regard the ideal as a module with a definite module basis and to treat each element in terms of coordinates. The other is to define ideals set theoretically by (1 to 3) with greater freedom from details of notation.

In favor of the second method it must be noted that even at the start the definition (1 to 3) would prove burdensome if it had to be related to the module basis $\mathfrak{a} = [a, b + c\omega_n]$, where ω_n is defined in Chapter IV, §10. Specifically, with

$$\alpha = ax + (b + c\omega_n)y,$$
$$\xi = r + s\omega_n, \qquad \alpha\xi = X + \omega_n Y,$$

we would have to write: For all integers r, s, x, y there exist integers X, Y such that

$$(r + s\omega_n)[ax + (b + c\omega_n)y] = aX + (b + c\omega_n)Y. \tag{4}$$

Yet the value of the module concept is not wholly computational. For example, an ideal, regarded as a module, contains rational integers such as j, the index of the module. (See Chapter IV, Exercise 7.) Actually, the module basis approach is required for calculations with quadratic forms but not for factorization calculations, as we shall see later on.

The reader might verify that the definition can be concisely expressed analogously to (4) as follows:

$$\text{If} \quad \alpha, \beta \in \mathfrak{a} \quad \text{and} \quad \xi, \eta \in \mathfrak{O}_n, \quad \text{then} \quad \alpha\xi + \beta\eta \in \mathfrak{a}. \tag{5}$$

We consider the set formed by the conjugates of the elements of an ideal $\mathfrak{a}$ in $\mathfrak{O}_n$. They are seen (in Exercise 3, below) to form an ideal in $\mathfrak{O}_n$, called the *conjugate ideal* and denoted by $\mathfrak{a}'$, like conjugate elements.

The ideal has the motivating property that a congruence is a more useful concept than for a module, For example, if $\mathfrak{M}$ is a *module* then a congruence can be subjected only to module operations:
If

$$\alpha \equiv \beta \pmod{\mathfrak{M}}, \qquad \alpha, \beta \in \mathfrak{O}_n$$

and

$$\gamma \equiv \delta \pmod{\mathfrak{M}}, \qquad \gamma, \delta \in \mathfrak{O}_n,$$

then

$$\alpha \pm \gamma \equiv \beta \pm \delta \pmod{\mathfrak{M}}.$$

For an ideal $\mathfrak{a}$, however, we also have the *multiplication*:
If

$$\alpha \equiv \beta \pmod{\mathfrak{a}}, \qquad \alpha, \beta \in \mathfrak{O}_n,$$

then

$$\omega\alpha \equiv \omega\beta \pmod{\mathfrak{a}}, \quad \text{for any } \omega \in \mathfrak{O}_n.$$

Thus (with $\beta = 0$) we see that "members of an ideal" serves as a generalization of "multiples of an integer" from rational to algebraic numbers.

EXERCISE 2. Verify in detail the equivalence of (1 to 3) and (5). Show that the set $\xi\mathfrak{O}_n$ for $\xi \in \mathfrak{O}_n$ forms an ideal. Show that the intersection of two ideals forms an ideal.

EXERCISE 3. Show that the set of conjugates of elements of an ideal form an ideal.

EXERCISE 4. If an ideal in $\mathfrak{O}_n$ contains two relatively prime rational integers, it consists of $\mathfrak{O}_n$.

3. Principal Ideals

Starting with α, a fixed element of $\mathfrak{O}_n$, we define the *principal ideal in* $\mathfrak{O}_n$

$$\mathfrak{a} = (\alpha) \tag{1}$$

as the set of $\alpha\xi$ where $\xi \in \mathfrak{O}_n$.

THEOREM 1. The ideal equality

$$(\alpha) = (\beta)$$

is valid if and only if α/β is a unit or $\alpha = \beta = 0$.

Proof. The proof is simple since $(\alpha) \ni \beta$; hence

$$\beta = \alpha\xi \text{ for some } \xi \in \mathfrak{O}_n; \tag{2}$$

likewise $(\beta) \ni \alpha$; hence

$$\alpha = \beta\eta \text{ for some } \eta \in \mathfrak{O}_n, \text{ whence, unless } \alpha = \beta = 0, \tag{3}$$

the substitution of (2) into (3) gives $\xi\eta = 1$. Q.E.D.

Otherwise expressed, the principal ideals generated by elements of an integral domain *identify all associates* of a given element with one another.

An integral domain in which all ideals are principal is called a *principal ideal integral domain.*

THEOREM 2. An integral domain with the Euclidean algorithm is a principal ideal integral domain.

Proof. We simply note that for any ideal $\mathfrak{a}$, we can define α as the element of $\mathfrak{a}$ with minimal positive norm. By the argument in Chapter VI, §10, $\mathfrak{a} = (\alpha)$. Q.E.D.

Thus for the fields listed in Chapter VI §8 $\mathfrak{O}_1$ has ideal factorization corresponding exactly to the ordinary factorizations. We shall see later that ideal factorizations are unique; hence we have another proof that ordinary factorizations into indecomposables[1] are unique when the Euclidean algorithm holds.

[1] In some elementary texts ideals are defined in the integral domain of rational integers for this purpose. For simplicity, we consider only ideals in $\mathfrak{O}_n$; thus (m) means $m\mathfrak{O}_n$.

As special cases $\mathfrak{O}_n = (1) = (\eta)$, where η is any unit of $\mathfrak{O}_n$ This is often called the *unit* ideal. Also the ideal (0) contains only 0, common to all ideals. Then $(0) \subseteq \mathfrak{a} \subseteq (1)$ for any ideal. The zero ideal is excluded from all consideration to simplify the discussion of factorization.

As a matter of practice, the symbols α and (α) might be confused and it may be necessary to distinguish the ideal $((7 - \sqrt{3})^3(7 - \sqrt{3}))$ from the product $(7 + \sqrt{3})^3(7 - \sqrt{3})$ by the use of additional parentheses (when the product is inconvenient to "write out").

As a further convenience, we shall speak of "ideal" rather than "ideal in $\mathfrak{O}_n$" when the context is clear.

4. Sum of Ideals, Basis

We define the sum of ideals as the set

$$\mathfrak{a} + \mathfrak{b} = \{\alpha + \beta\} \quad \text{where} \quad \alpha \in \mathfrak{a}, \qquad \beta \in \mathfrak{b}. \tag{1}$$

We must note first that the sum of two ideals is an ideal. To see this we use the definition (5) of §2 directly: let $\alpha_1 + \beta_1$ and $\alpha_2 + \beta_2$ be formed with $\alpha_i \in \mathfrak{a}$, $\beta_i \in \mathfrak{b}$; then (for the definition) form the quantity

$$\rho = \xi(\alpha_1 + \beta_1) + \eta(\alpha_2 + \beta_2) \quad \text{where} \quad \xi, \eta \in \mathfrak{O}_n.$$

But now $\rho = \alpha_3 + \beta_3$, where

$$\begin{cases} \alpha_3 = \xi\alpha_1 + \eta\alpha_2 \in \mathfrak{a}, \\ \beta_3 = \xi\beta_1 + \eta\beta_2 \in \mathfrak{b}. \end{cases}$$

Thus $\rho \in \mathfrak{a} + \mathfrak{b}$.

It is trivial to verify

$$\mathfrak{a} + \mathfrak{b} = \mathfrak{b} + \mathfrak{a}, \qquad \text{(Commutative law)} \tag{2}$$

$$\mathfrak{a} + (\mathfrak{b} + \mathfrak{c}) = (\mathfrak{a} + \mathfrak{b}) + \mathfrak{c}, \quad \text{(Associative law)} \tag{3}$$

$$\mathfrak{a} + \mathfrak{b} \supseteq \mathfrak{a}. \tag{4}$$

We next introduce the notations

$$\mathfrak{a} + \mathfrak{b} = (\mathfrak{a}, \mathfrak{b}),$$

or, in particular, if $\mathfrak{a} = (\alpha)$, $\mathfrak{b} = (\beta)$, we write

$$\begin{cases} \mathfrak{a} + \mathfrak{b} = (\alpha) + \mathfrak{b} = (\alpha, \beta), \\ \mathfrak{a} + \mathfrak{b} = (\alpha) + (\beta) = (\mathfrak{a}, \beta). \end{cases} \tag{5}$$

The variations of the notation for three or more addends are easily imagined and clearly consistent.

Thus the ideal

$$\mathfrak{a} = (\alpha_1, \alpha_2, \cdots, \alpha_s) = \alpha_1\mathfrak{O}_n + \alpha_2\mathfrak{O}_n + \cdots + \alpha_s\mathfrak{O}_n$$

consists of the aggregate

$$\{\xi_1\alpha_1 + \xi_2\alpha_2 + \cdots + \xi_s\alpha_s\}, \qquad \xi_i \in \mathfrak{O}_n.$$

We say that the algebraic integers $\alpha_1, \alpha_2, \cdots, \alpha_s$ form an *ideal basis* for $\mathfrak{a}$. This should not be confused with the *module* basis in Chapter IV, since ξ_i are algebraic (and not necessarily rational) integers, nor does the set of α_i need to be "minimal" (in the sense of unique representation of ideal elements through the ξ_i).

We should compare the notation in *rational* number theory, $(a, b) = g$ for g the gcd of a and b. It is consistent with the present ideal basis notation in that the basis $(a, b, \alpha, \cdots, \lambda)$, representing an ideal $\mathfrak{a}$ in $\mathfrak{O}_n$, can be replaced by $(g, \alpha, \cdots, \lambda)$. To see this result, consider the relations: $a\mathfrak{O}_n + b\mathfrak{O}_n \ni g$, and $g\mathfrak{O}_n \ni a$ and b, hence $a\mathfrak{O}_n + b\mathfrak{O}_n = g\mathfrak{O}_n$. In particular if $g = 1$, then $\mathfrak{a} = (1)$.

THEOREM 3. Every ideal in a quadratic integral domain has a finite basis.

Proof. Select any element $\alpha_1 \in \mathfrak{a}$, a given ideal. Then, if $\mathfrak{a} = (\alpha_1)$, the theorem is proved. If $\mathfrak{a} \neq (\alpha_1)$, select an $\alpha_2 \in \mathfrak{a}$, $\alpha_2 \notin (\alpha_1)$.

If $\mathfrak{a} = (\alpha_1, \alpha_2)$, the theorem is proved, otherwise let $\alpha_3 \in \mathfrak{a}$, $\alpha_3 \notin (\alpha_1, \alpha_2)$, etc.

We achieve what is called an *ascending chain* of ideals:

$$(6) \qquad (\alpha_1) \subset (\alpha_1, \alpha_2) \subset (\alpha_1, \alpha_2, \alpha_3) \subset \cdots \subset \mathfrak{a}.$$

No two consecutive elements of the chain are equal, since, generally, $\alpha_n \notin (\alpha_1, \alpha_2, \cdots, \alpha_{n-1})$. We need only show the so-called *ascending chain condition*, namely, *every ascending chain of ideals under inclusion is finite*; hence at some point $\mathfrak{a} = (\alpha_1, \alpha_2, \cdots, \alpha_s)$. In our context the condition is satisfied very cheaply by recalling that ideals are a special kind of module.

THEOREM 4. If $\mathfrak{M}$ is a given module in $\mathfrak{O}_n$, there is only a finite number of modules $\mathfrak{X}$ between $\mathfrak{M}$ and $\mathfrak{O}_n$:

$$\mathfrak{M} \subset \mathfrak{X} \subset \mathfrak{O}_n.$$

Proof. The index $[\mathfrak{O}_n/\mathfrak{X}] \leq$ index $[\mathfrak{O}_n/\mathfrak{M}]$, since in $\mathfrak{O}_n$ two elements α, β where $\alpha \not\equiv \beta \pmod{\mathfrak{X}}$ satisfy $\alpha \not\equiv \beta \pmod{\mathfrak{M}}$. The index fixes the value ac in the canonical basis, thereby restricting a, b, and c. Specifically,

$$\mathfrak{X} = [a, b + c\omega] \qquad 0 \leq b < a, \qquad 0 < c,$$

whence only a finite number of $\mathfrak{X}$ is possible. Q.E.D.

COROLLARY. There is only a finite number of modules with a fixed bound on the index.

To relate the ideal basis with the more general module basis, we note the following:

THEOREM 5. If the module $[\alpha, \beta]$ forms an ideal, then $[\alpha, \beta] = (\alpha, \beta)$, so that the module basis serves as an ideal basis.

Proof: We first note trivially that $(\alpha, \beta) \supseteq [\alpha, \beta]$, whether or not $[\alpha, \beta]$ is an ideal; but if $[\alpha, \beta]$ is an ideal, from the fact that $[\alpha, \beta] \ni \alpha$ and $[\alpha, \beta] \ni \beta$ it necessarily follows that $[\alpha, \beta] \supseteq (\alpha) + (\beta) = (\alpha, \beta)$. Q.E.D.

EXERCISE 5. Show that the module $[1 + \sqrt{2}, 1 - \sqrt{2}]$ is no ideal by (*a*) working directly from the definition of an ideal and by (*b*) showing that this module represents only those rational integers that are *even*, whereas the ideal $(1 + \sqrt{2}, 1 - \sqrt{2})$ is (1).

EXERCISE 6. Show that the elements of the ideals $(\alpha_1, \alpha_2, \cdots, \alpha_s)$ and $(\alpha_1', \alpha_2', \cdots, \alpha_s')$ are conjugates.

5. Rules for Transforming the Ideal Basis

It is easily seen that the *number* of elements in the *ideal* basis cannot be determined from the single fact that a quadratic module has *two* elements in the *module* basis. The greater flexibility of the ideal basis is emphasized by these simpler rules of transformation:

(a) The elements of an ideal basis may be rearranged.

(b) Repeated elements can be omitted.

(c) Any ideal element can be inserted in the basis as an additional element.

(d) Any basis element can be omitted if it is a member of the ideal determined by the other basis elements.

(e) In particular, a zero basis element can be omitted.

(f) A basis element can be replaced by its product with a unit.

We can verify that each law follows from definition in a manner very much like that of module theory (Chapter IV, §9). Rules (a), (c), and (d) easily are a minimal set of rules.

Indeed we can always reduce a quadratic ideal to two-element form by using basis operations for an ideal, since they include (among others) basis operations for a module. Yet it will not always be clear (as in the case of modules) when two *ideal bases* determine equal ideals unless each ideal is written as a module and reduced to canonical form by the method of Chapter IV, §9. For example, if $\mathfrak{O}_n = [1, \omega_n]$, then for any ideal: $(\alpha, \beta, \cdots) = \alpha\mathfrak{O}_n + \beta\mathfrak{O}_n + \cdots = [\alpha, \alpha\omega_n, \beta, \beta\omega_n, \cdots]$.

Yet, as the examples below will show, the *ideal* basis is generally more flexible than the module basis and is preferred for factorization problems.

EXERCISE 7. Let $\mathfrak{D}_n = \mathfrak{D}_1 = [1, \sqrt{3}]$. Verify step by step from the above laws:

Statement. $(33, 7 - 3\sqrt{3}) = (4 + 3\sqrt{3})$.

Proof. $(33, 7 - 3\sqrt{3}) = (33, 7 - 3\sqrt{3}, (7 - 3\sqrt{3})(7 + 3\sqrt{3})) = (33, 7 - 3\sqrt{3}, 22) = (33, 22, 7 - 3\sqrt{3}, 11) = (7 - 3\sqrt{3}, 11) = (4 + 3\sqrt{3}, 11) = ((4 + 3\sqrt{3}), (4 + 3\sqrt{3})(4 - 3\sqrt{3})) = (4 + 3\sqrt{3})$.

EXERCISE 8. Verify $(13, 7 + 5\sqrt{3}) = (4 + \sqrt{3})$. *Hint*. Solve $(7 + 5\sqrt{3})(x + y\sqrt{3}) = z + \sqrt{3}$, and verify $4 + \sqrt{3} \mid 7 + 5\sqrt{3}$.

EXERCISE 9. Verify $(1 + \sqrt{3}) = (1 - \sqrt{3})$.

EXERCISE 10. Verify $(4 + \sqrt{3}) \neq (4 - \sqrt{3})$. *Hint*. Show $(4 + \sqrt{3}, 4 - \sqrt{3}) = (1)$.

6. Product of Ideals, the Critical Theorem, Cancellation

We next define the *product* $\mathfrak{ab}$ of two ideals $\mathfrak{a}$ and $\mathfrak{b}$ as the ideal $\mathfrak{c}$ "generated by all products" $\alpha\beta$ where $\alpha \in \mathfrak{a}$, $\beta \in \mathfrak{b}$, or, more precisely, the aggregate of finite linear combinations ρ:

$$(1) \qquad \begin{cases} \rho = \sum_{i=1}^{q} \sum_{j=1}^{r} \alpha^{(i)} \beta^{(j)} \xi_{ij}, \\ \alpha^{(i)} \in \mathfrak{a}, \qquad \beta^{(j)} \in \mathfrak{b}, \qquad \xi_{ij} \in \mathfrak{D}_n. \end{cases}$$

It is clear that the set $\{\rho\}$ forms an ideal from the very definition of ideal. This definition of product is in no way dependent on any basis. We can write $\{\rho\} = \mathfrak{ab}$.

For convenience, we note that if

$$\mathfrak{a} = (\alpha_1, \cdots, \alpha_s)$$

$$\mathfrak{b} = (\beta_1, \cdots, \beta_t)$$

are bases then

$$\mathfrak{ab} = (\alpha_1\beta_1, \alpha_1\beta_2, \alpha_2\beta_1, \cdots, \alpha_s\beta_t).$$

If we call the right-hand ideal $\mathfrak{c}$, we find, easily, $\mathfrak{ab} \supseteq \mathfrak{c}$. On the other hand, any $\alpha^{(i)} = \sum_{k=1}^{s} \alpha_k \lambda_k^{(i)}$, $\beta^{(j)} = \sum_{l=1}^{t} \beta_l \mu_l^{(j)}$, where $\lambda_k^{(i)}, \mu_l^{(j)} \in \mathfrak{D}_n$. Thus any ρ of (1) satisfies

$$\rho = \sum_{i,j} \alpha^{(i)} \beta^{(j)} \xi_{ij} = \sum_{k,l} \alpha_k \beta_l \left(\sum_{i,j} \xi_{ij} \lambda_k^{(i)} \mu_l^{(j)} \right) \in \mathfrak{c},$$

or $\mathfrak{ab} \subseteq \mathfrak{c}$, and $\mathfrak{ab} = \mathfrak{c}$. Q.E.D.

As a special case, $(\alpha)(\beta) = (\alpha\beta)$, and the product of principal ideals is principal. The reader should show carefully (Exercise 11, below) that

$$\alpha\mathfrak{a} = (\alpha)\mathfrak{a}. \tag{2}$$

The parentheses on α could be omitted if desired, but we often leave the parentheses for emphasis. In particular, $(1)\ \mathfrak{a} = \mathfrak{a}$.

The existence of a quotient is more difficult to establish, and we shall first make some easier remarks.

LEMMA 1. If every element α of $\mathfrak{a}$ is divisible by a fixed nonzero element γ in $\mathfrak{O}_n$, then an ideal $\mathfrak{b}$ exists such that $\mathfrak{b} =$ the set $\{\alpha/\gamma\}$ and $\mathfrak{a} = (\gamma)\mathfrak{b}$.

LEMMA 2. If $\gamma\mathfrak{a} = \gamma\mathfrak{b}$ and $\gamma \neq 0$, then $\mathfrak{a} = \mathfrak{b}$.

It is further verified by proofs that we leave to the reader:

$$\mathfrak{a}\mathfrak{b} = \mathfrak{a}\mathfrak{b}, \qquad \text{(Commutative law)} \tag{3}$$

$$\mathfrak{a}(\mathfrak{b}\mathfrak{c}) = (\mathfrak{a}\mathfrak{b})\mathfrak{c}, \qquad \text{(Associative law)} \tag{4}$$

$$\mathfrak{a}(\mathfrak{b} + \mathfrak{c}) = \mathfrak{a}\mathfrak{b} + \mathfrak{a}\mathfrak{c}. \qquad \text{(Distributive law)} \tag{5}$$

We now say *ideal* $\mathfrak{a}$ *divides ideal* $\mathfrak{c}$ in $\mathfrak{O}_n$ (or $\mathfrak{a} \mid \mathfrak{c}$) if and only if an ideal $\mathfrak{b}$ exists in $\mathfrak{O}_n$ for which $\mathfrak{c} = \mathfrak{a}\mathfrak{b}$. Symbolically, we can write $\mathfrak{b} = \mathfrak{c}/\mathfrak{a}$.

From (1), if $\rho \in \mathfrak{a}\mathfrak{b}$, then $\rho \in \mathfrak{a}$. Thus $\mathfrak{a}\mathfrak{b} \subseteq \mathfrak{a}$, or *every divisor of an ideal contains the ideal.* This is like stating, in rational number theory, that all multiples of 6 are even (are contained among the multiples of 2) because $2 \mid 6$.

Analogously with Chapter VI, §2, we extend the definitions of indecomposable element in $\mathfrak{O}_n$ and prime element in $\mathfrak{O}_n$.

An *indecomposable ideal in* $\mathfrak{O}_n$ is an ideal $\mathfrak{q}$ in $\mathfrak{O}_n$ other than the (unit) ideal $\mathfrak{O}_n$, which has no ideal in $\mathfrak{O}_n$ as divisor other than $\mathfrak{q}$ and $\mathfrak{O}_n$.

A *prime ideal in* $\mathfrak{O}_n$ is an ideal $\mathfrak{p}$ in $\mathfrak{O}_n$ other than the (unit) ideal $\mathfrak{O}_n$, with the property that for any two ideals in $\mathfrak{O}_n$, $\mathfrak{a}$ and $\mathfrak{b}$, if $\mathfrak{p} \mid \mathfrak{a}\mathfrak{b}$, then $\mathfrak{p} \mid \mathfrak{a}$, or $\mathfrak{p} \mid \mathfrak{b}$.

Analogues of Lemma 3 and Theorems 1 and 2 of Chapter VI, §§2 and 3 hold (see Exercise 13 below). The situation is further simplified by the fact that under broad circumstances the indecomposable ideals become precisely the prime ideals. The following two theorems are critical:

THEOREM 6. If $\mathfrak{a}$ is an ideal (not zero) in $\mathfrak{O}_1$, the set of all integers of a quadratic field, then an ideal $\mathfrak{a}^*$ and $\alpha\ (\neq 0)$ in $\mathfrak{O}_1$ exist such that

$$\mathfrak{a}\mathfrak{a}^* = (\alpha).$$

THEOREM 7. If $\mathfrak{a}$ is an ideal (not zero) in the integral domain $\mathfrak{O}_n$ and if $\mathfrak{a}$ contains at least one element γ for which $N(\gamma)$ and n are relatively prime, then an ideal $\mathfrak{a}^*$ and an $\alpha \neq 0$ in $\mathfrak{O}_n$ exist such that

$$\mathfrak{a}\mathfrak{a}^* = (\alpha).$$

We shall assume these results for the present because of their critical nature, and from them we shall prove unique factorization of ideals. For ease of treatment (in §§7 to 10 below) we restrict ourselves to $\mathfrak{O}_1$, ($n = 1$), so that only Theorem 6 is relevant.

An easy consequence of Theorem 6 is cancellation.

THEOREM 8. If $\mathfrak{a}\mathfrak{b} = \mathfrak{a}\mathfrak{c}$, then $\mathfrak{b} = \mathfrak{c}$ (if $\mathfrak{a} \neq 0$).

Proof. Multiply by $\mathfrak{a}^*$. Then $\mathfrak{a}\mathfrak{a}^*\mathfrak{b} = \mathfrak{a}\mathfrak{a}^*\mathfrak{c}$; $(\alpha)\mathfrak{b} = (\alpha)\mathfrak{c}$. Lemma 2 gives the cancellation of α. Q.E.D.

EXERCISE 11. Prove (2) above.

EXERCISE 12. Prove Lemmas 1 and 2 and the distributive law (5) above.

EXERCISE 13. State and prove the analogues of Lemma 3 and Theorem 1 of Chapter VI, §2.

EXERCISE 14. Show that a module basis for the product of ideals $\mathfrak{a}$ and $\mathfrak{b}$ can be formed as follows: If $\mathfrak{a} = [\alpha_1, \cdots, \alpha_s]$ and $\mathfrak{b} = [\beta_1, \cdots, \beta_t]$, then $\mathfrak{a}\mathfrak{b} = [\alpha_1\beta_1, \cdots, \alpha_i\beta_j, \cdots, \alpha_s\beta_t]$. *Hint.* Show $\mathfrak{a}\mathfrak{b} = \alpha_1\mathfrak{b} + \cdots + \alpha_s\mathfrak{b}$ (and point out where ideal properties are needed).

EXERCISE 15. In $\mathfrak{O}_2$ for $R(\sqrt{-3})$ show that if $\mathfrak{a} = (2, 1 + \sqrt{-3})$ then $\mathfrak{a} \neq (2)$, whereas $\mathfrak{a}^2 = 2\mathfrak{a}$ (which contradicts cancellation). Obtain another such contradiction in $\mathfrak{O}_n$ for $n > 2$ by using $\mathfrak{a} = (n, n\sqrt{D_0}) \neq (n)$.

7.[1] "To Contain Is to Divide"

THEOREM 9. If $\mathfrak{a} \mid \mathfrak{c}$ (i.e., if a $\mathfrak{b}$ exists such that $\mathfrak{c} = \mathfrak{a}\mathfrak{b}$), then $\mathfrak{a} \supseteq \mathfrak{c}$. Conversely, if $\mathfrak{a} \supseteq \mathfrak{c}$, then $\mathfrak{a} \mid \mathfrak{c}$ (i.e., a $\mathfrak{b}$ exists such that $\mathfrak{c} = \mathfrak{a}\mathfrak{b}$).

Proof. The first part has been established. For the second part note that if $\mathfrak{a} \supset \mathfrak{c}$ then $\mathfrak{a}\mathfrak{a}^* \supset \mathfrak{c}\mathfrak{a}^*$ (using the terminology of Theorem 6) and $(\alpha) \supseteq \mathfrak{c}\mathfrak{a}^*$; it follows that every element of $\mathfrak{c}\mathfrak{a}^*$ is divisible by α. Thus $\mathfrak{c}\mathfrak{a}^* = (\alpha)\mathfrak{b}$, by Lemma 1. Then

$$\mathfrak{c}\mathfrak{a}^* = (\alpha)\mathfrak{b},$$

$$\mathfrak{c}\mathfrak{a}^*\mathfrak{a} = (\alpha)\mathfrak{a}\mathfrak{b},$$

$$\mathfrak{c}(\alpha) = (\alpha)\mathfrak{a}\mathfrak{b},$$

and by Lemma 2

$$\mathfrak{c} = \mathfrak{a}\mathfrak{b}.$$

[1] For simplicity in §7–10 the ideals are restricted to ideals of an $\mathfrak{O}_1$ type integral domain.

The last step consisted of dividing all elements of $\mathfrak{c}\alpha$ and $\alpha\mathfrak{a}\mathfrak{b}$ by α in obvious fashion. Q.E.D.

Hence the greatest common *divisor* of $\mathfrak{a}$ and $\mathfrak{b}$ becomes the smallest *containing ideal*, or $\mathfrak{a} + \mathfrak{b}$ by Exercise 16 (below). Theorem 4 can be rewritten as follows:

There is only a finite number of ideals dividing (or containing) a given ideal.

We say two ideals $\mathfrak{a}$, $\mathfrak{b}$ are *relatively prime* if there exists no ideal $\mathfrak{c} \neq (1)$ which divides (or contains) $\mathfrak{a}$ and $\mathfrak{b}$.

THEOREM 10. If two ideals $\mathfrak{a}$ and $\mathfrak{b}$ are relatively prime, then $\mathfrak{a} + \mathfrak{b} = (\mathfrak{a}, \mathfrak{b}) = (1)$.

Proof. Suppose $\mathfrak{a} + \mathfrak{b} = \mathfrak{c} \neq (1)$. Then $\mathfrak{c} \supseteq \mathfrak{a}$, $\mathfrak{c} \supseteq \mathfrak{b}$; hence $\mathfrak{c} \mid \mathfrak{a}$ and $\mathfrak{c} \mid \mathfrak{b}$, giving a contradiction. Q.E.D.

COROLLARY. If two ideals $\mathfrak{a}$, $\mathfrak{b}$ are relatively prime, there exist elements $\alpha \in \mathfrak{a}$ and $\beta \in \mathfrak{b}$ for which

$$\alpha + \beta = 1.$$

Proof. $\mathfrak{a} + \mathfrak{b}$ contains the element 1. Q.E.D.

This corollary can be regarded as a subtle version of the rational gcd algorithm: if there exists no t (except ± 1) which divides both the rational integers a and b, then we can solve $ax + by = 1$.

8. Unique Factorization

Unique factorization has two familiar steps: first we factor into indecomposables and then we show that the indecomposables are prime. Note that an ideal $\mathfrak{a}$ in $\mathfrak{O}_1$ which is not the zero ideal can have only a finite number of ideal divisors by Theorem 4, §4. Thus by continued decomposition of the ideal $\mathfrak{a}$ into factors and by the decomposition of these factors in turn, we find the following:

THEOREM 11. Any nonzero ideal in $\mathfrak{O}_1$ can be expressed as the product of a finite number of indecomposable ideals.

THEOREM 12. All indecomposable ideals in $\mathfrak{O}_1$ are prime ideals.

Proof. Let $\mathfrak{q}$ denote an indecomposable ideal, then show that if $\mathfrak{q} \mid \mathfrak{a}\mathfrak{b}$ and $\mathfrak{q} \nmid \mathfrak{a}$ then $\mathfrak{q} \mid \mathfrak{b}$. If $\mathfrak{q} \nmid \mathfrak{a}$, then $(\mathfrak{a}, \mathfrak{q}) = (1)$, or an $\alpha \in \mathfrak{a}$, $\pi \in \mathfrak{q}$ exist such that

$$\alpha + \pi = 1.$$

Then, if β is any element of $\mathfrak{b}$,

$$\alpha\beta + \pi\beta = \beta;$$

but if $\mathfrak{q} \mid \mathfrak{ab}$, $\mathfrak{q} \supseteq \mathfrak{ab}$, $\mathfrak{q} \ni \alpha\beta$, $\mathfrak{q} \ni \pi$, $\mathfrak{q} \ni \pi\beta$, then $\mathfrak{q} \ni \alpha\beta + \pi\beta = \beta$, and $\beta \in \mathfrak{q}$ for all $\beta \in \mathfrak{b}$. Thus $\mathfrak{q} \supseteq \mathfrak{b}$. Hence, necessarily, $\mathfrak{q} \mid \mathfrak{b}$. Q.E.D.

THEOREM 13. Unique factorization. Any nonzero ideal $\mathfrak{a}$ in $\mathfrak{O}_1$ has a unique decomposition into prime ideals.

We proceed as in elementary number theory (see Exercise 13 above). If all ideals are principal, the set of nonzero algebraic integers has unique factorization into prime algebraic integers if we identify associates and ignore the order of the factors. It is not immediate that the occurrence of nonprincipal ideals does indeed ruin the unique factorization into indecomposable algebraic integers, but we shall see this later on. The nature of prime ideals is also discussed in §10 below.

9. Sum and Product of Factored Ideals

We now know that any arbitrary nonzero ideal $\mathfrak{a}$ can be factored uniquely,

$$\mathfrak{a} = \mathfrak{p}_1^{e_1}\mathfrak{p}_2^{e_2} \cdots \mathfrak{p}_t^{e_t}, \qquad e_i \geq 0,$$

where the prime ideals are written $\mathfrak{p}_1, \mathfrak{p}_2, \cdots, \mathfrak{p}_t$. If we consider another nonzero ideal $\mathfrak{b}$, for convenience we can consider $\mathfrak{p}_1, \cdots \mathfrak{p}_t$ to include all prime ideal factors of $\mathfrak{a}$ and $\mathfrak{b}$:

$$\mathfrak{b} = \mathfrak{p}_1^{f_1}\mathfrak{p}_2^{f_2} \cdots \mathfrak{p}_t^{f_t}, \qquad f_i \geq 0,$$

although some e_i and f_i may be zero.

THEOREM 14. $\mathfrak{ab} = \mathfrak{p}_1^{e_1+f_1}\mathfrak{p}_2^{e_2+f_2} \cdots \mathfrak{p}_t^{e_t+f_t}$.

THEOREM 15. $\mathfrak{a} + \mathfrak{b} = \mathfrak{p}_1^{m_1}\mathfrak{p}_2^{m_2} \cdots \mathfrak{p}_t^{m_t}$,

$$m_1 = \min(e_i, f_i).$$

THEOREM 16. $\mathfrak{a} \cap \mathfrak{b} = \mathfrak{p}_1^{M_1}\mathfrak{p}_2^{M_2} \cdots \mathfrak{p}_t^{M_t}$,

$$M_i = \max(e_i, f_i).$$

THEOREM 17. $(\mathfrak{a} + \mathfrak{b})(\mathfrak{a} \cap \mathfrak{b}) = \mathfrak{ab}$.

Proof. Theorem 14 is a result of unique factorization. For Theorem 15 let $\mathfrak{c} = \mathfrak{p}_1^{m_1}\mathfrak{p}_2^{m_2} \cdots \mathfrak{p}_t^{m_t}$, where m_i is the smaller of e_i, f_i (or the common value if they are equal). Then $\mathfrak{a} = \mathfrak{ca}^*$, $\mathfrak{b} = \mathfrak{cb}^*$, and $\mathfrak{a}^*$ and $\mathfrak{b}^*$ are divisible only by primes $\mathfrak{p}_i$; but, if $\mathfrak{p}_i \mid \mathfrak{a}^*$, then $\mathfrak{p}_i \nmid \mathfrak{b}^*$, and vice versa. Thus $\mathfrak{a}^* + \mathfrak{b}^* = (1)$ and

$$\mathfrak{a} + \mathfrak{b} = \mathfrak{c}\mathfrak{a}^* + \mathfrak{c}\mathfrak{b}^* = \mathfrak{c}(\mathfrak{a}^* + \mathfrak{b}^*) = \mathfrak{c}.$$

For Theorem 16 note that if $\mathfrak{a} \ni \alpha$, $\mathfrak{b} \ni \alpha$ then $\mathfrak{a} \mid (\alpha)$, $\mathfrak{b} \mid (\alpha)$ and conversely. This is equivalent to $\mathfrak{q} \mid \alpha$, where

$$\mathfrak{q} = \mathfrak{p}_1^{M_1}\mathfrak{p}_2^{M_2} \cdots \mathfrak{p}_t^{M_t}$$

or "$\mathfrak{a} \ni \alpha, \mathfrak{b} \ni \alpha$" is equivalent to "$\mathfrak{q} \ni \alpha$," which means $\mathfrak{a} \cap \mathfrak{b} = \mathfrak{q}$. Theorem 17 will be recognized as the analogue[1] of "gcd $(a, b) \cdot$ lcm $(a, b) = ab$," and is proved from "$e_i + f_i = m_i + M_i$." Q.E.D.

A remarkable fact is that these theorems, true for *ideals*, are not all true for *integers*, even in a principal ideal domain. Trivially, if $a = 2^1 \cdot$ (odd number) and $b = 2^1 \cdot$ (odd number), $a + b$ is not necessarily $2^1 \cdot$ (odd number). For instance, if $a = 6$, $b = -14$, $a + b = -2^3$. Yet in ideal theory all is simpler:

$$2(3) + 2(-7) = (6) + (-14), \qquad (6, -14) = (2) = 2(1).$$

Theorem 15 also holds for any number of addends. If we define $\text{ord}_{\mathfrak{p}}\mathfrak{a} = e$, (the "order of $\mathfrak{p}$ in $\mathfrak{a}$") as the integer $e \geqq 0$ for which $\mathfrak{p}^e \parallel \mathfrak{a}$ (or $\mathfrak{p}^e \mid \mathfrak{a}$, $\mathfrak{p}^{e+1} \nmid \mathfrak{a}$), we have the following by induction on the number of addends.

THEOREM 18. If $\mathfrak{a}_1, \cdots, \mathfrak{a}_s$ are divisible only by $\mathfrak{p}_1, \cdots, \mathfrak{p}_t$ and no other prime ideals, then

$$\mathfrak{a}_1 + \cdots + \mathfrak{a}_s = \prod_{i=1}^{t} \mathfrak{p}_i^{m_i}, \qquad \text{where } \begin{cases} m_i = \min \text{ord}_{\mathfrak{p}_i}\mathfrak{a}_j, \\ 1 \leqq j \leqq s. \end{cases}$$

EXERCISE 16. Show that $\mathfrak{a} \cup \mathfrak{b}$ is not always an ideal but the smallest ideal containing it is $\mathfrak{a} + \mathfrak{b}$ by setting up the prime factors of each.

EXERCISE 17. Show that for any $\mathfrak{a}$ and any $\mathfrak{b}$ $(\neq (1))$ there exists an α such that $\alpha \in \mathfrak{a}$ but $\alpha \notin \mathfrak{a}\mathfrak{b}$. *Hint.* $\mathfrak{a}\mathfrak{b} \neq \mathfrak{a}$.

EXERCISE 18. Show that if $\mathfrak{a} \mid \mathfrak{c}$ and $\mathfrak{b} \mid \mathfrak{c}$ and $(\mathfrak{a}, \mathfrak{b}) = (1)$, then $\mathfrak{a}\mathfrak{b} \mid \mathfrak{c}$. Give two proofs: (*a*) using unique factorization and (*b*) using Theorem 17 directly.

10. Two-Element Basis, Prime Ideals

We now wish to consider bases and prime ideals further.

THEOREM 19. If $\mathfrak{a} \supseteq \mathfrak{b}$ and neither ideal is zero, then an element α of $\mathfrak{a}$ exists such that $\mathfrak{a} = (\mathfrak{b}, \alpha)$.

[1] Here "lcm" means "least common multiple" (and we assume $a > 0$, $b > 0$ for simplicity).

Proof. Let

$$\mathfrak{a} = \mathfrak{p}_1^{e_1}\mathfrak{p}_2^{e_2} \cdots \mathfrak{p}_t^{e_t}, \qquad e_i \geq 0,$$

$$\mathfrak{b} = \mathfrak{p}_1^{f_1}\mathfrak{p}_2^{f_2} \cdots \mathfrak{p}_t^{f_t}, \qquad f_i \geq 0,$$

using all $\mathfrak{p}_i$ which divide $\mathfrak{a}$ or $\mathfrak{b}$. Now since $\mathfrak{a} \supseteq \mathfrak{b}$, it follows that $\mathfrak{a} \mid \mathfrak{b}$ and $f_t \geq e_t$.

We write

$$\begin{aligned} \mathfrak{a}_1 &= \mathfrak{p}_1^{e_1}\mathfrak{p}_2^{e_2+1} \cdots \mathfrak{p}_t^{e_t+1} \\ \mathfrak{a}_2 &= \mathfrak{p}_1^{e_1+1}\mathfrak{p}_2^{e_2} \cdots \mathfrak{p}_t^{e_t+1}, \\ &\cdots\cdots\cdots \\ \mathfrak{a}_t &= \mathfrak{p}_1^{e_1+1}\mathfrak{p}_2^{e_2+1} \cdots \mathfrak{p}_t^{e_t}. \end{aligned}$$

Let α_1 be so chosen that $\alpha_1 \in \mathfrak{a}_1$ but $\alpha_1 \notin \mathfrak{a}_1\mathfrak{p}_1$. Then

$$(\alpha_1) = \mathfrak{a}_1\mathfrak{q}_1, \qquad \mathfrak{p}_1 \nmid \mathfrak{q}_1,$$

and likewise there exist $\alpha_i \in \mathfrak{a}_i$ such that

$$\begin{aligned} (\alpha_2) &= \mathfrak{a}_1\mathfrak{q}_2, \qquad \mathfrak{p}_2 \nmid \mathfrak{q}_2, \\ &\cdots\cdots\cdots \\ (\alpha_t) &= \mathfrak{a}_t\mathfrak{q}_t, \qquad \mathfrak{p}_t \nmid \mathfrak{q}_t. \end{aligned}$$

Now, by Theorem 18,

$$\mathfrak{j} = (\alpha_1) + (\alpha_2) + \cdots + (\alpha_t) = \mathfrak{p}_1^{e_1}\mathfrak{p}_2^{e_2} \cdots \mathfrak{p}_t^{e_t}\mathfrak{q},$$

where $\mathfrak{p}_i \nmid \mathfrak{q}$ for any i. Thus, if we call

$$\alpha = \alpha_1 + \alpha_2 + \cdots + \alpha_t \in \mathfrak{j},$$

then $\mathfrak{j} \mid (\alpha)$ or $(\alpha) = \mathfrak{p}_1^{g_1}\mathfrak{p}_2^{g_2} \cdots \mathfrak{p}_t^{g_t}\mathfrak{w}$, where $g_i \geq e_i$ and $\mathfrak{p}_i \nmid \mathfrak{w}$ for any i. Actually, each $g_i = e_i$. For example, if $g_1 > e_1$, then by writing

$$-\alpha_1 = \alpha_2 + \cdots + \alpha_t - \alpha$$

we find $\mathfrak{p}_1^{e_1+1} \mid \alpha_2, \cdots, \mathfrak{p}_1^{e_1+1} \mid \alpha_t, \mathfrak{p}_1^{e_1+1} \mid \mathfrak{p}_1^{g_1} \mid \alpha$. Thus $\mathfrak{p}_1^{e_1+1} \mid \alpha_1$ although $\mathfrak{p}_1 \nmid \mathfrak{q}_1$; this is a contradiction. Hence

$$(\alpha) = \mathfrak{p}_1^{e_1}\mathfrak{p}_2^{e_2} \cdots \mathfrak{p}_t^{e_t}\mathfrak{w}, \qquad (\mathfrak{p}_i \nmid \mathfrak{w} \text{ for any } i).$$

Finally

$$\mathfrak{b} + (\alpha) = \mathfrak{a}. \qquad \text{Q.E.D.}$$

THEOREM 20. For any arbitrary nonzero $\alpha_1 \in \mathfrak{a}$, a given ideal, there exists some specially selected α_2 for which

$$\mathfrak{a} = (\alpha_1, \alpha_2).$$

Proof. $\mathfrak{a} \supseteq (\alpha)$, hence Theorem 19 applies. (Note that if $\mathfrak{a} = (\alpha_1)$ then α_2 might equal α_1.) Q.E.D.

THEOREM 21. For any ideal $\mathfrak{a}$ we can find an integer $\alpha \in \mathfrak{a}$ such that

$$(\alpha) = \mathfrak{a}\mathfrak{c},$$

where $\mathfrak{c}$ is relatively prime to any preassigned ideal $\mathfrak{q}$.

Proof. Let $\mathfrak{b} = \mathfrak{a}\mathfrak{q}$ in Theorem 19 and find α so that $\alpha \in \mathfrak{a}$ (whence $\alpha = \mathfrak{a}\mathfrak{c}$) and

$$\mathfrak{a} = (\mathfrak{a}\mathfrak{q}, \alpha).$$

Using the distributive law, we see

$$\mathfrak{a} = (\mathfrak{a}\mathfrak{q}, \mathfrak{a}\mathfrak{c}) = \mathfrak{a}(\mathfrak{q}, \mathfrak{c}).$$

Hence $(\mathfrak{q}, \mathfrak{c}) = 1$. Q.E.D.

THEOREM 22. Every prime ideal $\mathfrak{p}$ belongs to a rational prime p determined uniquely by $\mathfrak{p} \mid (p)$.

Proof. Every ideal $\mathfrak{p}$ contains a rational integer $a \neq 0$ (see §2 above), or $\mathfrak{p} \mid a\mathfrak{O}_1$. Thus, if $a = \pm p_1^{e_1} \cdots p_t^{e_t}$ by the rational unique factorization, then, multiplying both sides by (1), the unit ideal in $\mathfrak{O}_1$, we see

$$(a) = (p_1)^{e_1}(p_2)^{e_2} \cdots (p_t)^{e_t}$$

and $\mathfrak{p}$ divides some prime (say p_i) by unique factorization. In fact, it divides only one p_i as we easily see; for, otherwise, if $\mathfrak{p} \mid p_i$ and $\mathfrak{p} \mid q\ (\neq p_i)$ for some prime q, then

$$\mathfrak{p} \mid (p_i), \quad \mathfrak{p} \mid (q),$$

$$\mathfrak{p} \supseteq (p_i, q) = (1).$$

This contradicts the definition of prime. Q.E.D.

THEOREM 23. Every prime ideal $\mathfrak{p}$ can be written as

$$\mathfrak{p} = (p, \pi), \qquad \text{where } N(\pi) \equiv 0 \pmod{p}.$$

Proof. Use Theorem 19 knowing $\mathfrak{p} \mid (p)$, hence $\mathfrak{p} \ni p$. Therefore, $\mathfrak{p} = (p, \pi)$ for some π. Now $\mathfrak{p} \ni \pi$ and consequently $\mathfrak{p} \ni \pi\pi' = N(\pi)$. Hence if $p \nmid N(\pi)$, $(1) = (p, N(\pi)) \subseteq \mathfrak{p}$, leading to a contradiction. Q.E.D.

COROLLARY. We can even select π so that the module basis is

$$\mathfrak{p} = [p, \pi] = (p, \pi), \qquad \text{where } p \mid N(\pi).$$

Proof. The rational integers a in $\mathfrak{p}$ are multiples of p (lest $\mathfrak{p} \supseteq (a, p) = (1)$). Therefore, when we construct the module basis of $\mathfrak{p}$ [as in Chapter IV, §9, (1)], the rational element is p and the other is (say) π. But by Theorem 5, §4 (above), $\mathfrak{p} = [p, \pi] = (p, \pi) \ni N(\pi)$. Hence $p \mid N(\pi)$.

Q.E.D.

The rational prime (p) has only a finite number of ideal divisors $\mathfrak{p}$ by unique factorization. (See Exercise 13 above.) Actually, in some cases $p \mid \pi$. Then $\mathfrak{p} = (p)$, and the prime ideal is the same as the rational prime as far as factorizations are concerned. In other cases $\pi \mid p$; then $\mathfrak{p} = (\pi)$.

It could still happen that the various possible values π in $\mathfrak{p} = (p, \pi)$ would not satisfy the condition $\pi \mid p$. For instance, in $R(\sqrt{-5})$ in Chapter VI, §6, if $p = 3$, there is no $\pi = a + b\sqrt{-5}$ for which $\pi \mid 3$ (as we saw). Thus a prime divisor of 3, namely $\mathfrak{p}$, is merely nonprincipal, e.g., $\mathfrak{p} = (3, 1 + 2\sqrt{-5})$. It will be seen that $3 = \mathfrak{p}\mathfrak{p}'$ where $\mathfrak{p}' = (3, 1 - 2\sqrt{-5})$ by very general results in Chapter VIII.

The integral domain $\mathfrak{O}$ is a principal ideal domain if and only if all *prime* ideals are principal. This is easily seen if we use unique factorization as well as the fact that the product of two principal ideals is principal. The following is less obvious.

THEOREM 24. The integral domain $\mathfrak{O}_1$ has unique factorization into indecomposables if and only if all ideals are principal.

Proof. If all ideals are principal, then the ideal factors can be identified with algebraic integers by ignoring units.

If some ideal is nonprincipal, then a prime ideal $\mathfrak{p}$ is nonprincipal. Write $\mathfrak{p} = (p, \pi)$, where $p \mid N(\pi)$. But then from $\pi\pi' = pq$ (say) it follows that if there is a unique decomposition there must be an indecomposable $\pi_1 \mid p$ such that $\pi_1 \mid \pi$ or $\pi_1 \mid \pi'$. In the first case $(\pi_1) \mid (p, \pi)$. Hence, since $\mathfrak{p} = (p, \pi)$ is prime, $(\pi_1) = (p, \pi) = \mathfrak{p}$. In the second case, by taking conjugates, $\pi_1{}' \mid p$ (since p is its own conjugate) and $\pi_1{}' \mid \pi$, where $(\pi_1{}') = \mathfrak{p}$, as before. Thus all prime ideals are principal from unique factorization.

Q.E.D.

11. The Critical Theorem and Hurwitz's Lemma

We are now prepared to prove the critical theorems (6 and 7) on which everything else depends. We start with $\mathfrak{O}_n = \mathfrak{O}_1$ the integral domain of all algebraic integers in the field $R(\sqrt{D})$, (D not a perfect square).

HURWITZ'S LEMMA[1]

If α and β are two algebraic integers in $\mathfrak{O}_1$ and if the rational integer g divides $\alpha\alpha'$, $\beta\beta'$ and the sum $\alpha\beta' + \beta\alpha'$, then g also divides the individual numbers $\alpha\beta'$ and $\beta\alpha'$.

[1] This lemma is a weak form of a result applicable to fields of arbitrary degree. The *stronger* result really bears the name Hurwitz's Lemma.

Proof. If $\xi = \alpha\beta'$, $\xi' = \alpha'\beta$, then ξ satisfies the quadratic equation

$$\xi^2 - A\xi + B = 0,$$

where the rational integers are defined as

$$A = \alpha\beta' + \beta\alpha' \qquad (= \xi + \xi'), \tag{1}$$

$$B = \alpha\beta'\beta\alpha' \qquad (= \xi\xi'). \tag{2}$$

Now $g \mid A$, $g^2 \mid B$. Hence ξ/g satisfies

$$\left(\frac{\xi}{g}\right)^2 - \frac{A}{g}\left(\frac{\xi}{g}\right) + \frac{B_2}{g} = 0 \tag{3}$$

with integral coefficients, whence ξ/g, ξ'/g are algebraic integers. Q.E.D.

We are now at the stage at which Dedekind's definition of integer becomes crucial. We need only observe (3). The reason that ξ/g belongs to $\mathfrak{O}_1$ is simply that it satisfies Dedekind's definition! For example, let

$$\mathfrak{O}_2 = [1, \sqrt{-3}], \qquad \alpha = 2, \qquad \beta = 1 + \sqrt{-3}, \alpha\beta' + \beta\alpha' = 4,$$
$$\alpha\alpha' = \beta\beta' = 4$$

yet 4 does not divide $\alpha\beta'$ or $\beta\alpha'$ *within* $\mathfrak{O}_2$.

For instance, $\alpha\beta'/4 = (1 + \sqrt{-3})/2$, which is an integer under Dedekind's definition (as a root of $\eta^2 - \eta + 1 = 0$) but not an element of $\mathfrak{O}_2$. Thus unique factorization succeeds in $\mathfrak{O}_1 = [1, (1 + \sqrt{-3})/2]$, although it fails in $\mathfrak{O}_2 = [1, \sqrt{-3}]$ one *counter example* being provided by $\beta\beta' = \alpha^2$.

We can prove Theorem 6 by showing for any ideal $\mathfrak{a}$ the ideal $\mathfrak{a}\mathfrak{a}' = (g)$ where $\mathfrak{a}'$ is the conjugate ideal and g is a rational integer. To prove this, recall by Exercise 6 that we can write the conjugates

$$\mathfrak{a} = (\alpha_1, \alpha_2, \cdots, \alpha_s), \tag{4}$$
$$\mathfrak{a}' = (\alpha_1', \alpha_2', \cdots, \alpha_s').$$

Then $\mathfrak{a}\mathfrak{a}' = (\alpha_1\alpha_1', \alpha_1\alpha_2', \alpha_2\alpha_1', \alpha_2\alpha_2', \cdots, \alpha_s\alpha_s')$. But we define

$$\mathfrak{c} = (\alpha_1\alpha_1', \alpha_1\alpha_2' + \alpha_2\alpha_1', \alpha_2\alpha_2', \cdots, \alpha_s\alpha_t' + \alpha_t\alpha_s', \alpha_s\alpha_s').$$

Now $\mathfrak{c}$ is an ideal whose basis consists wholly of rational integers [see (1)]. Hence $\mathfrak{c} = (g)$ for some rational integer $g \neq 0$, but $\mathfrak{a}\mathfrak{a}' \supseteq \mathfrak{c} = (g)$. On the other hand, by Hurwitz's lemma, g divides each basis element of $\mathfrak{a}\mathfrak{a}'$. Thus (g) contains all elements of $\mathfrak{a}\mathfrak{a}'$, and

$$(g) \supseteq \mathfrak{a}\mathfrak{a}' \supseteq (g)$$

or $(g) = \mathfrak{a}\mathfrak{a}'$, proving Theorem 6. The reader should check carefully to see that the theorems of §7 to §10 have not been used in the proof (which we deferred to the end only for emphasis).

To prove Theorem 7, we prove the following lemma.

LEMMA 3. If $\alpha \in \mathfrak{D}_n$ and g and n are relatively prime and if α/g is an algebraic integer, $(\alpha/g \in \mathfrak{D}_1)$, then it follows that $\alpha/g \in \mathfrak{D}_n$.

Proof. The module $\mathfrak{D}_n$ is the set of elements α of $\mathfrak{D}_1$ such that

$$\alpha \in \mathfrak{D}_1 \quad \text{and} \quad \alpha \equiv r \pmod{n}, \qquad (r \text{ rational}).$$

But if g and n are relatively prime, $1/g \equiv s \pmod{n}$ and $\alpha/g \equiv rs \pmod{n}$ a rational integer, whence $\alpha/g \in \mathfrak{D}_n$. Q.E.D.

We can prove Theorem 7 by choosing a basis (4) of $\mathfrak{a}$ in which (say) $N(\alpha_1) = \alpha_1\alpha_1'$ is prime to n. In the proof of Theorem 6, $\mathfrak{c} = (g)$, where g and n are relatively prime. Then Hurwitz's lemma is applicable, for ξ/g, ξ'/g belong to $\mathfrak{D}_n$ as well as to $\mathfrak{D}_1$.

Thus we could develop a unique factorization theory for $\mathfrak{D}_n$ *by considering only* $\mathfrak{a}$ *for which* $\mathfrak{a} \ni \alpha$, *where* $N(\alpha)$ *is prime to* n. Since every divisor of $\mathfrak{a}$ *contains* $\mathfrak{a}$ and α, the theory will carry over. A more convenient procedure, however, is to restrict the ideal theory to $\mathfrak{D}_1$, the set of all integers of $R(\sqrt{D})$, and to find the ideals of $\mathfrak{D}_n$ by a "projection" procedure afterwards. This procedure, in principle going back to Gauss's theory of quadratic forms, is outlined in Chapter XIII, §2.

For the present, we consider the ideal theory only in a quadratic integral domain $\mathfrak{D}_1$ of *all* integers, reserving only a small portion of Chapter XIII for the factorization theory in $\mathfrak{D}_n$.

For convenience, we speak of "ideals in $R(\sqrt{D})$" to mean "ideals in $\mathfrak{D}_1$ for $R(\sqrt{D})$" when the context is clear.

EXERCISE 19. A *maximal* ideal $\mathfrak{m}$ in $\mathfrak{D}_n$ is defined as an ideal $(\neq \mathfrak{D}_n)$ for which no ideal $\mathfrak{a}$ in $\mathfrak{D}_n$ satisfies $\mathfrak{m} \subset \mathfrak{a} \subset \mathfrak{D}_n$. Show that all maximal ideals are indecomposable in $\mathfrak{D}_n$ and state a sufficient condition for the converse. *Hint.* In $\mathfrak{D}_2$ for $R(\sqrt{-3})$, $(1 + \sqrt{-3}) \subset (1 + \sqrt{-3}, 1 - \sqrt{-3}) \subset \mathfrak{D}_2$.

chapter VIII

Norms and ideal classes[1]

1. Multiplicative Property of Norms

The definition of index was given for modules in Chapter IV, §8, and naturally extends to ideals (as submodules of $\mathfrak{O}_1$) in which the index is called the *norm*. Thus we write for the norm of an ideal $\mathfrak{a}$, $N[\mathfrak{a}] =$ index $[\mathfrak{O}_1/\mathfrak{a}]$. For $\mathfrak{a} = (0)$ we define $N[(0)] = 0$, but the zero ideal never really enters into the theory. Otherwise, the norm is always positive. Moreover, $N[\mathfrak{a}] = 1$ exactly when $\mathfrak{a} = (1)$, or $\mathfrak{O}_1$.

THEOREM 1. For any two ideals $\mathfrak{a}, \mathfrak{b}$

$$N[\mathfrak{a}]\,N[\mathfrak{b}] = N[\mathfrak{a}\mathfrak{b}]. \tag{1}$$

Proof. To see this result, we count residue classes. We let $\rho_1, \rho_2, \cdots, \rho_l$ be $l = N[\mathfrak{a}]$ different residue classes mod $\mathfrak{a}$ and we let $\pi_1, \pi_2, \cdots, \pi_m$ be $m = N[\mathfrak{b}]$ different residue classes mod $\mathfrak{b}$. Then consider the $lm = N[\mathfrak{a}]$ $N[\mathfrak{b}]$ quantities

$$\omega_{ij} = \rho_i + \alpha\pi_j, \qquad 1 \le i \le l, \qquad 1 \le j \le m, \tag{2a}$$

where α is selected so that $(\alpha) = \mathfrak{a}\mathfrak{c}$, $(\mathfrak{c}, \mathfrak{b}) = (1)$, by Theorem 21 in Chapter VII, §10.

[1] We recall the restriction to ideals in the integral domain $\mathfrak{O}_1$ for $R(\sqrt{d})$ in Chapters VIII to XIII except for Chapter XIII, §2.

Equation $2a$ is chosen by analogy with rational arithmetic. For example, if r is a variable residue class modulo a, i.e., $0 \leq r < a$, and if q is a variable residue class modulo b, i.e., $0 \leq q < b$, then, regardless of whether $(a, b) = 1$, the ab quantities

$$w = r + aq \tag{2b}$$

represent each residue classes modulo ab exactly once [for the two values $w' = r' + aq'$ and w in ($2b$) are congruent modulo ab exactly when $r = r'$ and $q = q'$]. We find the present proof harder only because the presence of nonprincipal ideals makes the selection of α become the crucial step.

First of all, the various numbers ω_{ij} are incongruent modulo $\mathfrak{ab}$. Let $\omega_{ij} \equiv \omega_{IJ} \pmod{\mathfrak{ab}}$ or

$$(\rho_i + \alpha\pi_j) - (\rho_I + \alpha\pi_J) \in \mathfrak{ab}. \tag{3}$$

Since $\mathfrak{a} \mid \mathfrak{ab}$,

$$(\rho_i + \alpha\pi_j) - (\rho_I + \alpha\pi_J) \in \mathfrak{a},$$

but, since $\alpha \in \mathfrak{a}$, then $\rho_i - \rho_I \in \mathfrak{a}$; hence $\rho_i = \rho_I$, otherwise different residue classes would be involved. Next we reduce (3) to

$$\alpha\pi_J - \alpha\pi_J \in \mathfrak{ab}; \tag{4}$$

thus $\mathfrak{ab} \mid (\alpha)(\pi_j - \pi_J)$, $\mathfrak{ab} \mid \mathfrak{ac}(\pi_j - \pi_J)$, and $\mathfrak{b} \mid \mathfrak{c}(\pi_j - \pi_J)$; but, since $(\mathfrak{c}, \mathfrak{b}) = 1$, $\mathfrak{b} \mid (\pi_j - \pi_J)$, and then $\pi_j - \pi_J \in \mathfrak{b}$, whence $\pi_j = \pi_J$; showing that all lm numbers ω_{ij} are incongruent mod $\mathfrak{ab}$.

We show that every ξ in $\mathfrak{O}$ is congruent to some ω_{ij} mod $\mathfrak{ab}$. First of all, $\xi \equiv \rho_i \pmod{\mathfrak{a}}$ for some i (by definition of the set ρ_i). We write $\xi - \rho_i = \alpha^*$, $\alpha^* \in \mathfrak{a}$, but $(\alpha, \mathfrak{ab}) = (\mathfrak{ac}, \mathfrak{ab}) = \mathfrak{a}(\mathfrak{c}, \mathfrak{b}) = \mathfrak{a}$. Thus α^* ($\in \mathfrak{a}$) is composed of an element of (α) plus an element of $\mathfrak{ab}$. Hence, for some integer θ

$$\alpha^* = \alpha\theta + \alpha^{(0)} \text{ where } \alpha^{(0)} \in \mathfrak{ab}.$$

But $\theta \equiv \pi_j \pmod{\mathfrak{b}}$ for some j, i.e., $\theta = \pi_j + \pi^{(0)}$ where $\pi^{(0)} \in \mathfrak{b}$. Finally, $\xi = \rho_i + \alpha^* = \rho_i + \alpha(\pi_j + \pi^{(0)}) + \alpha^{(0)} = \rho_i + \alpha\pi_j + \lambda$, where $\lambda = \alpha\pi^{(0)} + \alpha^{(0)} \in \mathfrak{ab}$. Q.E.D.

We can now relate the norm to more familiar concepts by showing no conflict in terminology.

THEOREM 2. If α is an algebraic or rational integer generating the principal ideal (α),

$$N[(\alpha)] = |N(\alpha)| .$$

Proof. First of all, if a is rational, then we shall see

$$N[(a)] = a^2, \tag{5}$$

To see this, note $(a) = a \cdot [1, \omega] = [a, a\omega]$. The index is therefore $a^2 = N[(a)]$. Next note that (α) and (α') have the same norm as ideals, since any two integers of $\mathfrak{O}$ congruent modulo (α) have conjugates congruent modulo (α') and vice versa, leading to a one-to-one correspondence of residue classes via conjugates. By Theorem 1 and Theorem 2 for rational ideals,

$$N[(\alpha)]^2 = N[(\alpha)]\,N[(\alpha')] = N[(\alpha)(\alpha')] = N[(N(\alpha))] = N(\alpha)^2.$$

Since $N[\mathfrak{a}] \geq 0$, necessarily, Theorem 2 follows.

We can further identify $N[\mathfrak{a}]$ by referring to the critical theorem in Chapter VII, §11; $\mathfrak{a}\mathfrak{a}' = (g)$. Here, by taking norms we find $N[\mathfrak{a}\mathfrak{a}'] = N[\mathfrak{a}]\,N[\mathfrak{a}'] = N[\mathfrak{a}]^2 = g^2$; hence $N[\mathfrak{a}] = |g|$. Thus

$$\mathfrak{a}\mathfrak{a}' = (N[\mathfrak{a}]). \tag{6}$$

Note that since $\mathfrak{a} \supseteq \mathfrak{a}\mathfrak{a}' = (g)$ the ideal of rational integers in $\mathfrak{a}$ is a divisor of the ideal $(N[\mathfrak{a}])$ (which it contains).

THEOREM 3. If the ideal $\mathfrak{a}$ is not divisible by any rational integral ideal in $\mathfrak{O}$ except (1), then the rational integers in $\mathfrak{a}$ are all the multiples of $N[\mathfrak{a}]$.

Proof. Since $\mathfrak{a}$ divides $(N[\mathfrak{a}])$ by (6), then $\mathfrak{a}$ contains $N[\mathfrak{a}]$. Let the rational integers in $\mathfrak{a}$ be given by the ideal (g) so that $N[\mathfrak{a}] = gk$. We show $k = 1$ and $N[\mathfrak{a}] = g$ as follows:

$$\mathfrak{a}\mathfrak{a}' = (gk),$$

and, since $\mathfrak{a} \supseteq (g)$, $\mathfrak{a} \mid (g)$ or $\mathfrak{a}\mathfrak{b} = (g)$ for an ideal $\mathfrak{b}$. Thus

$$\mathfrak{a}\mathfrak{a}' = \mathfrak{a}\mathfrak{b}k,$$
$$\mathfrak{a}' = \mathfrak{b}k,$$
$$\mathfrak{a} = \mathfrak{b}'k,$$

by taking conjugates; $k \mid \mathfrak{a}$ and $k = 1$. Q.E.D.

In particular, if we factor a rational prime $(p) = \Pi \mathfrak{p}_1^{e_i}$, taking norms we find $p^2 = N[\mathfrak{p}_1]^{e_1} N[\mathfrak{p}_2]^{e_2} \cdots$. Thus $N[\mathfrak{p}_i]$ is a power of p. This leads to several cases, namely

$$N[\mathfrak{p}_1] = p^2, \quad e_1 = 1; \quad N[\mathfrak{p}_1] = N[\mathfrak{p}_2] = p, \quad e_1 = e_2 = 1;$$
$$N[\mathfrak{p}_1] = p \quad e_1 = 2.$$

THEOREM 4. The quadratic-prime ideals $\mathfrak{p}$ are related to integers in the rational field in the following possible ways:

$(p) = (p)$, or (p) "does not factor", $N[(p)] = p^2$;

$(p) = \mathfrak{p}_1\mathfrak{p}_2$, or (p) "splits" into two different factors,

$$N[\mathfrak{p}_1] = N[\mathfrak{p}_2] = p;$$

$(p) = \mathfrak{p}_1^2$, or (p) "ramifies", $N[\mathfrak{p}_1] = p$.

Thus the norms provide the measure of the "size" of the ideal as a factor.

EXERCISE 1. Let $\mathfrak{a}$ be in $\mathfrak{O}_1$ and $n > 1$; show, if $(N[\mathfrak{a}], n) = 1$, that $\mathfrak{a}$ contains an element α for which $(N[\alpha], n) = 1$. Is a converse valid?

EXERCISE 2 (Chinese Remainder Theorem). Let ρ_i $(1 \leq i \leq N[\mathfrak{a}])$ and σ_j $(1 \leq j \leq N[\mathfrak{b}])$ represent the residue classes moduli $\mathfrak{a}$ and $\mathfrak{b}$, respectively, and let $(\mathfrak{a}, \mathfrak{b}) = (1)$. Show that the residue classes modulo $\mathfrak{ab}$ are determined by the pairs of residue classes (ρ_i, σ_j) in one-to-one manner. *Hint*. Use Exercise 18 of Chapter VII, §9.

EXERCISE 3. Consider the ideal $(a, b - \omega_0) = \mathfrak{a}$, where ω_0 is defined as usual (Chapter III, §7). Then, if $g = \gcd\ (a, N(b - \omega_0))$, show that the rational integers in $\mathfrak{a}$ are exactly the rational integral multiples of g. *Hint*. Set $a\alpha + (b - \omega_0)\beta = t$ and multiply by $b - \omega_0'$. Show $N[\mathfrak{a}] = g$ by noting that $x + y\omega_0 \equiv x + yb \pmod{\mathfrak{a}}$. Verify by actually calculating $\mathfrak{aa}'$.

EXERCISE 4. In Exercise 3, if $a \mid N(b - \omega_0)$, show the module $\mathfrak{A} = [a, b - \omega_0]$ equals the ideal $\mathfrak{a}$ by showing $\mathfrak{A} \leq \mathfrak{a}$, whereas index $[\mathfrak{O}/\mathfrak{A}] =$ index $[\mathfrak{O}/\mathfrak{a}]$. Verify $\mathfrak{A} = \mathfrak{a}$ by actually calculating the general term of each.

2. Class Structure

Once we see the failure of unique factorization of integers (without using ideals) we are led to measure the extent to which this "failure" prevails. For this purpose we say two ideals $\mathfrak{a}$, $\mathfrak{b}$ (not zero) fall into the same class, written $\mathfrak{a} \sim \mathfrak{b}$ if

$$\mathfrak{a}(\beta) = \mathfrak{b}(\alpha) \tag{1}$$

for integers α, β not zero. It is easily seen that equivalent ideals "form a class" in the logical sense. This means, if $\mathfrak{a} \sim \mathfrak{b}$, then $\mathfrak{b} \sim \mathfrak{a}$ and, if $\mathfrak{a} \sim \mathfrak{b}$, $\mathfrak{b} \sim \mathfrak{c}$, then $\mathfrak{a} \sim \mathfrak{c}$ and, finally, $\mathfrak{a} \sim \mathfrak{a}$. To see the second result, which is the least trivial, note that (1) and

$$\mathfrak{b}(\gamma) = \mathfrak{c}(\beta_0) \tag{2}$$

imply together

$$\mathfrak{a}(\beta\gamma) = \mathfrak{c}(\beta_0\alpha). \tag{3}$$

Now all principal ideals $(\alpha) \sim (1)$; hence, if there is but *one* class (the *principal* class), *unique* factorization prevails and conversely. The principal ideals are the identity class, since $(\alpha)\mathfrak{a} \sim \mathfrak{a}$. There is an inverse to the class of $\mathfrak{a}$, namely the class of $\mathfrak{a}^*$ where, by Theorem 6 in Chapter VII, $\mathfrak{aa}^* = (\alpha) \sim (1)$. Note that if $\mathfrak{a} \sim \mathfrak{b}$ and $\mathfrak{bb}^* = (\beta) \sim (1)$ then $\mathfrak{b}^* \sim \mathfrak{a}^*$. (For if $(\rho)\mathfrak{a} = (\sigma)\mathfrak{b}$ then $(\rho)\mathfrak{aa}^*\mathfrak{b}^* = (\sigma)\mathfrak{ba}^*\mathfrak{b}^*$ and $(\rho\alpha)\mathfrak{b}^* = (\sigma\beta)\mathfrak{a}^*$). In the quadratic case, of course, we can let $\mathfrak{a}^* = \mathfrak{a}'$.

We denote an ideal class of nonzero ideals by a capital roman letter, e.g.,

$$(\mathbf{I}):\quad (1) \sim (\alpha) \sim (\beta) \sim \cdots$$

$$(\mathbf{A}):\quad \mathfrak{a}_1 \sim \mathfrak{a}_2 \sim \mathfrak{a}_3 \sim \cdots$$

$$(\mathbf{B}):\quad \mathfrak{b}_1 \sim \mathfrak{b}_2 \sim \mathfrak{b}_3 \sim \cdots.$$

We define the product of two ideal classes $\mathbf{C} = \mathbf{AB}$ as the class belonging to any ideal $\mathfrak{c} = \mathfrak{a}_1\mathfrak{b}_1$ formed by multiplying representatives of each class. To make the definition valid, it must be verified that with symbols defined as above

$$\mathfrak{a}_1\mathfrak{b}_1 \sim \mathfrak{a}_2\mathfrak{b}_2,$$

but this is an easy consequence of definition. We need hardly add that commutativity and associativity, etc., follow from the ideals; e.g., we denote the *class* of $\mathfrak{a}^*$ by $\mathbf{A}^{-1}$ if $\mathbf{A}$ is the class of $\mathfrak{a}$ and $\mathfrak{a}\mathfrak{a}^*$ is principal.

$$(4) \qquad \begin{cases} \mathbf{AB} = \mathbf{BA}, & \text{(Commutative law)} \\ \mathbf{A(BC)} = \mathbf{B(AC)}, & \text{(Associative law).} \end{cases}$$

The sum $\mathbf{A} + \mathbf{B}$, however, is meaningless; e.g., if $\mathfrak{a}_1 \sim \mathfrak{a}_2$, $\mathfrak{b}_1 \sim \mathfrak{b}_2$, then $(\mathfrak{a}_1, \mathfrak{b}_1)$ is not necessarily equivalent to $(\mathfrak{a}_2, \mathfrak{b}_2)$. For instance, let $\mathfrak{a} = (\alpha, \beta)$ be nonprincipal. Then $(\alpha)\mathfrak{a} \sim \mathfrak{a}$ and $(\beta)\mathfrak{a} \sim \mathfrak{a}$, but $(\alpha)\mathfrak{a} + (\beta)\mathfrak{a}$ is not equivalent to $\mathfrak{a} + \mathfrak{a}$, for $\mathfrak{a} + \mathfrak{a}$ obviously $= \mathfrak{a}$, whereas $(\alpha)\mathfrak{a} + (\beta)\mathfrak{a} = ((\alpha) + (\beta))\mathfrak{a} = \mathfrak{a} \cdot \mathfrak{a}$. Hence we ask if $\mathfrak{a}\mathfrak{a} \sim \mathfrak{a}$. If $\mathfrak{a}\mathfrak{a}^* = (\alpha)$, $\mathfrak{a}\mathfrak{a}\mathfrak{a}^* \sim \mathfrak{a}\mathfrak{a}^*$, $\mathfrak{a}(\alpha) \sim (\alpha)$, and $\mathfrak{a} \sim (1)$, contrary to assumption.

Now we shall see why the class structure provides a "measure" of the remoteness of unique factorization once we show that the ideal classes form a *finite* (commutative) group.

THEOREM 5 (Minkowski). Every ideal $\mathfrak{a}$ contains an element α such that

$$0 < |N(\alpha)| \leq N[\mathfrak{a}]\sqrt{|d|}\,,$$

where d is the discriminant of the field.

This theorem, which is geometric in nature, is proved in the next two sections. We shall draw a few conclusions now.

COROLLARY 1. Every ideal class $\mathbf{A}$ contains an ideal $\mathfrak{a}$ such that $N[\mathfrak{a}] < \sqrt{|d|}$.

Proof. Let $\mathfrak{b}$ belong to $\mathbf{A}^{-1}$ and let $\mathfrak{b}$ contain an element β with the property that

$$0 < |N(\beta)| \leq N[\mathfrak{b}]\sqrt{|d|}.$$

Since $\beta \in \mathfrak{b}$, then $\mathfrak{b} \mid (\beta)$ and $\mathfrak{b}$ satisfies

$$(5) \qquad \mathfrak{b}\mathfrak{a} = (\beta) \text{ for some } \mathfrak{a}.$$

But $\mathfrak{a}$ belongs to class $\mathbf{A}$, since (5) indicates that it is inverse to $\mathbf{A}^{-1}$, the class of $\mathfrak{b}$. Hence taking norms of (5),

$$N[\mathfrak{b}]\, N[\mathfrak{a}] \leq N[\mathfrak{b}]\sqrt{|d|},$$

whence follows the conclusion. (The $\leq$ becomes a $<$, when $\sqrt{|d|}$ is not rational.) Q.E.D.

COROLLARY 2. The number of classes is finite.

Proof. Only a finite number of modules can have a norm (or index) less than $\sqrt{|d|}$. (See Theorem 4, Chapter VII, §4.) This is all the more true for ideals (since they are a special type of module). Q.E.D.

THEOREM 6. The ideal classes form a finite commutative group.

Proof. The finiteness was just shown. The group properties are (4) together with the inverse. Q.E.D.

COROLLARY. If h is the number of classes, then for any ideal $\mathfrak{a}$, $\mathfrak{a}^h$ is principal.

This follows from Lagrange's lemma (whereby the order of an element divides the order of the group).

THEOREM 7. Every ideal class $\mathbf{A}$ contains an ideal $\mathfrak{a}$ relatively prime to any preassigned ideal $\mathfrak{q}$.

Proof. Let $\mathfrak{b} \in \mathbf{A}^{-1}$. Then $\mathfrak{b} \supseteq \mathfrak{b}\mathfrak{q}$. Thus we can write $\mathfrak{b} = (\mathfrak{b}\mathfrak{q}, (\beta))$. Hence $(\beta) = \mathfrak{b}\mathfrak{a}$, where, by the definition of inverse class, $\mathfrak{a} \in \mathbf{A}$. But, since

$$\mathfrak{b} = (\mathfrak{b}\mathfrak{q}, \mathfrak{b}\mathfrak{a}) = \mathfrak{b}(\mathfrak{q}, \mathfrak{a}),$$

$(\mathfrak{q}, \mathfrak{a}) = (1)$. Q.E.D.

Thus it is *never* the case that all principal ideals have a common divisor with some fixed integer $m > 1$, for example.

The preceding theorems made no use of special properties of quadratic fields. For an elementary result, strictly true of the quadratic case, we note the following in retrospect:

THEOREM 8. In the quadratic case the conjugate of an ideal determines the reciprocal of its class.

3. Minkowski's Theorem

The proof of Minkowski's theorem requires a good deal of visualization. First of all, we consider a lattice parametrized by rational integral variables x and y.

$$(1) \qquad \begin{cases} \xi = \alpha x + \beta y, \\ \eta = \gamma x + \delta y, \end{cases}$$

or, in vector form,

(2) $$(\xi, \eta) = x(\alpha, \gamma) + y(\beta, \delta).$$

Here α, β, γ, δ are *real* numbers (rational or irrational, quadratic or otherwise). The lattice then consists of integral combinations of the basis vectors (α, γ) and (β, δ). In order that the basis vectors not be parallel, we say they form a *basis parallelogram* of nonzero area Δ. In analytic geometry

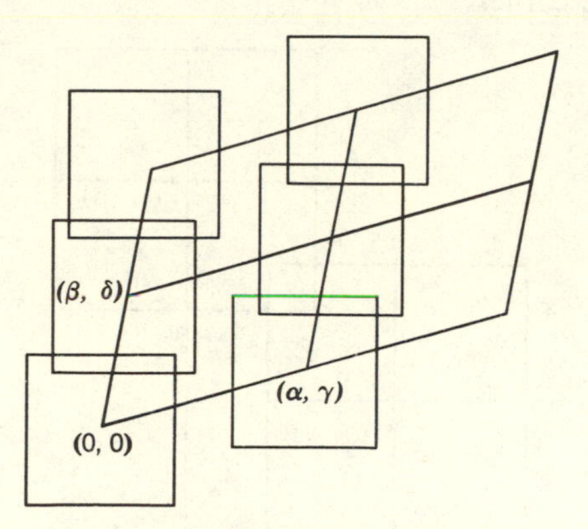

FIGURE 8.1

the area determined by the vectors (α,γ) and (β, δ) is given as the following absolute value of a determinant:

(3) $$\Delta = |\alpha\delta - \beta\gamma| > 0.$$

The parallelograms of area Δ cover the plane by translations (see Figure 8.1).

The vector closest to the origin need not be (α, β) nor (β, γ) but could be more difficult to obtain. For instance, if $(\alpha, \gamma) = (1, 1)$ and $(\beta, \delta) = (2 ,1)$, it is clear that $\pm(1, 0) = \mp(\alpha, \gamma) \pm (\beta, \delta)$ and $\pm(0, 1) = \pm 2(\alpha, \gamma) \mp (\beta, \delta)$ are the *closest* possible points (since all coordinates are integers).

We are ready to use this basic result of Minkowski.

THEOREM 9. In the foregoing lattice notation of (1) and (2), there is at least one $(\xi, \eta) \neq (0, 0)$ (for an integral $(x, y) \neq (0, 0)$) which satisfies

(4) $$\begin{cases} |\xi| \leq \sqrt{\Delta}, \\ |\eta| \leq \sqrt{\Delta}. \end{cases}$$

Proof. We construct, centered about each point of the lattice, a square of side $c\sqrt{\Delta}$ where c is a real constant $c > 1$ (and, for convenience, $c \leq 2$).

Thus we have squares of area $c^2\Delta(> \Delta)$ surrounding each point of the lattice. The basis parallelograms of area Δ completely cover the plane. It is now intuitively clear that some of the squares must *overlap*, since each has a larger area than the parallelogram.

To accomplish a rigorous proof, we must argue the following:

(a) In a large circle of radius R there are (approximately[1]) $N \sim \pi R^2/\Delta$ points (or parallelograms) if we associate each basis parallelogram with the lower right-hand corner.

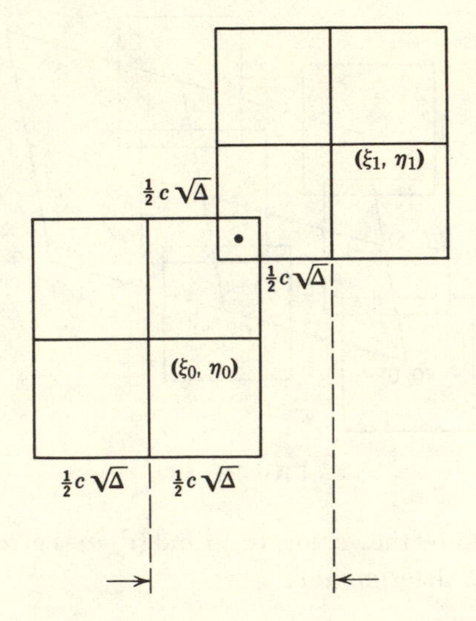

FIGURE 8.2. When overlapping occurs, distance is $<\frac{1}{2}c\sqrt{\Delta} + \frac{1}{2}c\sqrt{\Delta}$.

(b) The area of this circle would have to exceed $Nc^2\Delta \sim \pi R^2 c^2$ if the squares failed to overlap. This is impossible if $c > 1$.

Once we achieve overlapping, we can say for some two distinct lattice points

$$\begin{cases}(\xi_0, \eta_0) = x_0(\alpha, \gamma) + y_0(\beta, \delta), \\ (\xi_1, \eta_1) = x_1(\alpha, \gamma) + y_1(\beta, \delta),\end{cases} \qquad (x_0, y_0) \neq (x_1, y_1),$$

that (as in Figure 8.2),

$$\begin{cases}|\xi_0 - \xi_1| < \frac{1}{2}c\sqrt{\Delta} + \frac{1}{2}c\sqrt{\Delta}, \\ |\eta_0 - \eta_1| < \frac{1}{2}c\sqrt{\Delta} + \frac{1}{2}c\sqrt{\Delta},\end{cases} \tag{5}$$

[1] In calculus $A \sim B$ means $A/B \to 1$ as $R \to \infty$, not to be confused with "similar ideals." The rigorous argument is left to the student to complete. We merely wish to make clear the power of geometrical "existence" proofs scarcely like the constructional proofs of Euclid! A more rigorous argument is given on a related matter in Chapter X, §2.

Then, writing $(x_0 - x_1) = X$, $(y_0 - y_1) = Y$ and

$$(\xi, \eta) = X(\alpha, \gamma) + Y(\beta, \delta), \qquad (X, Y) \neq (0, 0),$$

we find that (5) becomes

$$\text{(6)} \qquad \begin{cases} |\xi| < c\sqrt{\Delta}, & (\xi, \eta) \neq (0, 0), \\ |\eta| < c\sqrt{\Delta}. \end{cases}$$

We have not quite achieved formula (4). We let $c = c_m = 1 + 1/m$, where the integer $m \to \infty$. For each m a solution exists for which

$$\text{(7)} \qquad \begin{cases} |\xi^{(m)}| < c_m\sqrt{\Delta}, & (\xi^{(m)}, \eta^{(m)}) \neq (0, 0), \\ |\eta^{(m)}| < c_m\sqrt{\Delta}. \end{cases}$$

But, since $1 < c_m \leq 2$, it is clear that only a *finite* number (say Q) of $(\xi^{(m)}, \eta^{(m)})$ is considered (namely, the number that lies in a square of side $4\sqrt{\Delta}$ centered at the origin).

By an adaptation of Dirichlet's boxing-in principle, if at most Q points are used in an infinitude of (7), as $m = 1, 2, 3, 4, \cdots$, at least one of these points (ξ^*, η^*) must be used in an infinite number for a special sequence of m. For this point

$$\text{(8)} \qquad \begin{cases} |\xi^*| < c_m\sqrt{\Delta}, \\ |\eta^*| < c_m\sqrt{\Delta}, \end{cases}$$

as $m \to \infty$ through a special set of values. Regardless of how $m \to \infty$, $c_m \to 1$. Hence (8) yields the desired result (4) on the limiting operation. [In retrospect, we can see (ξ^*, η^*) was valid in (8) for *all* m.] Q.E.D.

EXERCISE 5. In the lattice determined by $(\alpha, \gamma) = (1, 0)$ and $(\beta, \delta) = (m/n, 1/(n-1)^2)$ show that Theorem 9 leads to the solution of $mx + ny = 1$ if $(m, n) = 1$.

EXERCISE 6. In the lattice determined by $(\alpha, \gamma) = (1, 0)$ and $(\beta, \delta) = (\xi_2/\xi_1, 1/T^2)$ show that Theorem 9 leads to a variant of Lemma 5 of Chapter VI, §3.

EXERCISE 7. Show that the inequalities (4) of Theorem 9 can be replaced by the single inequality $|\xi| + |\eta| \leq \sqrt{2\Delta}$ by noting the area of the square determined thereby.

4. Norm Estimate

We shall actually prove the following theorem:

THEOREM 10. Every module $\mathfrak{M}$ in $\mathfrak{O}_1$ with different $\Delta(\mathfrak{M})$ contains an element α for which $0 < |N(\alpha)| \leq |\Delta(\mathfrak{M})|$.

Recalling the terminology of Chapter IV, §10, $\Delta^2(\mathfrak{M}) = j^2 d$ where $j = [\mathfrak{O}_1/\mathfrak{M}]$. If $\mathfrak{M}$ is the ideal $\mathfrak{a}$, then $j = N[\mathfrak{a}]$, and Theorem 5 (above) follows. (The different is not necessarily positive).

To prove Theorem 10, we first separate cases according to whether $d > 0$ or $d < 0$.

Let $d > 0$. The module $\mathfrak{M} = [\alpha_1, \alpha_2]$ in basis form. A general element for rational integral x and y is

$$(1) \qquad \begin{cases} \alpha = \alpha_1 x + \alpha_2 y, \\ \alpha' = \alpha_1' x + \alpha_2' y, \end{cases} \qquad \alpha \in \mathfrak{M},$$

where $|\Delta(\mathfrak{M})| = |\alpha_1\alpha_2' - \alpha_1'\alpha_2|$, which is precisely the parallelogram area Δ. Thus the desired result comes from Theorem 9. Q.E.D.

Let $d < 0$. The ideal $\mathfrak{M} = [\alpha_1, \alpha_2]$ is written as before.

$$(2) \qquad \begin{cases} \alpha = \alpha_1 x + \alpha_2 y, \\ \alpha' = \alpha_1' x + \alpha_2' y, \end{cases} \qquad \alpha \in \mathfrak{M},$$

Here

$$(3) \qquad \begin{cases} \alpha_1 = \beta_1 + i\gamma_1, & \alpha_1' = \beta_1 - i\gamma_1, \\ \alpha_2 = \beta_2 + i\gamma_2, & \alpha_2' = \beta_2 - i\gamma_2, \end{cases}$$

where $\beta_1, \gamma_1, \beta_2, \gamma_2$ are real. For example, if $\alpha_1 = a + b\sqrt{d}$, then $\beta_1 = a$, $\gamma_1 = b\sqrt{|d|}$. As before, by multiplying the determinant and taking absolute values, we can verify that

$$|\Delta(\mathfrak{M})| = \text{abs}\begin{vmatrix} \alpha_1 & \alpha_1' \\ \alpha_2 & \alpha_2' \end{vmatrix} = \text{abs}\begin{vmatrix} \beta_1 & \gamma_1 \\ \beta_2 & \gamma_2 \end{vmatrix} \text{abs}\begin{vmatrix} 1 & 1 \\ i & -i \end{vmatrix} = 2\,|\beta_1\gamma_2 - \gamma_1\beta_2|$$

Now, separating the real and imaginary parts in (2) and by substituting (3), we obtain components which form a lattice like that in Figure 4.2:

$$\begin{cases} \alpha = \rho + i\sigma, \\ \alpha' = \rho - i\sigma, \end{cases}$$

where

$$\begin{cases} \rho = \beta_1 x + \beta_2 y, \\ \sigma = \gamma_1 x + \gamma_2 y, \end{cases}$$

and

$$\Delta^* = |\beta_1\gamma_2 - \gamma_1\beta_2| = |\Delta(\mathfrak{M})|/2$$

is the parallelogram area. Hence, by Theorem 9, §3 (above), we determine a couple (x, y) or (ρ, σ) for which

$$|\rho| \le \sqrt{\Delta^*}, \qquad |\sigma| \le \sqrt{\Delta^*}.$$

Thus

$$\alpha\alpha' = \rho^2 + \sigma^2 \le 2\Delta^* = |\Delta(\mathfrak{M})|. \qquad \text{Q.E.D.}$$

It would be appropriate to remark that the depths of the norm estimation problem can be appreciated only in terms of further results which are not proved (or used) here.

THEOREM 11. Every ideal $\mathfrak{a}$ contains an element α of nonzero norm $\leq N[\mathfrak{a}]\sqrt{d/5}$ if d is positive or $\leq N[\mathfrak{a}]\sqrt{-d/3}$ if d is negative.

The extreme cases occur when $d = 5$ and $d = -3$, respectively, with $\mathfrak{a} = (1)$, $\alpha = 1$. Thus it is only a coincidence that in the weaker versions proved here the real and imaginary results seem to be the same. Geometrically, they are vastly different!

Minkowski's theorem can be *slightly* strengthened so that (4), §3, becomes

$$(4) \qquad |\xi| < \sqrt{\Delta}, \qquad |\eta| \leq \sqrt{\Delta},$$

Thus *one* specified inequality can be made strict. (See Exercise 9, below.) The "extreme" case is $(\alpha, \gamma) = (1, 0)$ and $(\beta, \delta) = (0, 1)$. Here $\Delta = 1$ and, when $x = 0$, $y = 1$, by (1) in §3, $\xi = 0$, $\eta = 1$. Thus beyond the use of one strict inequality no improvement can be made on Theorem 9. Other techniques are needed for a result like Theorem 11.

EXERCISE 8. Show that, starting with Exercise 7 and using the inequality $4|\xi\eta| \leq (|\xi| + |\eta|)^2$, we can improve Theorem 5 (slightly) to attain the inequality $0 < N(\alpha) \leq N[\mathfrak{a}]\sqrt{d/2}$ for real fields.

EXERCISE 9. Improve Minkowski's theorem to (4) by first solving $|\xi| \leq \sqrt{\Delta}\, m/(m + 1)$, $|\eta| \leq \sqrt{\Delta}\, (m + 1)/m$.

chapter IX

Class structure in quadratic fields

1. The Residue Character Theorem

As mentioned in Chapters VII and VIII, the prime ideals $\mathfrak{p}$, first of all, can arise only from rational primes (Chapter VII, Theorem 22) and, second, completely determine the class structure in that every equivalence class, say that of $\mathfrak{a} = \Pi \mathfrak{p}_i^{e_i}$, is determined by the equivalence classes of the $\mathfrak{p}_i$.

As a matter of fact, in 1882, Weber showed that a prime ideal exists in each equivalence class, but the result is deferred until Chapter X, §12. All we say is that we can build the class structure by using $\mathfrak{p}_i$ as generators. We therefore must know how to construct the $\mathfrak{p}_i$.

THEOREM 1. The rational prime p factors in the quadratic field $R(\sqrt{D})$ according to the following rules based on d, the discriminant of the field, and (d/p), the Kronecker symbol:

$$(1)\begin{cases}(p) = (p) \text{ or } p \text{ does not factor if and only if } (d/p) = -1; \\ (p) = \mathfrak{p}\mathfrak{p}' \text{ or } p \text{ splits into two different factors if and only if } (d/p) = +1; \\ (p) = \mathfrak{p}^2 \text{ (and } \mathfrak{p} = \mathfrak{p}') \text{ or } p \text{ ramifies if and only if } (d/p) = 0.\end{cases}$$

Here the rule is independent of whether $d > 0$ or $d < 0$ or whether $d = D \equiv 1 \pmod 4$ or $d/4 = D \not\equiv 1 \pmod 4$ (where D is a square-free integer).

Proof. The proof is wholly constructional, and we shall derive specific formulas for $\mathfrak{p}$ and $\mathfrak{p}'$.

We first show that if factors exist, hence $(p) = \mathfrak{p}\mathfrak{p}'$ or $\mathfrak{p}^2$, then $(d/p) = 1$ or 0. According to Theorem 23, Chapter VII, if $\mathfrak{p} \mid p$, then for some π, $\mathfrak{p} = (p, \pi)$ and $p \mid N(\pi)$. If $p \mid \pi$, then $\mathfrak{p} = p(1, \pi/p) = p$ and p does not factor. Therefore, we assume

$$(2) \qquad \mathfrak{p} = (p, \pi), \qquad p \mid N(\pi), \qquad p \nmid \pi.$$

First we take p odd. Then we write $\pi = a + b\sqrt{D}$ or $\pi = (a + b\sqrt{D})/2$, depending on whether $4D = d$ or $D = d$. In either case

$$N(\pi) = \begin{Bmatrix} a^2 - b^2D \\ (a^2 - b^2D)/4 \end{Bmatrix} \equiv 0 \pmod{p},$$

whereas $p \nmid b$ (for if $p \mid b$ then $p \mid a$ and $p \mid \pi$). Then

$$a^2 - b^2D \equiv 0 \pmod{p}$$

and, if $B \equiv b^{-1} \pmod{p}$, i.e., $bB \equiv 1 \pmod{p}$,

$$(aB)^2 \equiv D \pmod{p};$$

thus $(D/p) = 1$ or 0, as a result of the assumption in (2).

Next we take $p = 2$; then $(d/p) = -1$, just when $d = D \equiv 5 \pmod{8}$, according to Kronecker's symbol. If $\pi = (a + b\sqrt{D})/2$,

$$2 \mid N(\pi) = \frac{a^2 - b^2D}{4} \equiv 0 \pmod{2}, \qquad (a \equiv b \pmod{2})),$$

whereas $2 \nmid a$ and $2 \nmid b$, lest π be divisible by 2. Thus we can say that $a^2 - b^2D \equiv 0 \pmod{8}$. This contradicts the possibility that $D \equiv 5 \pmod{8}$, since, for odd a and b, $a^2 \equiv b^2 \equiv 1 \pmod{8}$. Again $(d/p) = 1$ or 0.

Now all we need show is the existence of prime divisors $\mathfrak{p}$, $\mathfrak{p}'$ to answer the requirements of the theorem when $(d/p) = 1$ or 0.

First let p be odd and $(d/p) = 0$. We can actually write out

$$(3) \qquad \mathfrak{p} = (p, \pi) \qquad \pi = \sqrt{D}.$$

Note $\mathfrak{p}' = \mathfrak{p}$, since $(p, \sqrt{D}) = (p, -\sqrt{D})$. Then $\mathfrak{p}^2 = (p, \pi)^2 = (p^2, p\pi, \pi^2) = p(p, \pi, D/p)$. But since D is square-free, $(p, D/p) = 1$; hence $\mathfrak{p}^2 = (p)$.

Next let p be odd and $(d/p) = 1$. There exists an a such that (since $d = D$ or $d = 2^2D$)

$$a^2 \equiv D \pmod{p} \quad \text{and} \quad (a, p) = 1.$$

Set $\pi = a + \sqrt{D}$ and define the ideals (not yet known to be prime):

$$\mathfrak{p} = (p, a + \sqrt{D}), \quad \mathfrak{p}' = (p, a - \sqrt{D}). \tag{4}$$

Now $\mathfrak{p} \neq \mathfrak{p}'$. For otherwise we could reason

$$\begin{aligned} \mathfrak{p} = \mathfrak{p}' = \mathfrak{p} + \mathfrak{p}' &= (p, a + \sqrt{D}, a - \sqrt{D}), \\ &= (p, a + \sqrt{D}, a - \sqrt{D}, 2a) = (1), \end{aligned}$$

since p is odd and $(p, a) = 1$. This is a contradiction by virtue of our next step that $(p) = \mathfrak{p}\mathfrak{p}' \neq (1)$. Observe

$$\begin{aligned} \mathfrak{p}\mathfrak{p}' &= (p^2, pa + p\sqrt{D}, pa - p\sqrt{D}, a^2 - D), \\ &= p(p, a + \sqrt{D}, a - \sqrt{D}, (a^2 - D)/p), \\ &= p(p, 2a, \cdots) = (p). \end{aligned}$$

Thus Theorem 4 of Chapter VIII, §1, applies and $\mathfrak{p}$ is prime.

We finally take care of p even. If $(d/2) = 1$, $d \equiv 1 \pmod 8$, and we write $2 = \mathfrak{p}\mathfrak{p}'$, where, again, we explicitly write out

$$\mathfrak{p} = (2, (1 + \sqrt{d})/2), \quad \mathfrak{p}' = (2, (1 - \sqrt{d})/2). \tag{5}$$

Once more $\mathfrak{p} \neq \mathfrak{p}'$, for otherwise

$$\begin{aligned} \mathfrak{p} = \mathfrak{p}' = \mathfrak{p} + \mathfrak{p}' &= [2, (1 + \sqrt{d})/2, (1 - \sqrt{d})/2] \\ &= [2, (1 + \sqrt{d})/2 + (1 - \sqrt{d})/2, \cdots] = (2, 1) = (1). \end{aligned}$$

This is false, since

$$\begin{aligned} \mathfrak{p}\mathfrak{p}' &= (4, 1 + \sqrt{d}, 1 - \sqrt{d}, 1 - d) = (4, 2, \cdots) \\ &= 2(2, 1, \cdots) = (2). \end{aligned}$$

The final case is $(d/2) = 0$, which means $D \equiv -1$ or $2 \pmod 4$. Here,

$$\pi = 1 + \sqrt{D} \quad \text{or} \quad \pi = \sqrt{D}. \tag{6}$$

with further details left as an exercise (below). Q.E.D.

For simplicity of application, we write the factorizations concisely according to the value of p:

$$(2) = \begin{cases} (2) & \text{if} \quad d = D \equiv 5 \pmod 8, \\ (2, (1 + \sqrt{D})/2)(2, (1 - \sqrt{D})/2) & \text{if} \quad d = D \equiv 1 \pmod 8, \\ (2, 1 + \sqrt{D})^2 & \text{if } d/4 = D \equiv -1 \pmod 4, \\ (2, \sqrt{D})^2 & \text{if } d/4 = D \equiv 2 \pmod 4; \end{cases}$$

and for $p > 2$,

$$(p) = \begin{cases} (p) & \text{if } p \nmid D \text{ and } x^2 \equiv D \pmod{p} \text{ is unsolvable,} \\ (p, x + \sqrt{D})(p, x - \sqrt{D}) & \text{if } p \nmid D \text{ and } x^2 \equiv D \pmod{p}, \\ (p, \sqrt{D})^2 & \text{if } p \mid D. \end{cases}$$

Another form of Theorem 1 is the following:

THEOREM 2. *If $\mathfrak{p}$ is any prime ideal factor of (p), then for $\omega = \sqrt{D}$, if $D \not\equiv 1 \pmod 4$ or $\omega = (1 + \sqrt{D})/2$ if $D \equiv 1 \pmod 4$ (D square-free), it follows that the equation (or equations)*

$$(7) \qquad \omega \equiv x \pmod{\mathfrak{p}}, \qquad \mathfrak{p} = \text{any prime divisor of } (p),$$

has $1 + (d/p)$ roots $x \pmod{p}$.

COROLLARY. *Under the conditions of the foregoing theorem, the equation*

$$N(x - \omega) \equiv 0 \pmod{p}$$

has $1 + (d/p)$ distinct roots $x \pmod{p}$.

EXERCISE 1. Verify that when $(d/2) = 0$ in (6), $(2) = \mathfrak{p}^2$ where $\mathfrak{p} = (2, \pi) = \mathfrak{p}'$.

EXERCISE 2. Verify Theorem 2 and corollary when p is odd. Do $p = 2$ separately.

EXERCISE 3 (Generalized Euler Φ-Function). Let $\Phi[\mathfrak{a}]$ denote the number of residue classes modulo $\mathfrak{a}$ which are relatively prime to $\mathfrak{a}$. Verify (*a*) $\Phi[\mathfrak{a}]\Phi[\mathfrak{b}] = \Phi[\mathfrak{a}\mathfrak{b}]$ if $(\mathfrak{a}, \mathfrak{b}) = (1)$ from Exercise 2 of Chapter VIII, §1; (*b*) if the prime $\mathfrak{p} \mid \mathfrak{a}$, then $\Phi[\mathfrak{a}\mathfrak{p}] = N[\mathfrak{p}]\Phi[\mathfrak{a}]$ from (2*a*) of Chapter VIII, §1; and (*c*) $\Phi(\mathfrak{p}) = N[\mathfrak{p}] - 1$. Next show $\Phi[\mathfrak{a}] = N[\mathfrak{a}]\Pi[1 - 1/N[\mathfrak{p}]]$ with the product extended over the primes $\mathfrak{p} \mid \mathfrak{a}$. Finally, show for a *rational* prime p, $\Phi[(p)] = [p - (d/p)][p - 1]$ and $\Phi[(a)] = \phi(a)a\Pi[1 - (d/p)/p]$, where the product extends over rational $p \mid a$ and $\phi(a)$ is the (ordinary) Euler ϕ-Function.

2. Primary Numbers

To every principal ideal (α) there corresponds an aggregate of generating associates β where $\beta = \alpha\eta$ and η is a unit. These ideals are indistinguishable, e.g., $(\alpha) = (\beta)$.

If $d < 0$ or the field is imaginary, then there are two units $+1$ and -1, except when $d = -3$, when there are six units $[\pm 1$ and $(\pm 1 \pm \sqrt{-3})/2]$, or when $d = -4$, when there are four units (± 1 and $\pm i$). In all these cases there is only a finite number of associates, and there is little purpose in distinguishing them one from another at the present stage in the theory (although Exercises 4 and 5 (below) are instructive in this fashion).

If $d > 0$, then a real *fundamental* unit η_1 exists such that $N(\eta_1) = \pm 1$ and any unit η is given by $\eta = \pm\eta_1{}^t$, $t = 0, \pm 1, \pm 2, \cdots$. There is now an *infinite* number of associates, and we are faced with a more acute problem of identifying some standard value.

We call the integer α of a real quadratic field *primary* when

$$(1) \qquad 1 \leq |\alpha/\alpha'| < \eta_1{}^2; \qquad \alpha > 0,$$

for η_1 (>1) the fundamental unit of the field.

THEOREM 3. Every real quadratic integer (except 0) has precisely one associate which is primary.

Proof. The most general associate of α_0 is $\pm\alpha_0\eta_1{}^t = \alpha$. If we write

$$\log |\alpha| = \log |\alpha_0| + t \log \eta_1,$$

then

$$\log |\alpha'| = \log |\alpha_0'| + t \log |\eta_1'|.$$

But $\log \eta_1 + \log |\eta_1'| = 0$, and, therefore, letting $\log |\alpha_0/\alpha_0'| = \xi$, we have, by subtraction,

$$\log |\alpha/\alpha'| = \xi + 2t \log \eta_1 = f(t).$$

Now $f(t)$ has only one value for which

$$0 \leq f(t) < 2 \log \eta_1,$$

namely for t the largest integer in $\xi/(2 \log \eta_1)$. For this t we choose the $\pm$ sign so that $\pm\alpha_0\eta_1{}^t > 0$. Q.E.D.

The term "primary" unfortunately tends to create confusion in view of the other meaning (power of a prime). The term is used because the uniquely chosen associate is of "primary" importance for ideal theory.

EXERCISE 4. Show that if α is in $R(\sqrt{-1})$ and $2 \nmid N(\alpha)$ then there is precisely one of the four associates β of α for which $\beta \equiv 1$ [mod $2(1 + \sqrt{-1})$].

EXERCISE 5. Show that if α is in $R(\sqrt{-3})$ and $3 \nmid N(\alpha)$ then there is precisely one of the six associates β of α for which $\beta \equiv 1 \pmod 3$.

EXERCISE 6. Show that the conditions for án algebraic integer α to be primary can be put into rational form by writing $\alpha = (x + y\sqrt{D})/2$, $\eta_1 = (a + b\sqrt{D})/2$. Then the condition for $1 \leq |\alpha/\alpha'| < |\eta_1/\eta_1'|$ is that x and y vanish or are positive and

$x/y > a/b$, if $N(\alpha) > 0$, $N(\eta_1) > 0$; $\quad x/y > bD/a$, if $N(\alpha) > 0$, $N(\eta_1) < 0$;
$x/y < bD/a$, if $N(\alpha) < 0$, $N(\eta_1) > 0$; $\quad x/y < a/b$, if $N(\alpha) < 0$, $N(\eta_1) < 0$.

3. Determination of Principal Ideals with Given Norms

We next show that the equation

$$N(\alpha) = n \tag{1}$$

can be solved for α in a finite number of steps.

THEOREM 4. The primary integers of norm n in a *real* quadratic field satisfy an equation of the type

$$\alpha^2 - A\alpha + n = 0, \tag{2}$$

where

$$|A| < \sqrt{|n|}\,(\eta_1 + 1). \tag{3}$$

Proof. By multiplying (1) in §2 with $|\alpha\alpha'| = |n|$, we find

$$|n| \leq \alpha^2 < |n|\,\eta_1{}^2,$$

$$\sqrt{|n|} \leq \alpha < \sqrt{|n|}\cdot\eta_1.$$

Thus

$$|\alpha + \alpha'| = |\alpha + n/\alpha| \leq |\alpha| + |n|/|\alpha| \leq \sqrt{|n|}\,\eta_1 + |n|/\sqrt{|n|}$$
$$= \sqrt{|n|}\,(\eta_1 + 1).$$

Q.E.D.

THEOREM 5. The integers of norm n (>0) in a complex quadratic field satisfy an equation of the type

$$\alpha^2 - A\alpha + n = 0, \tag{4}$$

$$|A| \leq 2n^{1/2}. \tag{5}$$

Proof. Here we note that $\alpha = r + si$, $\alpha' = r - si$,

$$A = \alpha + \alpha' = 2r \leq 2(r^2 + s^2)^{1/2} = 2(\alpha\alpha')^{1/2} = 2n^{1/2}.$$

Q.E.D.

LEMMA 1. If $d < 0$ then no number α of norm g exists in $R(\sqrt{D})$ if

$$g < |d|/4, \tag{6}$$

except if g is a perfect square and $\alpha = \pm\sqrt{g}$, a rational integer.

Proof. Observe that the relations (with $y \neq 0$)

$$|D|\,y^2 \leq x^2 - Dy^2 = g, \qquad D \not\equiv 1 \pmod 4, \qquad 4D = d < 0,$$

$$\left|\frac{D}{4}\right| y^2 \leq \frac{x^2 - Dy^2}{4} = g, \qquad D \equiv 1 \pmod 4, \qquad D = d < 0,$$

clearly contradict inequality (6).

Q.E.D.

As a matter of practice, the number of equations (2) or (4) can be very large. We would do well to make a first restriction to those A for which $A^2 - 4n$ is divisible by D, since α must belong to the field $R(\sqrt{D})$. As a further remark in the real case, note that since $|\eta_1|$ is smaller than any other (nonfundamental) unit (bigger than 1) we could not do any harm by using any unit $\eta \geq \eta_1 > 1$ instead of η_1 in (3) as long as we were merely interested in showing that (2) and (3) have *no* solution.

EXERCISE 7. Determine primary solutions, if any, of

$$N(\alpha) = 5 \text{ in } R(\sqrt{85}), \qquad \eta_1 = (9 + \sqrt{85})/2,$$
$$N(\alpha) = 2 \text{ in } R(\sqrt{7}), \qquad \eta_1 = 8 + 3\sqrt{7},$$
$$N(\alpha) = -2 \text{ in } R(\sqrt{7}), \qquad \eta_1 = 8 + 3\sqrt{7}.$$

EXERCISE 8. Show that if $d = g_1g_2 < 0$ and $(g_1/p) = (g_2/p) = -1$ then the divisors of p are nonprincipal. (Treat even p separately.)

EXERCISE 9. Show that if $d = g_1g_2 > 0$ and $(g_1/p) = (g_2/p) = -1$ and if all prime factors of d are $\equiv 1 \pmod 4$ then the divisors of p are nonprincipal.

4. Determination of Equivalence Classes

The most important problem in setting up class structures[1] consists of recognizing when two ideals $\mathfrak{a}$ and $\mathfrak{b}$ belong to the same class. Equivalently, since $\mathfrak{a}'$ lies in the reciprocal class, when we ask if $\mathfrak{a} \sim \mathfrak{b}$, we are asking if

$$\mathfrak{c} = \mathfrak{a}'\mathfrak{b} \sim 1.$$

Now the problem consists of taking an arbitrary ideal

$$\mathfrak{c} = (\gamma_1, \gamma_2, \cdots, \gamma_s) \tag{1}$$

in its basis form and asking if $\mathfrak{c}$ is principal (say) $= (\gamma)$. The norm $N[\mathfrak{c}]$ is either known from $N[\mathfrak{a}']N[\mathfrak{b}]$ or can be easily ascertained from the module

$$\mathfrak{c} = \mathfrak{M} = \gamma_1(1) + \gamma_2(1) + \cdots = \gamma_1[1, \omega] + \gamma_2[1, \omega] + \cdots,$$

where $[1, \omega]$ is the basis of the field. (The determination of a canonical basis, hence the norm or index of a module, was covered in Chapter IV, §7.) We then ask, by Theorems 4 and 5 (above), which primary numbers (γ), if any, have

$$|N(\gamma)| = N[\mathfrak{c}], \qquad \text{(the known value).} \tag{2}$$

[1] The theory of quadratic forms provides a faster algorithm for determining quadratic class structure (see Chapter XIII, §1), but the present methods are more easily generalized to fields of higher degree.

For these numbers γ we seek to determine if

$$\gamma \mid \gamma_1, \gamma \mid \gamma_2, \cdots, \gamma \mid \gamma_s, \tag{3}$$

for then $(\gamma) \mid \mathfrak{c}$, or $(\gamma)\mathfrak{q} = \mathfrak{c}$ for same ideal $\mathfrak{q}$. But (2) then tells us $N[\mathfrak{q}] = 1$, $\mathfrak{q} = (1)$. Thus $(\gamma) = \mathfrak{c}$ if and only if conditions (2) and (3) hold.

The procedure in determining class structure is clearly to factor each rational prime $p \leq \sqrt{|d|}$ and to see how many different equivalence classes can be built on the resulting prime ideals $\mathfrak{p}$ (in fact, the nonprincipal ones) by taking powers and products. The class number (or the number of elements in the equivalence class group) will be denoted by $h(d)$ for a field of discriminant d.

We shall denote factors of (p) by the use of *numerical* symbols,

$$\begin{cases} p = p \text{ if } p \text{ does not factor} & (p = p'), \\ p = p_1 p_2 \text{ if } p \text{ splits} & (p_1 = p_2', p_2 = p_1'), \\ p = p_1^2 \text{ if } p \text{ ramifies} & (p_1 = p_1'). \end{cases} \tag{4}$$

Thus we might write $3 = 3$, $3 = 3_1 3_2$, or $3 = 3_1^2$, as the case may be. This eliminates the parentheses and results in fewer symbols. The notation is due to Hasse. With it, (α) and α are used interchangeably.

Table III in the appendix, which shows the fantastic irregularity of $h(d)$, provides much additional useful information.

5. Some Imaginary Fields

First take $R(\sqrt{-1})$. Here $d = -4$ and only the prime $p = 2 \leq \sqrt{|d|}$. But $2 = 2_1^2$, where $2_1 = (1 + \sqrt{-1})$. In fact $(1 + \sqrt{-1}) = (1 - \sqrt{-1})$ since $1 + \sqrt{-1}$ and $1 - \sqrt{-1}$ differ by a factor of $\sqrt{-1}$. Hence all classes are principal, $h = 1$, and Theorem 1 shows

$$(p) = \pi\pi' \text{ if } (-4/p) = 1,$$
$$(p) \neq \pi\pi' \text{ if } (-4/p) = -1.$$

More precisely, by writing $\pi = x + \sqrt{-1}y$ we see, *for p odd*, that

$$p = x^2 + y^2 \tag{1}$$

if and only if $(-4/p) = (-1/p) = 1$, which means that $p \equiv 1 \pmod 4$ (by an elementary result in Chapter I). This is the famous Theorem of Fermat in the Introductory Survey.

More significantly, if $n = x^2 + y^2$, then if $(x, y) = 1$, all odd prime divisors of n are primes $\equiv 1 \pmod 4$ by the ideal factorization of $x + \sqrt{-1}y$ into primes π (which necessarily divide only $p \equiv 1 \pmod 4$).

We note in similar fashion when $D = -2, -3, -7, -11$, $h(d) = 1$.

When $D = -5$, $d = -20$, $p \le \sqrt{20}$; thus we must test $p = 2, 3$. Here all we need consider is $N(1 + \sqrt{-5}) = 6 = 2 \cdot 3$; hence $(1 + \sqrt{-5}) = 2_1 3_1$, whereas $2 = 2_1^2$. Hence all ideals are equivalent to 1 or 2_1, e.g., $3_2 \sim 2_1^{-1} \sim 2_1 \sim 3_1^{-1}$. Once we know that 2_1 is nonprincipal (by Lemma 1, §3), we know $h(-20) = 2$.

A more challenging case is $D = -14$, $d = -56$, where $h = 4$. We outline the essential steps for the reader to verify:

$$p < \sqrt{|d|} \text{ for } p = 2, 3, 5, 7,$$
$$(d/p) = +1 \text{ for } p = 3, 5,$$
$$(d/p) = 0 \text{ for } p = 2, 7,$$

(a) $$2 = 2_1^2, \quad 2_1 = (2, \sqrt{-14}),$$
$$7 = 7_1^2, \quad 7_1 = (7, \sqrt{-14}),$$

(b) $$2_1 \cdot 7_1 = (14, 2\sqrt{-14}, 7\sqrt{-14}, 14)$$
$$= \sqrt{-14}(\cdots, 2, 7, \cdots) = \sqrt{-14}.$$

(c) $$3 = 3_1 \cdot 3_2; \quad 3_1 = (3, 1 + \sqrt{-14}), \quad 3_2 = (3, 1 - \sqrt{-14}),$$
$$15 = N(\alpha), \quad \alpha = 1 + \sqrt{-14},$$
$$5 = 5_1 5_2; \quad 5_1 = (5, 1 + \sqrt{-14}), \quad 5_2 = (5, 1 - \sqrt{-14}).$$

(d) $$3_1 \cdot 5_1 = (15, 3 + 3\sqrt{-14}, 5 + 5\sqrt{-14}, (1 + \sqrt{-14})^2)$$
$$= (15, 3 + 3\sqrt{-14}, 5 + 5\sqrt{-14},$$
$$6 + 6\sqrt{-14}, (1 + \sqrt{-14})^2)$$
$$= (15, 1 + \sqrt{-14}, (1 + \sqrt{-14})^2) = (1 + \sqrt{-14}) = \alpha'.$$
$$3_1^2 = (9, 3 + 3\sqrt{-14}, -13 + 2\sqrt{-14})$$
$$= (9, 3 + 3\sqrt{-14}, -13 + 2\sqrt{-14}, 16 + \sqrt{-14})$$
$$= (9, -2 + \sqrt{-14}).$$

(e) $$3_1^2 2_1 = (18, -14 - 2\sqrt{-14}, 9\sqrt{-14}, -4 + 2\sqrt{-14})$$
$$= (18, 16 + \sqrt{-14}) = (18, -2 + \sqrt{-14})$$
$$= (-2 + \sqrt{-14}).$$

Hence, if $3_1 \sim \mathbf{J}$, then $\mathbf{J}^4 = \mathbf{I}$ and $2_1 \sim \mathbf{J}^2$ by (a), (e). Furthermore, $3_2 \sim \mathbf{J}^3$ by (c); likewise $7_1 \sim 2_1$, $5_2 \sim 3_1$, $5_1 \sim 3_2$, and 3_1^2 is nonprincipal by Lemma 1, §3, since $N(3_1^2) = 9$ and $3_1^2 \neq (3)$.

EXERCISE 10. Work out the class structure for $D = -21$, $D = -31$.

6. Class Number Unity

It was conjectured by Gauss that the only fields of class number 1 for $D < 0$ are

$$D = -1, -2, -3, -7, -11, -19, -43, -67, -163.$$

It is seen that conditions for an imaginary field of class number 1 are increasingly complicated as $|D|$ increases. For example, if $p < |d/4|$, then p cannot have a nontrivial ideal factor that is principal (by Lemma 1, §3). Since $|d/4| > \sqrt{|d|}$ for $|d| > 16$, we see that for $h(d) = 1$ and $|d| > 16$ it is necessary and (easily) sufficient that

$$(d/p) = -1 \text{ for } p < \sqrt{d}.$$

First of all, d is prime and therefore $d \equiv -1 \pmod 4$; otherwise, d has a prime divisor, $p_1 < \sqrt{d}$ for which $(d/p_1) = 0$. Thus

$$\begin{array}{lll}
(d/2) = -1, & \text{so } d \equiv 5 & \pmod 8, \text{ if } |d| > 16, \\
(d/3) = -1, & \text{so } d \equiv -1 & \pmod 3, \text{ if } |d| > 16, \\
(d/5) = -1, & \text{so } d \equiv \pm 2 & \pmod 5, \text{ if } |d| > 25 \\
(d/7) = -1, & \text{so } d \equiv 3, 5, 6 & \pmod 7, \text{ if } |d| > 49 \\
\vdots & \vdots & \vdots
\end{array}$$

Thus we see that d becomes subjected to an increasing number of restrictions if $h(d) = 1$. Eventually, a conclusive proof that there is only a finite number of $d < 0$ for which $h(d) = 1$ was given in 1934 by Heilbronn and Linfoot. It seems "very certain" that the last one is $d = -163$ (as Gauss conjectured), on the basis of numerical evidence of Lehmer showing $|d| > \frac{1}{2}10^9$, if $d < -163$ and $h(d) = 1$.

In the case of real fields, incidentally, Gauss conjectured that $h(d) = 1$ infinitely often (and has not been contradicted or justified).

7. Units and Class Calculation of Real Quadratic Fields

The same procedures are valid in the case of real quadratic fields. The difficulty is always that there is no easy way to tell if an algebraic integer exists with given norm, even if the unit is known, except by labored trial and error.

We shall discuss another procedure readily applicable to the real and complex case for obtaining both class structure and the fundamental unit (when required).

Let the field $R(\sqrt{D})$ have basis $[1, \omega]$. Then call $f(m) = N(m - \omega)$. Thus, with D square-free,

(1)
$$f(m) = \begin{cases} m(m-1) - (D-1)/4, & D \equiv 1 \pmod 4, \quad \omega = (1 + \sqrt{D})/2, \\ m^2 - D, & D \not\equiv 1 \pmod 4, \quad \omega = \sqrt{D}. \end{cases}$$

We take a range of m,

(2)
$$0 \leq m < \sqrt{|d|}.$$

Next we calculate the values $f(m)$ in the range and factor each answer. Those p for which $(d/p) = 1$ or 0 will appear as factors of $f(m)$ for each x_i such that (see Theorem 2, §1)

$$x_i \equiv \omega \ (\text{mod } \mathfrak{p}_i), \ \mathfrak{p}_i \mid p, \text{ if } p < \sqrt{|d|}.$$

No other p (i.e., for which $(d/p) = -1$) appears as a factor of any $N(\omega - m) = f(m)$, for then $p \mid (\omega - m)(\omega' - m)$, and since (p) is prime, p divides one factor (say) $\omega - m$, which contradicts the basis $\mathfrak{O}_1 = [1, \omega]$ by Theorem 3, Chapter IV, §6.

We therefore have taken into account all $\mathfrak{p}$ of norm $< \sqrt{|d|}$. The class structure and units can be deduced from the fact that each $(\omega - m)$ is a principal ideal not divisible by any rational prime. Hence $m - \omega$ is never divisible by both $\mathfrak{p}$ and $\mathfrak{p}'$. Perhaps a somewhat difficult example can indicate how to "play by ear." Let

$$D = 79, \quad d = 79 \cdot 4, \quad \omega = \sqrt{79}, \quad p \leq \sqrt{316} = 17, \cdots,$$
$$f(m) = m^2 - 79.$$

Function		Values		Factors		Norms
$f(0)$	=	-79	=	-79	=	$N(0 - \omega)$
$f(1)$	=	-78	=	$-2 \cdot 3 \cdot 13$	=	$N(1 - \omega)$
$f(2)$	=	-75	=	$-3 \cdot 5^2$	=	$N(2 - \omega)$
$f(3)$	=	-70	=	$-2 \cdot 5 \cdot 7$	=	$N(3 - \omega)$
$f(4)$	=	-63	=	$-3^2 \cdot 7$	=	$N(4 - \omega)$
$f(5)$	=	-54	=	$-2 \cdot 3^3$	=	$N(5 - \omega)$
$f(6)$	=	-43	=	-43	=	$N(6 - \omega)$
$f(7)$	=	-30	=	$-2 \cdot 3 \cdot 5$	=	$N(7 - \omega)$
$f(8)$	=	-15	=	$-3 \cdot 5$	=	$N(8 - \omega)$
$f(9)$	=	2	=	2	=	$N(9 - \omega)$
$f(10)$	=	21	=	$3 \cdot 7$	=	$N(10 - \omega)$
⋮				⋮		⋮

We should really test all m up to 16 to be sure of $p \leq 17$, but let us stop right here for a moment at $m = 10$. In the range $0 \leq m \leq 10$ we have all residues modulo 2, 3, 5, 7, or 11. Hence the factors prove

$$(d/p) = 0 \text{ for } p = 2 \quad \text{and} \quad (d/p) = 1 \text{ for } p = 3, 5, 7$$
$$\text{(which are present as factors),}$$
$$(d/p) = -1 \text{ for } p = 11 \text{ (which is absent as a factor).}$$

We have yet to test $p = 13$ and $p = 17$, but $(p =)$ 13, by a stroke of *good luck*, occurs in that $13 \mid N(1 - \omega)$. We might suspect, however, $(4 \cdot 79/17) = -1$, and rather than calculate $f(11), \cdots, f(16)$ we note $(79/17) = (-6/17) = (6/17) = (2/17)(3/17) = 1 \cdot (17/3) = (-1/3) = -1$ by reciprocity.

We next look for generators of the ideal classes. Again we are *lucky*: $2 = 2_1^2$ where $2_1 = 2_2 = (9 - \omega)$ is principal, or $2_1 \sim 1$. Next we can write, in ideal factors,

$$\mathbf{I} \sim (1 - \omega) = 2_1 3_1 13_1 \sim 3_1 13_1.$$

Hence 13_1 is in the cycle generated by 3_1 (in fact, the inverse). From now on 3_1 and 3_2 are *labeled* by the residue classes. For instance, if $q \equiv 1 \pmod 3$, then $q - \omega = 3[(q - 1)/3] + (1 - \omega)$ and

$$3_1 \mid (q - \omega) \quad \text{whereas} \quad 3_2 \nmid (q - \omega),$$

since $3_1 \neq 3_2$ and $3_1 3_2 = (3) \nmid (q - \omega)$. Likewise $3 \mid N(2 - \omega)$; thus, if $q \equiv 2 \pmod 3$,

$$3_2 \mid (q - \omega) \quad \text{whereas} \quad 3_1 \nmid (q - \omega).$$

We therefore write

$$\mathbf{I} \sim (8 - \omega) = 3_2 5_1$$

(which henceforth labels 5_1 and 5_2 according to residues of m mod 5). It is clear that 5_1 is in the cycle generated by 3_2. Likewise from $(4 - \omega) = 3_1^2 7_1$ it is clear that 7_1 also lies in the cycle generated by 3_1. Thus, using conjugates as inverses, it is clear that powers of 3_1 generate all ideal classes. Finally,

$$\mathbf{I} \sim (5 - \omega) = 2_1 3_2^3 \sim 3_2^3.$$

Thus 3_2 *is of order* 3 *or principal.*

We now digress: to find a unit, note that since $2_1 = 2_2$

$$(9 - \omega) = 2_1 = 2_2 = (9 - \omega').$$

Thus $(9 - \omega)/(9 - \omega')$ is a unit $= \eta$.

$$\eta = \frac{(9 - \omega)^2}{N(9 - \omega)} = \frac{81 - 18\omega + 79}{2} = 80 - 9\sqrt{79}.$$

Another unit can be found by setting

$$7 - \omega = 2_1 3_1 5_2$$

$$8 - \omega = 3_2 5_1$$

Thus $(7 - \omega)/(9 - \omega)(8 - \omega')$ is a unit $= \eta^*$.

With somewhat more laborious calculation, using conjugates,

$$\eta^* = \frac{(7 - \sqrt{79})(9 + \sqrt{79})(8 - \sqrt{79})}{(9 - \sqrt{79})(9 + \sqrt{79})(8 - \sqrt{79})(8 + \sqrt{79})} = -1.$$

This unit is not necessary for the problem, but it is inserted to remind the reader to not *always* expect a lucky result. We can actually prove from the procedures of Chapter VI that $80 + 9\sqrt{79}$ is the fundamental unit.

Now to prove 3_1 is nonprincipal or $N(a + b\omega) \neq \pm 3$ for integral a, b we refer to Theorem 5. We must simply show the root of equation

$$\alpha^2 - A\alpha \pm 3 = 0$$

never belongs to $R(\sqrt{79})$ when $|A| < \sqrt{n}(1 + \eta_0) = \sqrt{3}(80 + 9\sqrt{79}) = 278 \cdots$. But $A^2 \pm 4n = A^2 \pm 12 = 79g^2$ for some integer g, in order that α belong to $R(\sqrt{79})$. By the power residue tables, we note the solvability of

$$A^2 \equiv -12 \equiv 3^{48} \pmod{79}.$$

(Since $79 \equiv -1 \pmod 4$, $+12$ is *consequently* a nonresidue.) Thus

$$A \equiv \pm 3^{24} \equiv \pm 15 \pmod{79}.$$

We now try $A = 79k \pm 15$, for $k = 0, 1, 2, \cdots$.

A	$A^2 + 12$	
$A = 15$	$A^2 + 12 = 237$	$= 3 \cdot 79$
$= 64$	$= 4108$	$= 2^2 \cdot 13 \cdot 79$
$= 94$	$= 8848$	$= 2^4 \cdot 7 \cdot 79$
$= 143$	$= 20461$	$= 7 \cdot 37 \cdot 79$
$= 173$	$= 29941$	$= 379 \cdot 79$
$= 222$	$= 49296$	$= 2^4 \cdot 3 \cdot 13 \cdot 79$
$= 252$	$= 63516$	$= 2^2 \cdot 3 \cdot 67 \cdot 79$

In no case where $A < 279$ is $A^2 \pm 12 = 79g^2$ for an integer g. Thus 3_1 is nonprincipal and $h(316) = 3$.

EXERCISE 11. Complete the table with $f(11)$ through $f(16)$ and list the ideal factors of each $q - \omega$, $0 \leq q \leq 16$.

EXERCISE 12. Find another unit (not ± 1) from the information in the extended table. Compare with η_1.

EXERCISE 13. From the following data alone deduce the class number of $R(\sqrt{-79})$ by justifying the factorizations on the right:

$$f(x) = x^2 + x + 20 = N(x + \omega)$$
$$\omega = (+1 + \sqrt{-79})/2$$

$f(0) = 20$	$(\omega) = 2_1{}^2 5_1$
$f(1) = 22$	$(1 + \omega) = 2_2 11_1$
$f(2) = 26$	$(2 + \omega) = 2_1 13_1$
$f(3) = 32$	$(3 + \omega) = 2_2{}^5$
$f(4) = 40$	$(4 + \omega) = 2_1{}^3 5_2$
$f(5) = 50$	$(5 + \omega) = 2_2 5_1{}^2$
$f(6) = 62$	$(6 + \omega) = 2_1 31_1$

EXERCISE 14. Find the class structure of $R(\sqrt{-31})$ and $R(\sqrt{31})$ by the method of this section.

EXERCISE 15. Find a nontrivial unit of $R(\sqrt{31})$ from the table for Exercise 14. *Hint*. Factor $6 - \sqrt{31}$ and $9 - \sqrt{31}$ and recall $2_1{}^2 = 2$.

EXERCISE 16. Show that if g and p are primes and $r \geq 0$ is an odd integer then the field generated by a square free (negative) number

$$r^2 - 4g^p = d < 0.$$

has a class number divisible by p if $|d| > 4g$. (Note in Exercise 13, $r = 7, g = 2, p = 5, d = -79$.)

EXERCISE 17. Invent some fields of class number divisible by 2, 3, 5, 7, 11 by experimenting with the preceding problem. Try $g = 2$ and see which suitable r exists. Also try $g = 3$, etc.

*8. The Famous Polynomials $x^2 + x + q$

Euler discovered that the polynomials for certain positive values of q

$$f_q(x) = x^2 + x + q \tag{1}$$

take on only prime values when $0 \leq x \leq q - 2$. The values of q and the polynomials are listed as follows:

$$\begin{array}{llll} & x^2 + x + 3; & q = 3, & 1 - 4q = -11; \\ & x^2 + x + 5; & q = 5, & 1 - 4q = -19; \\ (2) & x^2 + x + 11; & q = 11, & 1 - 4q = -43; \\ & x^2 + x + 17; & q = 17, & 1 - 4q = -67; \\ & x^2 + x + 41; & q = 41, & 1 - 4q = -163. \end{array}$$

The values of $1 - 4q = d$, coincidentally, are precisely those for which the field $R(\sqrt{d})$ has class number 1. In fact, for $1 - 4q = -7, -3$ or $q = 2, 1$, which are not listed, the polynomials $f_q(x)$ still do assume only prime values for $0 \leq x \leq q - 2$, although in a trivial sense.

THEOREM 6. The polynomial $f_q(x)$ will assume prime values for $0 \leq x \leq q - 2$ if and only if for $d = 1 - 4q$, $R(\sqrt{d})$ has class number 1.

Proof. Let us assume $f_q(a)$ is *composite* for some a in the range $0 \leq a \leq q - 2$:

$$f_q(a) \leq (q-2)^2 + (q-2) + q = q^2 - 2q + 2 < q^2.$$

Now we can conclude that at least one prime factor of $f_q(a)$, namely p, is $\leq (q-1)$. Thus

$$p \leq q - 1 = (1-d)/4 - 1 = (|d| - 3)/4 < |d|/4,$$

whereas, if we define

$$\omega = \frac{1+\sqrt{d}}{2},$$

then

$$f_q(a) = N(a+\omega) \equiv 0 \pmod{p}.$$

Now, easily, $(p, a+\omega) = \mathfrak{p} \mid (p)$, yet $\mathfrak{p}$ is not principal, since no prime exists of norm p (by Lemma 1, §3). Thus the class number is unequal to 1.

Let us assume, conversely, that $f_q(x)$ is always prime for $0 \leq x \leq q - 2$. We shall prove that for all p in the range $2 \leq p \leq \sqrt{|d|}$ we would have the relation $(d/p) = -1$. (Hence the class number is 1 for want of primes to split or ramify!) First we note that $q - 2 \geq \sqrt{|d|}$, since $(q-2)^2 \geq 4q - 1$ when $q \geq 11$. (This means that the theorem must be verified separately for $q = 3, 5, 7$.) If any prime $p \leq \sqrt{|d|}$ exists for which $(d/p) = +1$ or 0, then some integer a exists for which (by Theorem 2)

$$f_q(a) = N(a+\omega) \equiv 0 \pmod{p}.$$

Since each such a is determined only modulo p, we write

$$0 \leq a < p \leq \sqrt{|d|} \leq q - 2. \tag{3}$$

Thus, if $f_q(a)$ is always prime, $f_q(a) = p$, which is a contradiction, since $p = f_q(a) \geq f_q(0) = q$, contrary to (3). Q.E.D.

PART 3

APPLICATIONS OF IDEAL THEORY

*chapter X

Class number formulas and primes in arithmetic progression

1. Introduction of Analysis into Number Theory

In this chapter we derive a formula for $h(d)$, the class number of a quadratic field of discriminant d, which makes use of infinite processes such as series and limits. The purpose of the formula is to enable us to calculate $h(d)$ directly from d but without gaining any group-theoretic knowledge of the class structure.

The real value of this formula historically is that it enabled Dirichlet to prove the following famous result (1837):

THEOREM 1. If a and m are relatively prime positive integers, then there exists an infinitude of primes in the arithmetic progression

$$a, a + m, a + 2m, a + 3m, a + 4m, \cdots.$$

The result was monumental for many reasons. First, as we shall see, it required infinite series, convergence, limits, logarithms, etc., and any number of concepts seemingly alien to the theory of integers. From this point forward it became an increasingly acceptable procedure to use limiting processes in number theory. Second, the fact that class structure should be relevant to arithmetic progressions is still largely unexplored, and results, as good as they are, lie in an esoteric synthesis of number theory, analysis, and algebra called "class-field theory."

The present chapter is therefore the one of greatest mathematical scope in the book, but it can scarcely do justice to the subject matter because of its brevity.

2. Lattice Points in Ellipse

We begin with the problem, due to Gauss, of finding the number of lattice points with integral coordinates in a family of ellipses. We consider the set of ellipses

$$Ax^2 + Bxy + Cy^2 = T \tag{1}$$

in the (x, y) plane, where A, B, C are fixed integers with

$$-B^2 + 4AC = \Delta > 0 \tag{2}$$

and T is a positive integer which is to approach infinity. It would seem intuitively clear that the number of integral lattice points inside the ellipse is approximately the area.

Before considering this more closely, we show that the exact value of the area is $2\pi T/\sqrt{\Delta}$. A rotation of axis is known which enables us to write

$$A'x'^2 + C'y'^2 = T, \quad (\text{or } B' = 0), \tag{3}$$

where x', y' is a new coordinate system at angle arctan $B/(A - C)$ with the old one. The area inside the ellipse (3), by an elementary calculation, is $\pi T/\sqrt{A'C'}$, but by the rotational invariance property of a conic $B^2 - 4AC = B'^2 - 4A'C'$; thus the area is $2\pi T/\sqrt{4A'C' - 0^2} = 2\pi T/\sqrt{\Delta}$. The major axis of the ellipse is by similarity exactly $\sqrt{T} \cdot k$, where k is a constant, namely, the major axis of the ellipse,

$$Ax^2 + Bxy + Cy^2 = 1, \tag{4}$$

which we need not find explicitly.

We now consider lattice points inside any smooth convex curve of width M.

There is little difficulty in showing the geometric relationship between area and the number of lattice points. All we need do is to surround each integral[1] lattice point (x, y) by a square of side 1, centered at this point, e.g., bounded by the lines of ordinates $y \pm \frac{1}{2}$ and abscissas $x \pm \frac{1}{2}$. The (x, y) *interior* to the curve determine a set of squares $\mathfrak{S}$. These are shown in Figure 10.1. (Note *vertical* shading in those squares of $\mathfrak{S}$ which intersect the boundary of the curve. If we shade *horizontally* all squares outside $\mathfrak{S}$

[1] To avoid a burdensome notation, we shall use the same symbols (x, y) for lattice points as for the "general" point of the plane.

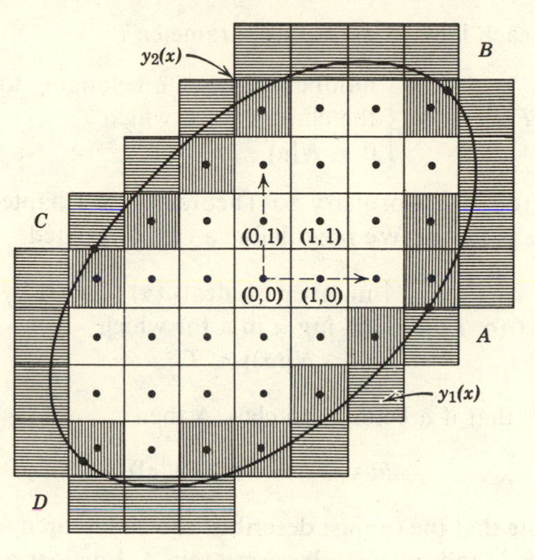

FIGURE 10.1

which intersect the boundary of the curve, we find that the difference between the number of squares in $\mathfrak{S}$ (or the number of lattice points interior to the curve) and the area of the curve is less in value than the number of shaded squares in Figure 10.1. It can then be shown (see Exercise 1, below) that the number of shaded squares is less than $8 + 8M$.

If $T > 1$, then $8 + 8k\sqrt{T} < (8 + 8k)\sqrt{T}$, leading to the following result (in terms of constants k and k' determined by the shape of the ellipse):

THEOREM 2 (Gauss). The number of lattice points inside ellipse (1) is given by

$$N = 2\pi T/\sqrt{\Delta} + \text{error}, \tag{5}$$

where the error is bounded by $k'\sqrt{T}$ as $T \to \infty$. (We say that the error has "order of magnitude" $\sqrt{T}$.)

EXERCISE 1. Prove that the number of shaded squares in Figure 10.1 is less than $8 + 8M$ by selecting four points A, B, C, D on the curve for which the slope is ± 1. Show, for example, that if on the arc DA the slope lies between -1 and $+1$ then the shaded squares covering that portion DA of the arc are fewer than 2 plus twice the x-projection of the curve.

3. Ideal Density in Complex Fields

In order to obtain an ideal-theoretic interpretation of Gauss's lattice-point result, we consider a complex quadratic field of discriminant $d < 0$.

We define for each ideal class **A** and parameter T

$$(1) \qquad F(\mathbf{A}, T) = \begin{cases} \text{number of ideals } \mathfrak{a} \text{ belonging to} \\ \text{the class } \mathbf{A}^{-1} \text{ for which} \\ 0 < N[\mathfrak{a}] \leq T. \end{cases}$$

It is evident from the Corollary to Theorem 4 in Chapter VII, §4 that $F(\mathbf{A}, T)$ is always finite. We now define a closely related

$$(2) \qquad G(\mathfrak{a}, T) = \begin{cases} \text{number of ideals } (\alpha) \text{ formed by} \\ \text{taking } \alpha \text{ in } \mathfrak{a} \text{ for which} \\ 0 < N[(\alpha)] \leq T. \end{cases}$$

It then follows that if $\mathfrak{a}$ belongs to class **A** then

$$(3) \qquad F(\mathbf{A}, T) = G(\mathfrak{a}, TN[\mathfrak{a}])$$

To see this, note that the (α) just described can be factored $(\alpha) = \mathfrak{a}\mathfrak{b}$, where $\mathfrak{b}$ belongs to $\mathbf{A}^{-1}$, and, conversely, every $\mathfrak{b}$ in $\mathbf{A}^{-1}$ defines a principal ideal $(\alpha) = \mathfrak{a}\mathfrak{b}$. Thus every $\mathfrak{b}$ in $\mathbf{A}^{-1}$ with $N[\mathfrak{b}] \leq T$ corresponds uniquely to an ideal (α) in $\mathfrak{a}$ with $N[(\alpha)] \leq TN[\mathfrak{a}]$.

The function $G(\mathfrak{a}, TN[\mathfrak{a}])$ is found from the inequality

$$(4a) \qquad 0 < N[(\alpha)] \leq TN[\mathfrak{a}].$$

Specifically, if $\mathfrak{a} = [\alpha_1, \alpha_2]$ then, in coordinate form, for α in $\mathfrak{a}$

$$(5a) \qquad \begin{cases} \alpha = \alpha_1 x + \alpha_2 y \\ \alpha' = \alpha_1' x + \alpha_2' y \\ N[(\alpha)] = \alpha\alpha' = (\alpha_1\alpha_1')x^2 + (\alpha_1\alpha_2' + \alpha_2\alpha_1')xy + \alpha_2\alpha_2' y^2. \end{cases}$$

We are therefore concerned with lattice points inside the ellipse,

$$(4b) \qquad Ax^2 + Bxy + Cy^2 = TN[\mathfrak{a}],$$

where

$$(5b) \qquad A = \alpha_1\alpha_1', \qquad B = \alpha_1\alpha_2' + \alpha_2\alpha_1', \qquad C = \alpha_2\alpha_2',$$

and

$$\Delta = 4(\alpha_1\alpha_1')(\alpha_2\alpha_2') - (\alpha_1\alpha_2' + \alpha_2\alpha_1')^2,$$

$$0 < \Delta = -(\alpha_1\alpha_2' - \alpha_2\alpha_1')^2 = -(\sqrt{d}N[\mathfrak{a}])^2,$$

$$(6) \qquad |\Delta| = |d|\, N[\mathfrak{a}]^2$$

by the index formula (6) of Chapter IV, §10.

We must, however, ask if each α inside the ellipse (4*b*) corresponds to a different ideal. Generally α and $-\alpha$ represent the same ideal. If we define

w as the number of units of the field of $\sqrt{d}$, then by the result in Chapter VI

$$\begin{cases} w = 6 \text{ for } d = -3, \\ w = 4 \text{ for } d = -4, \\ w = 2 \text{ for other } d. \end{cases}$$

Hence each *ideal* (α) satisfying $(4a)$ corresponds to w *points* α in the ellipse.

Therefore, from the number of lattice points in ellipse $(4a)$

$$\begin{aligned} F(\mathbf{A}, T) &= G(\mathfrak{a}, TN[\mathfrak{a}]) \\ &= \frac{1}{w} \cdot \frac{2\pi TN[\mathfrak{a}]}{\sqrt{|d|N[\mathfrak{a}]^2}} + \text{error}. \end{aligned}$$

$$\text{(7)} \qquad \begin{cases} F(\mathbf{A}, T) = T\kappa + \text{error} \\ \kappa = \dfrac{2\pi}{w\sqrt{|d|}} . \end{cases}$$

Here κ is the (complex) *Dirichlet structure constant*. The error has the order of magnitude of $\sqrt{T}$.

If we go further and define

$$\text{(8)} \qquad F(T) = \begin{cases} \text{number of ideals of any class with} \\ 0 < N[\mathfrak{a}] \leq T, \end{cases}$$

we see that

$$\text{(9)} \qquad F(T) = \sum_{\mathbf{A}} F(\mathbf{A}, T),$$

where the summation is over h different classes $\mathbf{A}$. But for each class (7) is valid; thus

$$\text{(10)} \qquad F(T) = h\kappa T + \text{error},$$

where the error still has the order of magnitude of $\sqrt{T}$. A quicker way of stating the essential result is that the ideals have *norm density* given by

$$\text{(11)} \qquad \lim_{T \to \infty} F(T)/T = h\kappa.$$

4. Ideal Density in Real Fields

To extend the result of §3 to real quadratic fields, we start with the function $G(\mathfrak{a}, T)$ defined precisely as in (2), §3. The difference is that the locus [corresponding to $(4b)$, §3] is derived from

$$\text{(1)} \qquad |N(\alpha)| \leq TN[\mathfrak{a}].$$

Using substitutions analogous with (5*b*), §3, we write

$$\alpha = \alpha_1 x + \alpha_2 y, \quad \alpha' = \alpha_1' x + \alpha_2' y$$

and

$$|(\alpha_1\alpha_1')x^2 + (\alpha_1\alpha_2' + \alpha_2\alpha_1')xy + \alpha_2\alpha_2' y^2| \leq TN[\mathfrak{a}]. \tag{2}$$

As we shall see, the inequality (2) is satisfied by an infinite number of pairs of integers, as T becomes large enough.

We therefore select from the lattice points satisfying (2) only those which correspond to *primary* numbers (Chapter IX, §2). For these numbers α the conjugates α and α' are related by

$$1 \leq |\alpha/\alpha'| < \eta_1^2, \qquad \alpha > 0, \tag{3}$$

where η_1 is the fundamental unit. Thus, if we supplement inequality (2) by

$$1 \leq \left|\frac{\alpha_1 x + \alpha_2 y}{\alpha_1' x + \alpha_2' y}\right| < \eta_1^2, \qquad \alpha_1 x + \alpha_2 y > 0. \tag{4}$$

We then obtain *two* sectors of a hyperbola. In finding the number of lattice points, once more we use the area as an approximation. This time it is convenient to change coordinates:

$$\begin{cases} \xi = \alpha_1 x + \alpha_2 y, \\ \xi' = \alpha_1' x + \alpha_2' y; \end{cases} \tag{5}$$

then it can be shown that the upper one of the sectors, defined by the inequalities analogous to (1) and (4), is

$$\begin{cases} \qquad |\xi\xi'| \leq U = TN[\mathfrak{a}] \\ 1 \leq \xi/\xi' < \eta_1^2, \qquad \xi > \xi' > 0. \end{cases} \tag{6}$$

Its area is precisely the following combination of an integral and equal triangles:

$$\int_{\sqrt{U}}^{\sqrt{U}\eta_1} U d\xi/\xi + \Delta OAC - \Delta OBD = U \log \eta_1,$$

as seen in Figure 10.2. The ratio of areas from the $\xi\xi'$ to the xy plane is the determinant $|\alpha_1\alpha_2' - \alpha_2\alpha_1'| = \sqrt{d}\, N[\mathfrak{a}]$. Hence, using two sectors of the hyperbola,

$$G(\mathfrak{a}, TN[\mathfrak{a}]) = \frac{2TN[\mathfrak{a}] \log \eta_1}{\sqrt{d}\, N[\mathfrak{a}]} + \text{error}; \tag{7}$$

and we define $F(\mathbf{A}, T)$, as in (1), §3, by

$$F(\mathbf{A}, T) = T\kappa + \text{error} \tag{8}$$

$$\kappa = \frac{2 \log \eta_1}{\sqrt{d}},$$

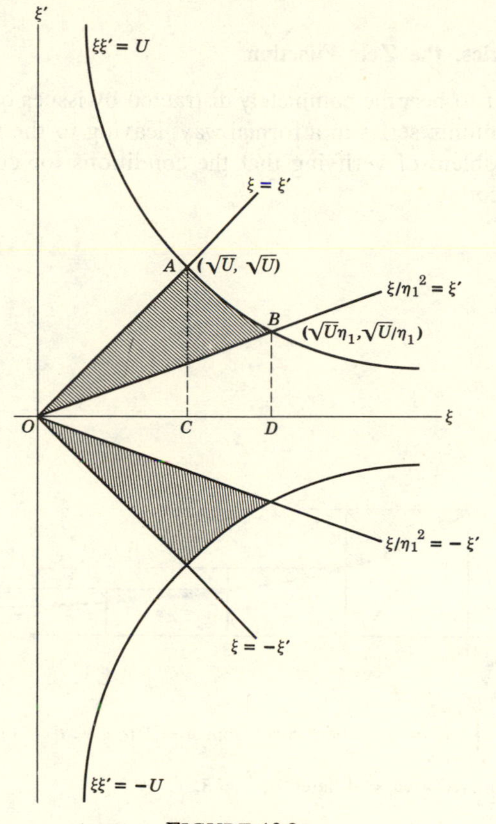

FIGURE 10.2

where κ is the (real) *Dirichlet structure constant*. Finally, with $F(T)$ defined [as in (8), §3] by $\sum_{\mathbf{A}} F(\mathbf{A}, T)$,

$$F(T) = h\kappa T + \text{error}, \tag{9}$$

where the errors are all of "order of magnitude" $\sqrt{T}$. The *norm density* again is $h\kappa$.

EXERCISE 2. Verify that the required sectors of the hyperbola (containing primary solutions) for $|x^2 - 2y^2| \leq 7$ are limited by $x > 2y > 0$ and $y > x > 0$ by applying Exercise 6 of Chapter IX, §2. (Here $\alpha_1 = 1$, $\alpha_2 = \sqrt{2}$, $\eta_1 = 1 + \sqrt{2}$.) Plot the primary solutions in the (x, y) plane.

EXERCISE 3. Extend the argument in §2 to show that the error term in approximating lattice points by area of a hyperbolic sector is still of the order $\sqrt{T}$.

EXERCISE 4. Tabulate $F(T)$ for $d = -4$ and $d = 8$ for $T \leq 25$ by considering the factorizations of all algebraic integers of the respective fields of norm ≤ 25. (Here the ideals are, of course, all principal.) Observe the ratio $F(T)/T$.

5. Infinite Series, the Zeta-Function

In order not to become completely distracted by issues of real analysis, we shall use infinite series in a formal way, leaving to the more energetic reader the problem of verifying that the conditions for convergence are taken into account.

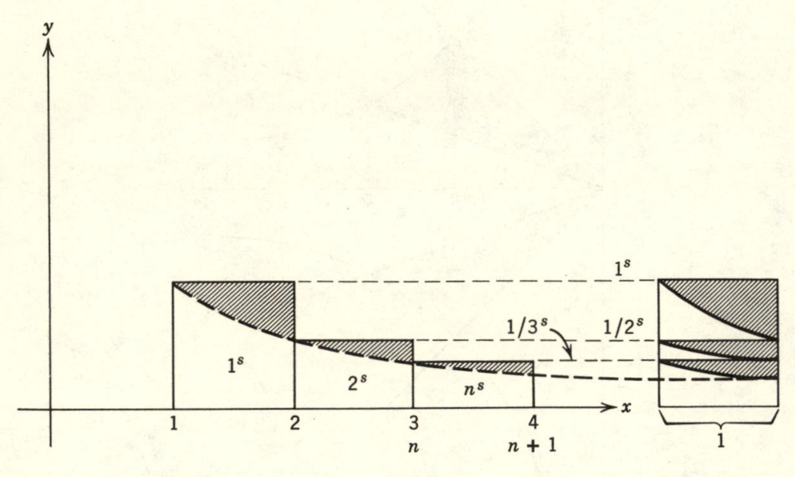

FIGURE 10.3. $\int_1^{n+1}$ = area under curve from $x = 1$ to $x = n + 1$; $\sum_1^n$ = area under step-function; shaded area = difference; $n = 3$.

The main type of infinite series that we shall use is the so-called zeta series:

$$\zeta(s) = 1 + \frac{1}{2^s} + \frac{1}{3^s} + \frac{1}{4^s} + \cdots. \tag{1}$$

Here s is a real continuous variable, $s \geq 1$. By the experience of elementary calculus, we know that the series converges for $s > 1$ by comparison with the area

$$\int^{\infty} \frac{dx}{x^s} = \frac{x^{1-s}}{1-s}\Big]^{\infty} = \text{"finite integral."} \tag{2}$$

More important is the result that

$$\zeta(s)(s-1) \to 1, \tag{3}$$

as $s \to 1$. To see this, note by comparison of areas in Figure 10.3 that

$$\left| \sum_{x=1}^{n} \frac{1}{x^s} - \int_1^{n+1} \frac{dx}{x^s} \right| < 1, \qquad \left| \sum_1^n \frac{1}{x^s} - \frac{1}{s-1}\left(1 - \frac{1}{(n+1)^{s-1}}\right) \right| < 1.$$

If $n \to \infty$, we find

$$\left| \zeta(s) - \frac{1}{s-1} \right| < 1,$$

or multiplying

$$|(s-1)\zeta(s) - 1| < (s-1),$$

and as $s \to 1$ the conclusion (3) holds. Hence $\zeta(s) \to \infty$ as $s \to 1$. Actually, the only interesting values of s in the whole chapter are those s that are near 1.

6. Euler Factorization

We now note a form of the unique factorization theorem. Consider the infinite product of series, multiplied over all primes, namely,

$$(1) \quad \left(1 + \frac{1}{2^s} + \frac{1}{2^{2s}} + \frac{1}{2^{3s}} + \cdots\right)\left(1 + \frac{1}{3^s} + \frac{1}{3^{2s}} + \frac{1}{3^{3s}} + \cdots\right) \cdot \left(1 + \frac{1}{5^s} + \frac{1}{5^{2s}} + \cdots\right) \cdots \left(1 + \frac{1}{p^s} + \frac{1}{p^{2s}} + \frac{1}{p^{3s}} + \cdots\right) \cdots .$$

The formal multiplication must yield all products of

$$\frac{1}{p_i^{a_i s}}$$

for primes p_i and exponents a_i. The unique factorization theorem tells us that every positive integer n occurs once to make up

$$(2) \quad \sum_1^\infty \frac{1}{n^s} = \zeta(s).$$

But

$$1 + \frac{1}{p^s} + \left(\frac{1}{p^s}\right)^2 + \left(\frac{1}{p^s}\right)^3 + \cdots + = \frac{1}{1 - (1/p^s)}.$$

Thus

$$(3) \quad \sum_1^\infty \frac{1}{n^s} = \zeta(s) = \prod_p (1 - p^{-s})^{-1},$$

where the product is extended over all primes.

Actually, Euler attached a great significance to the representation (3). He noted, on the left, that as $s \to 1$, $\zeta(s) \to \infty$, whereas on the right a product over primes occurred. Hence there must be an *infinite* number of primes; otherwise $\zeta(s) \to \Pi(1 - p^{-1})$ (over a finite set) or $\zeta(s) \not\to \infty$.

A related statement is the following:

THEOREM 3. If $f(a)$ is a completely[1] multiplicative function of positive integers a, b,

$$f(a)f(b) = f(ab),$$

then

$$\sum_{a=1}^{\infty} \frac{f(a)}{a^s} = \prod_p \left(1 - \frac{f(p)}{p^s}\right)^{-1}, \tag{4}$$

provided the sum on the left is absolutely convergent.

A word of explanation is in order concerning the use of infinite series. It is ordinarily assumed that care concerning the use of series is reserved for the "ultrafastidious" personalities in mathematics. This was generally believed until Riemann (1859) attempted to find a distribution formula for the nth prime, using the $\zeta(s)$ function. He made many "plausible" statements (and some incorrect ones), believing that the analysis was only a matter of detail, but it took almost 50 years before enough details could be supplied by his successors to prove the most basic result, that p_n, the nth prime, satisfies the limit

$$p_n/(n \log n) \to 1 \quad \text{as} \quad n \to \infty.$$

As is well known, all of Riemann's statements have *not* been settled, and, as a result, analysis, as applied to number theory, is treated with caution bordering on suspicion. The use of infinite series, *here*, gives no cause for suspicion, and the reader can supply any missing details.

As an example of what is involved in the analysis, let us introduce logarithms. We write in accordance with calculus

$$\log(1 + t) = t - \frac{t^2}{2} + \frac{t^3}{3} - \cdots \tag{5}$$

$$\log\left(1 - \frac{1}{p^s}\right)^{-1} = \frac{1}{p^s} + \frac{1}{2p^{2s}} + \frac{1}{3p^{3s}} + \cdots + \tag{6}$$

$$= \frac{1}{p^s} + \text{error}$$

where the error is less than

$$\frac{1}{2p^{2s}} + \frac{1}{2p^{3s}} + \frac{1}{2p^{4s}} + \cdots + = \frac{1}{2p^{2s}}\left(1 - \frac{1}{p^s}\right)^{-1} < \frac{1}{p^{2s}},$$

since $p^s > 2$ and $[1 - (1/p^s)]^{-1} < [1 - (1/2)]^{-1} = 2$.

[1] An (ordinary) multiplicative function is one for which the indicated relationship is imposed only when $(a, b) = 1$, such as $f(a) = \phi(a)$, the Euler function of elementary number theory.

Thus, taking logarithms of relation (3), we find

$$(7)\quad \begin{cases} \log \zeta(s) = \sum \log \left(1 - \dfrac{1}{p^s}\right)^{-1} \\ \qquad\quad = \sum_p \dfrac{1}{p^s} + \text{error.} \end{cases}$$

The error is less than $\Sigma\,(1/p^{2s}) < \Sigma\,(1/p^2) < \zeta(2)$, for $s > 1$. Hence since $\zeta(s)(s-1) \to 1$,

$$\log \zeta(s) = \log\,[\zeta(s)(s-1)] + \log \frac{1}{s-1}$$

and

$$(8)\quad \sum_p \frac{1}{p^s} = \log \frac{1}{(s-1)} + \text{bounded error,}$$

as $s \to 1$, providing additional evidence that there is an infinite number of primes, according to Euler's technique.

EXERCISE 5. Using Hilbert's example of "nonunique factorization," consider

$$\sum_n \frac{1}{n^s} \neq \prod_p \left(1 - \frac{1}{p^s}\right)^{-1},$$

where the sum is over positive $n \equiv 1 \pmod 4$ and the product over "indecomposible" $p \equiv 1 \pmod 4$, with no factor $< p$ of the same type. (See Chapter III, §5.) Single out the first term g/n^s for which both sides will fail to check.

7. The Zeta-Function and L-Series for a Field

We next consider the series called the zeta-function for a field

$$(1)\quad \zeta(s; d) = \sum \frac{1}{N[\mathfrak{a}]^s}$$

summed over the norms of all ideals (excluding zero) in the field of discriminant d. Rather than concern ourselves with convergence immediately, we note "formally," i.e., assuming convergence,

$$(2)\quad \zeta(s; d) = \prod \left(1 - \frac{1}{N[\mathfrak{p}]^s}\right)^{-1}$$

over all prime ideals by virtue of unique factorization, according to the method of Euler.

The methods of Chapter IX tell us how to decompose further the product

$$(3)\quad \zeta(s; d) = \textstyle\prod_1 \prod_2 \prod_3.$$

Here $\prod_1$ is the product over all primes $\mathfrak{p}$ $(= q)$ for which $(d/q) = -1$; for these $N[(q)] = q^2$. Furthermore, $\prod_2$ is the product over all primes $\mathfrak{p}$ for which $\mathfrak{p} \mid p$ and $(d/p) = +1$. There are *two* factors for $\mathfrak{p}_1$ and $\mathfrak{p}_2$ and $N[\mathfrak{p}_1] = N[\mathfrak{p}_2] = p$. Finally, $\prod_3$ is the product over all primes $\mathfrak{r} \mid r \mid d$, for which $(d/r) = 0$, $N[\mathfrak{r}] = r$. Hence

$$\zeta(s; d) = \prod\nolimits_1 \left(1 - \frac{1}{q^{2s}}\right)^{-1} \prod\nolimits_2 \left(1 - \frac{1}{p^s}\right)^{-2} \prod\nolimits_3 \left(1 - \frac{1}{r^s}\right)^{-1}, \tag{4}$$

or

$$\begin{aligned} \zeta(s; d) = &\prod\nolimits_1 \left(1 - \frac{1}{q^s}\right)^{-1}\left(1 + \frac{1}{q^s}\right)^{-1} \\ &\cdot \prod\nolimits_2 \left(1 - \frac{1}{p^s}\right)^{-1}\left(1 - \frac{1}{p^s}\right)^{-1} \\ &\cdot \prod\nolimits_3 \left(1 - \frac{1}{r^s}\right)^{-1}. \end{aligned} \tag{5}$$

We can easily recognize $\zeta(s)$ in the product, since categories p, q, r exhaust all primes. Thus

$$\zeta(s; d) = \zeta(s)L(s; d), \tag{6a}$$

where we define

$$L(s; d) = \prod_q\nolimits_1 \left(1 + \frac{1}{q^s}\right)^{-1} \prod_p\nolimits_2 \left(1 - \frac{1}{p^s}\right)^{-1} \prod_r\nolimits_3 (1). \tag{6b}$$

Now, combining all cases (p, q, and the "missing" r) into one symbol p, we find

$$L(s; d) = \prod_p \left(1 - \frac{(d/p)}{p^s}\right)^{-1}, \tag{7}$$

taken over all primes p. By Theorem 2, since $f(x) = (d/x)$ is completely multiplicative for $x > 0$ (by properties of the Kronecker symbol),

$$L(s; d) = \sum_1^{\infty} \frac{(d/n)}{n^s}. \tag{8}$$

This function $L(s; d)$ is known as *Dirichlet's L-series*. In all cases the convergence is no problem when $s > 1$.

EXERCISE 6. Write out the first 25 terms of $\zeta(s; d)$ for $d = -4$ and $d = 8$, noting that some values of $N(\mathfrak{a})$ occur more than once, some not at all. (Use Exercise 4, §4 above.)

8. Connection with Ideal Classes

To tie the zeta-function for a field with a class number we return to the function $F(T)$ defined in (8), §3. We note that if in the series we collect all terms with equal values of $N(\mathfrak{a})$ we find

$$(1) \quad \zeta(s; d) = \sum \frac{1}{N[\mathfrak{a}]^s} = \frac{F(1)}{1^s} + \frac{F(2) - F(1)}{2^s} + \frac{F(3) - F(2)}{3^s} + \cdots,$$

since $F(T) - F(T-1)$ is the *precise* number of ideals $\mathfrak{a}$ for which $T = N[\mathfrak{a}]$. If we rearrange the series, we find

$$\zeta(s; d) = F(1)\left(\frac{1}{1^s} - \frac{1}{2^s}\right) + F(2)\left(\frac{1}{2^s} - \frac{1}{3^s}\right) + F(3)\left(\frac{1}{3^s} - \frac{1}{4^s}\right) + \cdots +,$$

$$(2) \qquad \zeta(s; d) = \sum_{T=1}^{\infty} F(T)\left(\frac{1}{T^s} - \frac{1}{(T+1)^s}\right).$$

It is easily seen that

$$(3) \qquad \frac{1}{T^s} - \frac{1}{(T+1)^s} = \frac{1}{T^s}\left[1 - \left(1 + \frac{1}{T}\right)^{-s}\right] = \frac{1}{T^s}\left[1 - \left(1 - \frac{s}{T} + \frac{s(1-s)}{T^2} \cdots\right)\right] = \frac{s}{T^{s+1}} + \text{error}.$$

It can therefore be seen that for all $s > 1$ the error is $< C_1/T^{s+2}$ for C_1 constant. Thus, substituting into (2) and summing, we see

$$(4) \qquad \zeta(s; d) = s \sum_{1}^{\infty} \frac{F(T)}{T} \cdot \frac{1}{T^s} + \text{finite error.}^1$$

But $F(T)/T \to h\kappa$ for the various values of κ and $h(d)$ in §§3 and 4 (above). The error in the limit is less than $c_2/\sqrt{T}$ for c_2, a constant. Thus

$$(5) \qquad \zeta(s; d) = sh\kappa \sum_{1}^{\infty} \frac{1}{T^s} + \text{finite error}$$

or

$$(6) \qquad (s-1)\zeta(s; d) = s(s-1)\zeta(s)h\kappa + (\text{finite error})(s-1),$$

as $s \to 1$. If we refer to (6*a*) in §7 and (3) in §5;

$$(7) \qquad L(s; d) \to h\kappa,$$

as $s \to 1$. We can go further and *set* $s = 1$, from (8) in §7. Then

$$(8) \qquad h\kappa = L(1; d) = \sum_{1}^{\infty} \frac{(d/n)}{n}.$$

[1] We shall use the expression "finite error" to mean a variable that remains finite as $s \to 1$.

We have thus asserted that the limit of $L(s; d)$ as $s \to 1$ is merely $L(1; d)$, found by substituting $s = 1$. This proposition of continuity of the series in the parameter s is not trivial, although it involves well-known procedures of analysis.[1] We shall merely rewrite $L(s; d)$,

$$L(s; d) = \sum_{n=1}^{\infty} \frac{(d/n)}{n^s}, \tag{9}$$

in a form that will make convergence manifest for $s \geq 1$.

We first define

$$E(n) = (d/1) + (d/2) + \cdots + (d/n); \qquad E(0) = 0, \tag{10}$$

so that, analogous with the rearrangement performed earlier,

$$(d/n) = E(n) - E(n-1), \qquad (n \geq 1). \tag{11}$$

But if $n = kd + r$, e.g., n/d, has quotient k and remainder r,

$$E(n) = \sum_{x=1}^{kd} (d/x) + \sum_{x=kd+1}^{kd+r} (d/x). \tag{12}$$

Now $(d/kd + r) = (d/r)$ by the conductor properties of the discriminant d (see Chapter II, §6). Furthermore,

$$\sum_{x=1}^{d} (d/x) = 0,$$

since the orthogonality relation (16) of Chapter II, §2, now applies with $\chi(x) = (d/x)$ and $\chi_0(x)$, the two different characters modulo d. Thus the first sum in (12) is 0 and

$$|E(n)| \leq \sum_{x=kd+1}^{kd+r} |(d/x)| \leq \sum_{x=1}^{r} 1 = r < d. \tag{13}$$

Using the rearrangement of (2),

$$L(s; d) = \sum_{n=1}^{\infty} \frac{E(n) - E(n-1)}{n^s} = \sum_{n=1}^{\infty} E(n)\left(\frac{1}{n^s} - \frac{1}{(n+1)^s}\right); \tag{14}$$

and inequality (13) yields on a term-by-term comparison [see (3)]

$$|L(s; d)| \leq \sum_{n=1}^{\infty} \left(\frac{d \cdot s}{n^{s+1}} + \text{error}\right), \tag{15}$$

where the error is $< C_2/n^{s+2}$. This shows convergence for $s \geq 1$.

[1] The reader, familiar with the sufficiency of *uniform* convergence for continuity, will recognize the series (14), below, to be uniformly convergent for $s \geq 1$, whence the same will hold for the rearranged series (9).

9. Some Simple Class Numbers

As our first example, take $d = -4$. Here

$$L(1; -4) = 1 - \tfrac{1}{3} + \tfrac{1}{5} - \tfrac{1}{7} + \tfrac{1}{9} + \cdots,$$

since $(-4/x) = \pm 1$ according as $x \equiv \pm 1 \pmod 4$, (x odd). It is easy to see that $L(1; -4) = \pi/4$. In fact,

$$\frac{\pi}{4} = \int_0^1 \frac{dx}{1 + x^2} = \int_0^1 (1 - x^2 + x^4 - x^6 + \cdots +)\, dx = L(1; -4),$$

integrating term by term. But $\kappa = 2\pi/4\sqrt{|-4|} = \pi/4$. Thus $h = 1$.

As a second example, take $d = 5$. Here

$$L(1; 5) = (1 - \tfrac{1}{2} - \tfrac{1}{3} + \tfrac{1}{4}) + (\tfrac{1}{6} - \tfrac{1}{7} - \tfrac{1}{8} + \tfrac{1}{9}) + \cdots,$$

the plus sign going with residue and the minus sign with nonresidues modulo 5.

$$\begin{aligned} L(1; 5) &= \int_0^1 dx[(1 - x - x^2 + x^3) + (x^5 - x^6 - x^7 + x^8) + \cdots] \\ &= \int_0^1 dx(1 - x - x^2 + x^3)[1 + x^5 + x^{10} + x^{15} + \cdots] \\ &= \int_0^1 dx \frac{(1 - x - x^2 + x^3)}{1 - x^5} = \int_0^1 \frac{(1 - x^2)\, dx}{(1 + x + x^2 + x^3 + x^4)}. \end{aligned}$$

It is well-known that any rational function may be integrated by the method of (complex) partial fractions. We can avail ourselves, however, of the following trick. Let $x + 1/x = y$, $(1 - 1/x^2)\, dx = dy$;

$$\begin{aligned} L(1; 5) &= -\int_0^1 \frac{(1 - 1/x^2)\, dx}{1/x^2 + 1/x + 1 + x + x^2} = \int_2^\infty \frac{dy}{y^2 + y - 1} = \int_{5/2}^\infty \frac{dz}{z^2 - 5/4} \\ &= \frac{1}{2\sqrt{5/4}} \log \frac{5/2 + \sqrt{5}/2}{5/2 - \sqrt{5}/2} = \frac{2}{\sqrt{5}} \log \frac{\sqrt{5} + 1}{2} \end{aligned}$$

by well-known formulas. But for $d = 5$, $\eta_1 = (\sqrt{5} + 1)/2$; hence $\kappa = L(1; 5)$ and $h = 1$.

The evaluation of other integrals is a more complicated matter. It is not hard to see that in general we shall obtain

$$L(1; d) = \int_0^1 \frac{f_d(x)\, dx}{1 - x^{|d|}},$$

where $f_d(x) = \sum_{t=1}^{|d|} x^{t-1}(d/t)$. By the theory of partial fractions, we can decompose

$$\frac{f_d(x)}{1 - x^{|d|}} = \sum \frac{A + Bx}{x^2 - ax + b}$$

by *first* reducing the left-hand fraction to lowest terms and then factoring the new *reduced* denominator into quadratic factors. For example, when $d = 8$, using the indicated partial fractions

$$\frac{f_d(x)}{1 - x^{|d|}} = \frac{(1 - x^2)(1 - x^4)}{1 - x^8} = \frac{1 - x^2}{1 + x^4},$$

$$\frac{1 - x^2}{1 + x^4} = \frac{A + Bx}{x^2 - \sqrt{2}x + 1} + \frac{A' - B'x}{x^2 + \sqrt{2}x + 1},$$

we find, comparing coefficients, that $A = A' = \frac{1}{2}$, $B = B' = -\frac{1}{4}\sqrt{2}$.

It will be no surprise that the class number can be explicitly evaluated for all quadratic fields by evaluating the integral for $L(1; d)$. The manipulations involved are of no further interest,[1] at present.

We shall make a more startling use of the formula for $L(1; d)$ even without being able to evaluate it explicitly.

EXERCISE 7. Carry out the calculation of h when $d = -3$.

EXERCISE 8. Do the same when $d = 8$.

EXERCISE 9. If $xf_d(x) = g(x)$, show $g(1/x) = g(x)$ or $-g(x)$ according to the sign of d, by Exercise 9, Chapter II, §6.

10. Dirichlet L-Series and Primes in Arithmetic Progression

The historical consummation of the class-number formula has not been the numerical usage of the formula but the application to primes in arithmetic progression. We must first digress to define the Dirichlet L-series for a character $\chi(x)$ modulo m as

$$L(s, \chi) = \sum_{1}^{\infty} \frac{\chi(n)}{n^s}. \tag{1}$$

The product formula (4) of §6 yields

$$L(s, \chi) = \prod_{p} \left(1 - \frac{\chi(p)}{p^s}\right)^{-1}, \tag{2}$$

and the logarithm formula (5) of §6 yields similarly for $s > 1$

$$\log L(s, \chi) = \sum_{p} \chi(p)/p^s + \text{error}; \tag{3}$$

where the "error" again remains bounded as $s \to 1$, similarly to (7) of §6. In all that precedes, of course, we note $|\chi(p)| = 1$ if $p \nmid m$ or 0 if $p \mid m$, hence all estimates on the error in §8 (above) are certainly still valid here.

[1] See formula (17) in the Concluding Survey. There may interest in the fact that a logarithm occurs in $L(1; d)$ only if $d > 0$ which is partially explained by Exercise 9 (below).

We must interpret the logarithm as the complex logarithm, in general, using the MacLaurin series

$$\log(1+z) = z - z^2/2 + z^3/3 + \cdots + (-1)^{n+1}(z^n/n) + \cdots \tag{4}$$

valid for real or complex z if $|z| < 1$.

The connection with primes in arithmetic progressions is brought out if we take some definite relatively prime residue class a modulo m. If we multiply (3) by $\chi^{-1}(a)$ and sum over χ in $\mathbf{X}$ which denotes the finite set of $\phi(m)$ characters,

$$\begin{aligned}\sum_{\chi \text{ in } \mathbf{X}} \chi^{-1}(a) \log L(s, \chi) &= \sum_{\chi}\sum_{p} \frac{\chi^{-1}(a)\,\chi(p)}{p^s} + \text{error} \\ &= \sum_{p} \frac{1}{p^s} \sum_{\chi} \chi^{-1}(a)\,\chi(p) + \text{error}.\end{aligned} \tag{5}$$

Hence, by (17) in §2, Chapter II, the inner sum

$$\sum_{\chi} \chi^{-1}(a)\,\chi(p)$$

equals $\phi(m)$ if $p \equiv a \pmod m$ and 0 otherwise. Thus

$$\sum_{\chi} \chi^{-1}(a) \log L(s, \chi) = \phi(m) \sum_{p \equiv a \pmod m} \frac{1}{p^s} + \text{error}, \tag{6}$$

where the "error" remains finite as $s \to 1$. We shall show that the left-hand term approaches ∞ as $s \to 1$. This will prove Dirichlet's theorem (for the sum on the right would have to contain an infinite number of primes).

To see a simple case, let us take $m = 4$, where our purpose is to show that there is an infinitude of primes of type $p \equiv 1 \pmod 4$ and of type $q \equiv 3 \pmod 4$. Analogously, with Euler's proof in §3, we would wish to show that the following two quantities approach ∞ as $s \to 1$:

$$\begin{cases} \Pi_1 = \prod_{p} (1 - 1/q^s)^{-1}, & (s > 1), \\ \Pi_2 = \prod_{q} (1 - 1/p^s)^{-1}, & (s > 1). \end{cases} \tag{7}$$

Here we note (for $m = 4$) that χ can be two functions:

$$\chi_1(y) = (4/y) = 1,$$
$$\chi_4(y) = (-4/y) = (-1)^{(y-1)/2}, \quad (y > 0),$$

for y odd, whereas each $\chi(y) = 0$ when y is even. Symbolically,

$$\chi_1(p) = \chi_4(p) = 1; \qquad \chi_1(q) = -\chi_4(q) = 1.$$

We also introduce, in accordance with $\chi_4(q)$,

$$\Pi_1' = \prod_{q} (1 + 1/q^s)^{-1}, \qquad (s > 1),$$

so that if we define $\prod_3 = \prod_1 \prod_1'$ we obtain the bounds

$$(8) \qquad 1 \le \prod_3 = \prod_q (1 - 1/q^{2s})^{-1} \le \zeta(2s).$$

Now the L-series gives us essentially two expressions of "unknown behavior," $\prod_1 \prod_2$ and $\prod_2/\prod_1$,

$$(8a) \qquad L(s, \chi_1) = (\prod_1 \prod_2) = \zeta(s)(1 - \tfrac{1}{2}^s)$$

$$(8b) \qquad L(s, \chi_4) = \prod_2 \prod_1' = (\prod_2/\prod_1) \cdot \prod_3.$$

By multiplying and dividing these equations, we obtain

$$(9a) \qquad \prod_2{}^2 = \zeta(s)(1 - \tfrac{1}{2}^s)[L(s, \chi_4)/\prod_3]$$

$$(9b) \qquad \prod_1{}^2 = \zeta(s)(1 - \tfrac{1}{2}^s)[\prod_3/L(s, \chi_4)].$$

But as $s \to 1$, $\zeta(s) \to \infty$, $L(s, \chi_4) \to \pi/4$ (by §9), whereas $1 \le \prod_3 \le \zeta(2)$. Hence $\prod_1 \to \infty$ and $\prod_2 \to \infty$. Q.E.D.

The reader will note that adding logarithms is simpler, as a matter of notation, than handling products, but log $[L(s, \chi_1)\, L(s, \chi_4)]$ and log $[L(s, \chi_1)/L(s, \chi_4)]$ correspond to $\sum_\chi \chi^{-1}(1) \log L(s, \chi)$ and $\sum_\chi \chi^{-1}(3)$ log $L(s, \chi)$ required in (5). The replacement of $\prod_1'$ by $1/\prod_1$ is easily justified by the use of logarithms and the neglect of higher order terms in (3).

11. Behavior of the L-Series, Conclusion of Proof

We first distinguish three types of L-series; taken modulo m:

TYPE I SERIES

Here $\chi = \chi_1$, the unit character only. Then

$$\chi_1(x) = 1 \text{ if } (x, m) = 1 \quad \text{and} \quad \chi_1(x) = 0 \text{ if } (x, m) > 1.$$

Thus

$$L(s, \chi_1) = \prod_{(p,m)=1} \left(1 - \frac{1}{p^s}\right)^{-1}$$

$$= \zeta(s) \cdot \prod_{p \mid m} \left(1 - \frac{1}{p^s}\right).$$

The second product $\prod_{p \mid m} \cdots$ has a finite number of factors and approaches

$$\prod_{p \mid m} \left(1 - \frac{1}{p}\right),$$

as $s \to 1$. But $\log \zeta(s) = -\log(s-1) +$ finite error; thus

$$\log L(s, \chi_1) = -\log(s-1) + \text{finite error} \tag{1}$$

as $s \to 1$.

TYPE II SERIES

Here χ is real and $\chi \neq \chi_1$. We use the theorem that every real character modulo m satisfies, by Dirichlet's Lemma,

$$\chi(a) = (M/a), \tag{2}$$

where M is a suitably chosen positive integer, not a perfect square (see Chapter II, §7). Thus we can write $M = g^2 d$ where d is the discriminant of some quadratic field. It is easily shown that M, m, d have (ignoring sign) the same square-free kernel, but the exact relationship is irrelevant. Thus for a Type II series

$$\begin{aligned} L(s,\chi) &= \prod \left(1 - \frac{\chi(p)}{p^s}\right)^{-1} = \prod \left(1 - \frac{(M/p)}{p^s}\right)^{-1} \\ &= \prod_{\text{all } p} \left(1 - \frac{(g^2 d/p)}{p^s}\right)^{-1} = \prod_{p \nmid g}^{(1)} \prod_{p \mid g}^{(2)}. \end{aligned} \tag{3}$$

It is clear that if $p \nmid g$, then $(g^2 d/p) = (d/p)$; otherwise, if $p \mid g$, $(g^2 d/p) = 0$. Hence $\prod^{(2)} = 1$, and if we compare

$$L(s; d) = \prod_{\text{all } p} \left(1 - \frac{(d/p)}{p^s}\right)^{-1}, \tag{4}$$

then, supplying the factors for $\prod^{(1)}$ in $L(s, \chi)$, we see

$$L(s; d) = L(s, \chi) \cdot \prod_{p \mid g} \left(1 - \frac{(d/p)}{p^s}\right)^{-1}. \tag{5}$$

Now, since only a finite number of primes divides g, the factor $\prod_{p \mid g}$ approaches the finite limit ($\neq 0$)

$$\prod_{p \mid g} \left(1 - \frac{(d/p)}{p}\right)^{-1}$$

as $s \to 1$. We have only to observe that the class number $h(d) \geq 1$; hence $L(s; d)$ and $L(s, \chi)$ *approach nonzero limits* as $s \to 1$ by formula (8) of §8 (above) for χ of Type II.

TYPE III SERIES

Here χ is complex. *We show, likewise, that* $L(s, \chi) \to L(1, \chi) \neq 0$ as $s \to 1$. To see this, let us first note that a continuous derivative

$$L'(s, \chi) = -\sum_{1}^{\infty} \chi(n) \log n/n^s \tag{6}$$

exists for $s \geq 1$. The expression given here is a "term-by-term" derivative.[1]

Hence, if, to the contrary of the assumption of the non-vanishing of the L-series,

$$L(1, \chi_*) = 0, \qquad (\chi_* \text{ of Type III}). \tag{7}$$

then we could conclude by the mean value theorem of calculus

$$L(s, \chi_*) = L(s, \chi_*) - L(1, \chi_*) = L'(s_0, \chi_*)(s - 1) \tag{8}$$

for $1 < s_0 < s$. (We henceforth keep s real.) This would all be true even if χ_* were real, but the *complex* character χ_* has the property that for its complex conjugate $\bar{\chi}_*$ it necessarily follows that

$$L(1, \bar{\chi}_*) = 0 \tag{9}$$

by taking a term-by-term conjugate of (7). Then

$$L(s, \bar{\chi}_*) = L'(s_0, \bar{\chi}_*)(s - 1). \tag{10}$$

Now if we sum (3) of §10 over all characters (or take (6) of §10 with $a = 1$), we find

$$\sum_{\chi} \log L(s, \chi) = \phi(m) \sum_{p \equiv 1 \pmod{m}} \frac{1}{p^s} + \text{finite error}. \tag{11}$$

The sum on the right is ≥ 0, as $s \to 1$. It may (and actually does) approach ∞, but it certainly will not approach $-\infty$. If we examine the sum on the left in (11), we will find, by (8) and (10), *at least two* different χ, (χ_* and $\bar{\chi}_*$) for which a term of order $\log(s - 1)$ is contributed by a vanishing $L(1, \chi)$ but only one $\chi = \chi_1$ for which $-\log(s - 1)$ is contributed to this left-hand sum. The net result is that the left-hand side of (11) still approaches $-\infty$, (even more so, if some other $L(s, \chi) \to 0$ or even if, in (8), $L'(s_0, \chi_*) \to 0$ as $s \to 1$). This makes the left-hand side of (11) approach $-\infty$, contradicting the assumption (7).

[1] The derivative is valid when series (6) is shown to be uniformly convergent for $s \geq 1$. In Exercise 10 we perform a rearrangement analogous to §8 to show convergence.

Thus, as $s \to 1$

$$\text{(12)} \qquad \log L(s, \chi) = \begin{cases} -\log(s-1) + \text{finite terms if } \chi = \chi_1 \\ \text{merely finite terms if } \chi \neq \chi_1. \end{cases}$$

Returning to (6) in §10, we see from the result (12) in §11 that as $s \to 1$ only χ_1 matters in the sum, or

$$\text{(13)} \qquad \frac{1}{\phi(m)} \log\left(\frac{1}{s-1}\right) = \sum_{p \equiv a \bmod m} \frac{1}{p^s} + \text{finite error}.$$

Thus there is an infinite number of primes in the arithmetic progression as required.

As an outgrowth of the comparison, we note by (8) in §6

$$\text{(14)} \qquad \frac{\displaystyle\sum_{p \equiv a \pmod{m}} \frac{1}{p^s}}{\displaystyle\sum_{\text{all } p} \frac{1}{p^s}} \to \frac{1}{\phi(m)}, \qquad (\text{as } s \to 1).$$

Thus "in some sense" each of the various $\phi(m)$ relatively prime residue classes modulo m contains an "equal density" of primes if we measure density by the ratio on the left. This mode of measurement has been subsequently termed "Dirichlet density." To conclude the same linear density in the sense of a *count* is not at all easy; e.g., if $\prod(x; a, m)$ is the number of primes $\leq x$ in the given arithmetic progression (and if $\prod(x)$ is the total number of primes $\leq x$), it is by no means immediate that

$$\text{(15)} \qquad \frac{\prod(x; a, m)}{\prod(x)} \to \frac{1}{\phi(m)}, \qquad (\text{as } x \to \infty),$$

although this difficult result is true.

EXERCISE 10. In a grouping analogous to §8, (9) to (15), show the convergence of the series in (6) for $s \geq 1$.

**12. Weber's Theorem on Primes in Ideal Classes

On looking back, we note that quadratic field theory entered the theorem on primes in arithmetic progression *at only one spot*, in showing $L(1, \chi) \neq 0$ if χ is a character of Type II. To see this curious fact in greater perspective, let us consider a very closely related theorem of Weber (1882).

THEOREM 4. Every ideal class of a quadratic field contains an infinite number of primes.

Proof. Let d be the field discriminant as usual. First of all, the primes p, for which $(d/p) = -1$, do not split and therefore are in the principal class.

There is an infinite number of such p, for instance, taken from each arithmetic progression $dx + a$ where a is chosen so that $(d/a) = -1$. We shall show an infinitude of primes $\mathfrak{p}$ to exist in each class such that $\mathfrak{p} \mid p$ and $(d/p) = 1$. (This even strengthens the result for the principal class.)

In the earlier proofs we considered (in effect) a set of $\phi(d)$ characters $\chi(n)$ defined for the $\phi(d)$ residue classes n for which $(n, d) = 1$. For this proof we introduce the characters $\psi(\mathbf{A})$ defined on the class group with h elements symbolized by $\mathbf{A}$. There are h such characters, and the unit character $\psi_1(\mathbf{A})$ is 1 for each $\mathbf{A}$. If $\mathfrak{a} \in \mathbf{A}$, we extend the definition $\psi(\mathfrak{a}) = \psi(\mathbf{A})$. The multiplicative property of characters, of course, is $\psi(\mathfrak{a})\psi(\mathfrak{b}) = \psi(\mathfrak{ab})$.

Then, analogously with (1) and (2), §10, we define the *modified zeta function*:

$$\text{(1)} \qquad \zeta(s, d; \psi) = \sum_{\mathfrak{a}} \psi(\mathfrak{a})/N[\mathfrak{a}]^s = \prod_{\mathfrak{p}} (1 - \psi(\mathfrak{p})/N[\mathfrak{p}]^s)^{-1},$$

where the sum is over all nonzero ideals $\mathfrak{a}$ and the product is over all prime ideals $\mathfrak{p}$. Then, analogous with (3) and (6), §10,

$$\text{(2)} \qquad \log \zeta(s, d; \psi) = \sum_{\mathfrak{p}} \psi(\mathfrak{p})/N[\mathfrak{p}]^s + \text{finite error},$$

where "finite error" refers again to the limit $s \to 1$. Then, for $\mathbf{A}$, a fixed ideal-class, we take sums over all h characters ψ, using orthogonality:

$$\text{(3)} \qquad \sum_{\psi} \psi^{-1}(\mathbf{A}) \log \zeta(s, d; \psi) = h \sum_{\mathfrak{p} \in \mathbf{A}} 1/N[\mathfrak{p}]^s + \text{finite error}.$$

As before, we define the modified zeta-functions $\zeta(s, d; \psi)$ of three types: Type I, for which $\psi = \psi_1(\mathbf{A})$ is the unit character, Type II, for which $\psi = \psi(\mathbf{A})$ is real for all $\mathbf{A}$, and Type III, for which ψ has complex values for some $\mathbf{A}$. The proof of the theorem then consists in showing that as $s \to 1$

$$\text{(4)} \qquad \begin{cases} (s-1)\ \zeta(s, d; \psi_1) \to (\text{nonzero limit}), \text{ for Type I}, \\ \qquad\qquad \zeta(s, d; \psi) \to (\text{nonzero limit}), \text{ for Type II or III}. \end{cases}$$

We shall focus our attention on Type II, using an analogue of Dirichlet's Lemma of Chapter II, §7, whose proof is deferred until Exercise 20 in §3, Chapter XIII.

DIRICHLET-WEBER LEMMA

The only real characters on the class structure group have the form given by the so-called generic (*Jacobi*) *character*

$$\text{(5)} \qquad \psi(\mathbf{A}) = \left(\frac{d_\psi}{N[\mathfrak{a}]}\right)$$

where d_ψ represents a fixed integer (actually some divisor of 4d) and $\mathfrak{a}$ is any ideal in **A** *selected to have a norm relatively prime to 2d. (The definition of $\psi(\mathbf{A})$ is independent of $\mathfrak{a}$ in* **A** *for these values of d_ψ.)*

Then, for $\zeta(s, d; \psi)$, a Type II function, we remove the factors $\mathfrak{p}$ for which $(N[\mathfrak{p}], 2d) \neq 1$ and obtain

$$(6) \quad \prod_{\mathfrak{p} \mid 2d} \left(1 - \frac{\psi(\mathfrak{p})}{N[\mathfrak{p}]^s}\right) \cdot \zeta(s, d; \psi) = \prod_{\mathfrak{p} \nmid 2d} \left(1 - \frac{(d_\psi/N[\mathfrak{p}])}{N(\mathfrak{p})^s}\right)^{-1} = \prod\nolimits_1 \prod\nolimits_2;$$

as we next separate the product into two of the three types of §7 above. For those p satisfying $(d/p) = -1$, $N[(p)] = p^2$,

$$(7a) \qquad \prod\nolimits_1 = \prod_{p \nmid 2d} \left(1 - \frac{1}{p^{2s}}\right)^{-1},$$

and for those p satisfying $\mathfrak{p}\mathfrak{p}' = (p)$, $(d/p) = 1$, $N[\mathfrak{p}] = N[\mathfrak{p}'] = p$,

$$(7b) \qquad \prod\nolimits_2 = \prod_{p \nmid 2d} \left(1 - \frac{(d_\psi/p)}{p^s}\right)^{-2}.$$

We then see that the seemingly different types of factors can be unified as

$$(8) \qquad \prod\nolimits_1 \prod\nolimits_2 = \prod_{p \nmid 2d} \left(1 - \frac{(d_\psi/p)}{p^s}\right)^{-1} \prod_{p \nmid 2d} \left(1 - \frac{(d/p)(d_\psi/p)}{p^s}\right)^{-1}.$$

according to the values $(d/p) = -1$ and $+1$. But this expression is essentially the product of two L-series, the L-series for $\chi^{(1)}(n) = (d_\psi/n)$ and the L-series for $\chi^{(2)}(n) = (dd_\psi/n)$, each with the restriction $(n, 2d) = 1$. Each are Type II, whence from (6) the ratio

$$(9) \qquad \zeta(s, d; \psi)/[L(s, \chi^{(1)})L(s, \chi^{(2)})]$$

consists of only a *finite* product over $\mathfrak{p} \mid 2d$, with nonzero limit as $s \to 1$. Thus, finally, for Type II,

$$(10) \qquad \zeta(s, d; \psi) \to \text{(nonzero limit)}.$$

(The reader can supply the details as Exercise 11 below.) Q.E.D.

Although Weber's theorem confirmed the significance of Type II series, after Dirichlet's original proof various shorter proofs of Theorem 1 were developed to circumvent quadratic field theory. In fact, in retrospect, the use of quadratic fields seemed to be like "burning down a house to roast a pig." More recently, however, mathematicians have come to regard the general connection between prime decomposition in fields and primes in arithmetic progression as very deep and certainly still not completely explored. It might suffice to note only that for d a discriminant the primes in a single arithmetic progression modulo d [e.g., $p \equiv a \pmod{d}$ $(a, d) = 1$], all factor or stay unfactored alike, as $(d/a) = 1$ or $(d/a) = -1$. It is this circumstance that ultimately guaranteed the occurrence of an infinitude of primes in arithmetic progression!

EXERCISE 11. Complete the proof of Theorem 4 by showing (*a*) the "finite error" statement is valid in (3), (*b*) the factors $\prod_1$ and $\prod_2$ are unified by (8), and (*c*) the ratio (9) has a nonzero limit as $s \to 1$. Supply other details needed.

chapter XI

Quadratic reciprocity

1. Rational Use of Class Numbers

Quadratic reciprocity will be familiar to the reader as probably one of the culminating theorems of elementary texts in number theory. The theorem was conjectured experimentally by Euler (1760) and "almost" proved several times. The first actual proof was given by Gauss about 1796. Gauss indeed gave seven proofs, and by 1915 there were 56 proofs! Subsequently, at least a dozen more proofs were discovered. Obviously, there is something intrinsically appealing in such proofs to bring out new viewpoints so often.[1]

We shall give an ideal-theoretic proof, which is a variant of one due to Kummer (1861) using quadratic forms.

It is desirable to ask to what extent reciprocity has been invoked in ideal theory. First of all, we used reciprocity to show that $\chi(n) = (a/n)$ is determined by n modulo $f(a)$, the conductor. We also used reciprocity to evaluate some otherwise tiresome symbols (a/b). We did *not* use reciprocity, however, to prove Theorem 1 in Chapter IX on the factorization of p in the field of $\sqrt{D}$, (where D, as usual, is square-free). From this theorem we shall prove reciprocity now.

Let us first note a result concerning forms and ideals, which indicates the role of class number in *rational* number theory, typified in this chapter.

[1] We follow the count made by Bachmann (see Special References by Chapter below). Included in his list is a proof by Cauchy related to heat conduction theory!

THEOREM 1. If the field of $\sqrt{D}$ has class number h, then, whenever $(d/p) = 1$ for d the discriminant and p, an odd prime,

$$(1) \qquad \pm p^h = \begin{cases} (x^2 - Dy^2), & d/4 = D \not\equiv 1 \pmod 4 \\ (x^2 - Dy^2)/4, & d = D \equiv 1 \pmod 4 \end{cases}$$

has a solution in x and y relatively prime to p.

Proof. If $(d/p) = 1$, $p = \mathfrak{p}\mathfrak{p}'$ and $N[\mathfrak{p}] = p$. But (by the corollary to Theorem 6, Chapter VIII, §2) $\mathfrak{p}^h = (\pi_0)$, a principal ideal not divisible by $\mathfrak{p}'$, hence not divisible by p. Thus $\pi_0 = x + \sqrt{D}y$ or $(x + \sqrt{D}y)/2$ and we obtain (1) by taking norms:

$$N[\mathfrak{p}]^h = N[(\pi_0)] = |N(\pi_0)|.$$

Furthermore, $(x, p) = (y, p) = 1$; otherwise, from (1), $(x, y, p) = p$ and $p \mid (\pi_0)$. Q.E.D.

Note that if $D < 0$ then only the $+$ sign (in the $\pm$ symbol) applies as $N[(\pi_0)] = N(\pi_0) > 0$.

EXERCISE 1. When $d = -20$, $D = -5$, $h = 2$. For $(-20/p) = 1$ and $p < 30$, verify, by numerical work,

$$p = u^2 + 5v^2 \qquad (u, p) = (v, p) = 1$$

in some cases; but, regardless of this,

$$p^2 = x^2 + 5y^2 \qquad (x, p) = (y, p) = 1$$

in all cases.

EXERCISE 2. Under what circumstances will a fixed k exist, $1 \leq k < h$, such that whenever $(d/p) = 1$ we can use p^k instead of p^h in (1)? For example, if the class group is $\mathbf{Z}(p^a) \times \mathbf{Z}(p^b)$ and $a \geq b > 0$, then $h = p^{a+b}$. What would be a suitable $k(<h)$?

EXERCISE 3. Show that if h is odd and $q \mid D$ then $(\pm p/q) = 1$, according to the sign in (1).

2. Results on Units

We have in the past used only algebraic integers in the field, $R(\sqrt{D})$, and never fractions. In this chapter only, we use fractions α/β symbolically where α and β are algebraic integers such that $\beta \neq 0$. Then

$$(1) \qquad \frac{\alpha}{\beta} = \frac{\gamma}{\delta}$$

with $\delta \neq 0$, means precisely the result of "cross multiplication"

$$(2) \qquad \alpha\delta = \beta\gamma,$$

and conversely. The usual laws of cancellation are consistent, e.g.,

$$\frac{\alpha}{\beta} = \frac{\alpha\delta}{\beta\delta}, \tag{3}$$

if $\delta \neq 0$, as cross multiplication easily justifies.

THEOREM 2. *If $N(\epsilon) = 1$ for some unit ϵ, then for some algebraic integer γ, with conjugate γ',*

$$\epsilon = \frac{\gamma}{\gamma'}. \tag{4}$$

Proof. We have only to let

$$\begin{cases} \gamma = 1 + \epsilon \text{ if } \epsilon \neq -1, \\ \gamma = \sqrt{D} \quad \text{if } \epsilon = -1. \end{cases} \tag{5}$$

Q.E.D.

COROLLARY. *If α and β are algebraic integers such that $N(\alpha) = N(\beta)$, then for some algebraic integer γ*

$$\frac{\alpha}{\beta} = \frac{\gamma}{\gamma'}. \tag{6}$$

The theorem (and corollary) are referred to as Hilberts's Theorem 90 (of the famous Zahlbericht). They have an exalted place in the advanced theory, since condition (4) is trivially sufficient and very profoundly necessary for $N(\epsilon) = 1$. The proof of the corollary is left as Exercise 4.

THEOREM 3. *If the discriminant d of a field is positive and has only one prime divisor, then the fundamental unit η_1 satisfies*

$$N(\eta_1) = -1. \tag{7}$$

Proof. First of all, $d = 8$, for $R(\sqrt{2})$, and prime values $d = p \equiv 1 \pmod 4$, for $R(\sqrt{p})$, are the only cases in the hypothesis. If $N(\eta_1) = +1$, we could write (for integral γ),

$$\eta_1 = \frac{\gamma}{\gamma'}, \tag{8}$$

by Theorem 2. We could also remove any rational common factor from either γ or γ', which we consider henceforth to have been done. Then, in ideal terminology,

$$(\gamma) = (\gamma'). \tag{9}$$

This means that if $\mathfrak{q}$ is a prime ideal dividing γ then

$$\mathfrak{q} \mid (\gamma'). \tag{10}$$

Taking conjugates,

$$\mathfrak{q}' \mid (\gamma). \tag{11}$$

Thus both $\mathfrak{q}$ and $\mathfrak{q}'$ divide (γ). If $\mathfrak{q} \neq \mathfrak{q}'$, then, by ideal factorization, $\mathfrak{q}\mathfrak{q}' = (q)$ divides (γ), leading to a contradiction, and $\mathfrak{q} = \mathfrak{q}'$. But the only self-conjugate prime ideals are divisors of q for which $(d/q) = 0$. Now, $d = p$ or $d = 8$, hence $p = q$ or $2 = q = p$ and, since $p = (\sqrt{p})^2$, clearly

$$\mathfrak{q} = (\sqrt{p}) \tag{12}$$

for *any* $\mathfrak{q}$ that divide γ. Thus either

$$\gamma = \epsilon \tag{13}$$

(which means *no* prime divides γ) or

$$\gamma = \epsilon\sqrt{p}, \tag{14}$$

where ϵ is a unit. In these two cases, if $N(\epsilon) = \pm 1$,

$$\eta_1 = \gamma/\gamma' = \epsilon/\epsilon' = \epsilon^2/\epsilon\epsilon' = \pm\epsilon^2, \tag{13a}$$

$$\eta_1 = \gamma/\gamma' = \epsilon\sqrt{p}/(-\epsilon'\sqrt{p}) = -\epsilon^2/\epsilon\epsilon' = \mp\epsilon^2. \tag{13b}$$

In neither case is η_1 a fundamental unit. Hence by contradiction,

$$N(\eta_1) = -1. \qquad \text{Q.E.D.}$$

To see a corresponding result stated independently of algebraic number theory, consider the following:

THEOREM 4. If $p \equiv 1 \pmod 4$ is a positive prime, then the equation

$$x^2 - py^2 = -1 \tag{14}$$

has a solution in integers x, y.

Proof. As a direct consequence of Theorem 3, equation

$$\frac{u^2 - pv^2}{4} = -1 \tag{15}$$

has a solution in integers u, v, where $u \equiv v \pmod 2$. In fact, $\eta_1 = (u + v\sqrt{p})/2$. We next recall that by Exercise 19, Chapter VI, §9, $\eta_1{}^3 \equiv 1 \pmod 2$ and, consequently,

$$\eta_1{}^3 = x + y\sqrt{p}, \tag{16}$$

where x and y are necessarily integers but $N(\eta_1{}^3) = -1$. Q.E.D.

Incidentally, it is easy to recall the following [by reference to Chapter VI, §4]:

THEOREM 5. The Pell equation,

$$x^2 - my^2 = -1, \quad \text{or} \quad -4, \tag{17}$$

has no solution in integers if m has a prime factor $q \equiv -1 \pmod 4$.

If m has only prime factors which are $\equiv 1$ modulo 4 (or equal to 2), there is no decisive result. For example, from the fundamental units,

$$\begin{cases} x^2 - 10y^2 = -1 \text{ is solvable } (3, 1), & (10 = 2 \cdot 5), \\ x^2 - 34y^2 = -1 \text{ is unsolvable}, & (34 = 2 \cdot 17), \end{cases}$$

$$\begin{cases} x^2 - 65y^2 = -1 \text{ is solvable } (8, 1), & (65 = 5 \cdot 13), \\ x^2 - 221y^2 = -1 \text{ is unsolvable}, & (221 = 17 \cdot 13). \end{cases}$$

Once more we can admire the unpredictability of algebraic number theory!

EXERCISE 4. Prove the corollary to Theorem 2 (above). *Hint.* First try $\gamma = 1 + \alpha/\beta$.

EXERCISE 5. Verify the unsolvability of $x^2 - 34y^2 = -1$ and $x^2 - 221y^2 = -1$ by making use of the units in Table III (appendix) and $(15 + \sqrt{221})/2$.

3. Results on Class Structure

THEOREM 6. If the discriminant of a field is positive or negative and contains only one prime factor, then the class number of the field is odd.

Proof. We first note that if a group **G** *is of even order it has an element* **A**$(\neq$ **I**$)$ *such that* $\mathbf{A}^2 = \mathbf{I}$. Otherwise, let us remove the identity **I** and pair off different elements of the group $\mathbf{A}, \mathbf{A}^{-1}$; $\mathbf{B}, \mathbf{B}^{-1}; \cdots$ until the group is exhausted. But this process yields an *odd* number of elements in **G** (hence a contradiction) unless for some **A**, $\mathbf{A} = \mathbf{A}^{-1}$.

If the class number of the field of $\sqrt{D}$ is even, there exists a class **K** for which

$$\mathbf{K} \neq \mathbf{I}, \tag{1}$$

$$\mathbf{K}^2 = \mathbf{I}.$$

Now multiplying by $\mathbf{K}^{-1} = \mathbf{K}'$, the conjugate

$$\mathbf{K} = \mathbf{K}' \neq \mathbf{I}. \tag{2}$$

An ideal $\mathfrak{j}$ (in **K**) exists such that

$$\mathfrak{j} \sim \mathfrak{j}'. \tag{3}$$

We cannot conclude $\mathfrak{j} = \mathfrak{j}'$, as examples will indicate later on. At best, we see

$$\alpha\mathfrak{j} = \beta\mathfrak{j}'. \tag{4}$$

Thus $N[(\alpha)] = N[(\beta)]$, since $N[\mathfrak{j}] = N[\mathfrak{j}']$ and

$$N(\alpha) = \pm N(\beta). \tag{5}$$

If $N(\alpha) = N(\beta)$, it follows from the Corollary to Theorem 2, for some integer γ,

$$\frac{\alpha}{\beta} = \frac{\gamma}{\gamma'}. \tag{6}$$

Multiplying (4) by γ' on both sides and using $\alpha\gamma' = \beta\gamma'$, we see that β cancels and

$$\gamma\mathfrak{j} = \gamma'\mathfrak{j}', \tag{7}$$

or, if $\gamma\mathfrak{j} = \mathfrak{k}$,

$$\mathfrak{k} = \mathfrak{k}'. \tag{8}$$

If, however, $N(\alpha) = -N(\beta)$, we rewrite (4) as

$$\alpha\mathfrak{j} = \beta\eta\mathfrak{j}', \tag{9}$$

where $N(\eta) = -1$ by Theorem 2, and the same result (8) follows.

We now consider the ideal factors of $\mathfrak{k}$ and $\mathfrak{k}'$. As before, we eliminate rational factors and find that only ramified primes remain. Thus

$$\mathfrak{k} = (1) \quad \text{or} \quad (\sqrt{p}), \tag{10}$$

and $\mathfrak{k} \in \mathbf{I}$, contradicting condition (1). Q.E.D.

THEOREM 7. For two different positive primes $p \equiv q \equiv -1 \pmod 4$ the field $R(\sqrt{pq})$ has the property that the factors $\mathfrak{p}$, $\mathfrak{q}$ of $(p) = \mathfrak{p}^2$, $(q) = \mathfrak{q}^2$ are principal.

Proof. By Theorem 5 the fundamental unit of $R(\sqrt{pq})$ or η_1 has $N(\eta_1) = +1$. Thus

$$\eta_1 = \frac{\alpha}{\alpha'} \tag{11}$$

by Theorem 2 and

$$(\alpha) = (\alpha'). \tag{12}$$

Reducing by rational factors, we see the ideal factors of α or α' divide the discriminant pq ($\equiv 1 \bmod 4$). Therefore (α) can be only

(12*a*) $(\alpha) = (1)$, hence $\alpha = \eta$.

(12*b*) $(\alpha) = \mathfrak{p}$,

(12*c*) $(\alpha) = \mathfrak{q}$,

(12*d*) $(\alpha) = \mathfrak{pq}$, hence $\alpha = \eta\sqrt{pq}$.

where η is a unit (and, incidentally, $\mathfrak{pq} = (\sqrt{pq})$, regardless of whether $\mathfrak{p}$ and $\mathfrak{q}$ are principal). Now, under hypothesis (12*a*),

$$\eta_1 = \frac{\alpha}{\alpha'} = \frac{\eta}{\eta'} = \frac{\eta^2}{\eta\eta'} = \eta^2,$$

and, under hypothesis (12*d*), similarly,

$$\eta_1 = \frac{\alpha}{\alpha'} = \frac{\eta\sqrt{pq}}{\eta'(-\sqrt{pq}} = -\eta^2,$$

whence η_1 is not fundamental. This leaves hypothesis (12*b*) or (12*c*), where $\mathfrak{p}$ or $\mathfrak{q}$ appears as principal. Since $\mathfrak{pq} \sim 1$ both $\mathfrak{p}$ and $\mathfrak{q}$ are principal.

Q.E.D.

In terms of forms, if

$$\mathfrak{p} = \left(\frac{x + y\sqrt{pq}}{2}\right),$$

by taking norms we see

$$p = \pm\left(\frac{x^2 - y^2pq}{4}\right).$$

Clearly, $x = pX$; hence, in new notation, we obtain the following result:

COROLLARY. If $p \equiv q \equiv -1 \pmod 4$ for distinct positive primes p, q, equation

(13) $$\pm 4 = pX^2 - qY^2$$

is solvable in X and Y for some choice of sign. (We shall later see that both the $+$ and $-$ signs cannot be admissible for a given p and q.)

EXERCISE 6. An *ambiguous* ideal is defined as an ideal $\mathfrak{a}$ for which $\mathfrak{a} = \mathfrak{a}'$, whereas no rational integer (except ± 1) divides $\mathfrak{a}$. An *ambiguous* ideal-class is defined as a class $\mathbf{K}$ for which $\mathbf{K} = \mathbf{K}'$. Re-examining Theorem 6, show that a real field $R(\sqrt{D})$ can have an ambiguous ideal-class containing no ambiguous ideal only if the fundamental unit of the field has positive norm. (Observe that, equivalently $\mathfrak{a}^2 \sim (1)$, $\mathbf{K}^2 = \mathbf{I}$.)

EXERCISE 7. Observe that for $R(\sqrt{34})$ the ideals $\mathfrak{3}_1 = (3, 1 + \sqrt{34})$, $\mathfrak{3}_2 = (3, 1 - \sqrt{34})$ satisfy the relation $\mathfrak{3}_1{}^2 = (5 - \sqrt{34})$, whereas for no integer α can one have $\mathfrak{3}_1\alpha = \mathfrak{3}_2\alpha'$; since then $(\alpha(5 - \sqrt{34})) = (3\alpha')$ leading to a contradiction when we recall $N(\eta_1) = +1$ for $R(\sqrt{34})$.

EXERCISE 8. Generalizing the preceding exercise, show that if $D = a^2 + b^2$ and $N(\eta_1) = +1$ in $R(\sqrt{D})$ then an ambiguous ideal-class exists which contains no ambiguous ideals. *Hint*. If $\mathfrak{a} = (a, b + \sqrt{D})$ for odd a, show that $\mathfrak{a}\mathfrak{a}' = (a)$, whereas $\mathfrak{a}^2 = (b + \sqrt{D})$.

EXERCISE 9. Show that only at most one ideal-class can be ambiguous without possessing an ambiguous ideal. *Hint*. If $\mathfrak{a} \sim \mathfrak{a}'$ and $\mathfrak{b} \sim \mathfrak{b}'$, then either $\mathfrak{a}$ or $\mathfrak{b}$ or $\mathfrak{a}\mathfrak{b}$ is an ambiguous ideal ignoring rational factors (see Theorem 6).

EXERCISE 10. Show if a field of discriminant $d > 0$ possesses an ideal-class without ambiguous ideal then no prime divisor of d is $\equiv -1 \pmod 4$, whence $d = a^2 + b^2$ (by Chapter IX, §5). *Hint*. From the relation $N(\alpha) = -N(\beta)$ in the proof of Theorem 6, deduce the solvability in ξ of $N(\xi) = -m^2$.

EXERCISE 11. Show that the only ambiguous ideals are the unit ideal and those whose prime ideal factors divide the discriminant. *Hint*. Compare Theorem 7.

EXERCISE 12. Find the γ, for which $\eta_1 = \gamma/\gamma'$ in accordance with Theorem 2, for the fundamental units of $R(\sqrt{21})$ and $R(\sqrt{34})$. Note the relationship with Theorem 7 and with Exercise 7 (above).

EXERCISE 13. Show that the only ambiguous principal ideals can be (1), $(\sqrt{D})$, (γ), and $(\gamma\sqrt{D})$ (possibly divided by a rational number), where D is the square-free generator of the field and γ refers to Exercise 12 when $D > 0$ and $N(\eta_1) = +1$. *Hint*. First show that when $D < -3$, $(\alpha) = (\alpha')$ only when $\alpha = \pm\alpha'$, whence α or $\alpha/\sqrt{D}$ is rational (the cases $D = -1$ and -3 being trivial). Next take the case $D > 0$ and $N(\eta_1) = +1$. If $\alpha = \eta_1{}^t\alpha'$, set $\rho = \alpha/\gamma^t$, whereas if $\alpha = -\eta_1{}^t\alpha'$ set $\rho = \alpha/(\gamma^t\sqrt{D})$, and note $\rho = \rho'$. Finally, if $D > 0$ and $N(\eta_1) = -1$, then on the assumption that $\alpha = \pm\alpha'\eta_1{}^t$ show that t is even and specify under which conditions $\alpha/\eta^{t/2}$ or $\alpha/(\eta^{t/2}\sqrt{D})$ is rational.

4. Quadratic Reciprocity Preliminaries

The quadratic reciprocity theorem consists of three statements:

$$(1) \qquad (-1/p) = (-1)^{(p-1)/2}$$

$$(2) \qquad (2/p) = (-1)^{(p^2-1)/8}$$

$$(3) \qquad (q/p) = (p/q)(-1)^{(p-1)/2\cdot(q-1)/2}$$

where p, q are different odd positive primes. The third statement is, understandably, called the "main" result, whereas the first two are called "completion" theorems.

It would be wise to dispose of the completion theorems first, particularly since (1) is needed as a tool in the proof of the main theorem.

THEOREM 8. If p is an odd prime,

$$x^2 \equiv -1 \pmod{p} \tag{4}$$

is solvable in x precisely when

$$p \equiv +1 \pmod{4}. \tag{5}$$

[This constitutes (1).]

Proof. If we review Theorem 1 in Chapter IX, we find, without help of any reciprocity theorem, that (4) is solvable precisely when we have

$$(p) = \mathfrak{p}\mathfrak{p}' \text{ in } R(\sqrt{-1}). \tag{6}$$

Thus the solvability of (4) is interchangeable with that of (6).

We now assume $p \equiv 1 \pmod{4}$. Then, since $R(\sqrt{p})$ has a unit $\eta_1 = (x + y\sqrt{p})/2$ of norm -1 (by Theorem 3, §2), it follows that

$$\frac{x^2 - py^2}{4} = -1 \tag{7}$$

and easily $x^2 \equiv -4 \pmod{p}$, whence $x/2$ (if x is even) or $(x + p)/2$ (if x is odd) satisfies condition (4).

Conversely, if $x^2 \equiv -1 \pmod{p}$ or, equivalently, $p = \mathfrak{p}\mathfrak{p}'$, then, since the class number is unity, in $R(\sqrt{-1})$

$$p = N(\mathfrak{p}) = x^2 + y^2, \tag{8}$$

whence it is clear, taking cases modulo 4, that $p \equiv 1$. Q.E.D.

THEOREM 9. If p is an odd prime,

$$x^2 \equiv 2 \pmod{p} \tag{9}$$

is solvable in x if and only if

$$p \equiv \pm 1 \pmod{8}. \tag{10}$$

[This constitutes (2).]

Proof. This theorem is somewhat more involved. First of all, by reviewing Theorem 1 of Chapter IX, §1, we see that this time (10) is precisely the condition that

$$2 = \mathfrak{t}\mathfrak{t}' \text{ in } R(\sqrt{p^*}),$$

where $p^* = p$ if $p \equiv 1 \pmod{4}$ and $p^* = -p$ if $p \equiv -1 \pmod{4}$. Note (10) is equivalent to $p^* \equiv 1 \pmod{8}$.

Now assume (10) or, equivalently, $p^* \equiv 1 \pmod 8$. Then, since $R(\sqrt{p^*})$ has an odd class number h, $\mathfrak{t}^h = \omega$ is principal, or, as in Theorem 1 (above), since $N[\mathfrak{t}] = 2$,

$$2^h = \pm N(\omega), \qquad \omega \text{ in } R(\sqrt{p^*}), \tag{11}$$

the sign can be taken as $+$, for the fundamental unit η_1 having norm -1 can multiply ω, yielding $+2^h = N(\eta_1\omega)$ otherwise. At any rate,

$$2^h = \frac{x^2 - p^*y^2}{4}, \quad h \text{ odd}, \tag{12}$$

whence, eventually, we see (9) is solvable modulo p^* or modulo p.

Conversely, assume that (9) is solvable modulo p. Then

$$(p) = \mathfrak{p}\mathfrak{p}' \text{ in } R(\sqrt{2}), \tag{13}$$

where $\mathfrak{p} = (p, x + \sqrt{2}) = (u + v\sqrt{2})$, since $R(\sqrt{2})$ has only the principal class, or

$$\pm p = N(u + v\sqrt{2}) = u^2 - 2v^2. \tag{14}$$

Once more we check cases modulo 8 and find that $p \equiv \pm 1$. Q.E.D.

On considering the last two proofs carefully, we note that the field $R(\sqrt{-1})$ or $R(\sqrt{2})$ was used to prove the result "in one direction" and $R(\sqrt{p})$ or $R(\sqrt{p^*})$ "in the other." Although Euler's criterion and Gauss's criterion in elementary number theory may seem easier, these proofs are much more meaningful.

5. The Main Theorem

For this section we specify our notation with the following positive integers:

$$r, r_1 = \text{any odd primes},$$
$$p = \text{any odd prime} \equiv 1 \pmod 4,$$
$$q, q_1 = \text{any odd primes} \equiv -1 \pmod 4.$$

Once more we let $r^* = \pm r \equiv 1 \pmod 4$. Thus $p^* = p$; $q^* = -q$. We must show

$$(r/p) = (p/r), \tag{1}$$
$$(q/q_1) = -(q_1/q). \tag{2}$$

This includes all cases in (3) in §4.

THEOREM 10. If $(r^*/r_1) = 1$, then $(r_1/r) = 1$.

Proof. From the hypothesis we can factor r_1 in $R(\sqrt{r^*})$, setting $r_1 = \mathfrak{r}_1\mathfrak{r}_1'$. Now, $R(\sqrt{r^*})$ has odd class number h (and the fundamental unit η_1 of norm -1 if $r^* > 0$). Hence, by the method of Theorem 1, (above),

$$\mathfrak{r}_1^h = (\omega) \text{ for } \omega = (x + y\sqrt{r^*})/2; \tag{3}$$

and taking norms, we find

$$r_1^h = +N(\omega), \tag{4}$$

the sign of the norm being taken into account by using $(\eta\omega)$ instead of (ω). Then

$$4r_1^h = x^2 - y^2r^* \equiv x^2 \pmod{r}, \tag{5}$$

and, since h is odd, then $(r_1/r) = 1$. Q.E.D.

COROLLARY 1. If $(p/r_1) = 1$, then $(r_1/p) = 1$.

Proof. Let $r = p = r^*$ in Theorem 10. Q.E.D.

COROLLARY 2. If $(r/p) = 1$, then $(p/r) = 1$.

Proof. Letting $p = r_1$, we see $(r/p) = (r^*/p) = 1$, as $(-1/p) = 1$. Thus by Theorem 10, $(p/r) = 1$. Q.E.D.

Thus statement (1) follows, as the Legendre symbols here have only two values $+1$ and -1 (not zero).

THEOREM 11. If $(q/q_1) = 1$, then $(q_1/q) = -1$, and conversely.

Proof. From Theorem 7 at least one of the two equations

$$\pm 4 = qX^2 - q_1Y^2 \tag{6}$$

is solvable. If the sign is $+$, then $4 \equiv qX^2 \pmod{q_1}$ and $(q/q_1) = +1$, whereas $4 \equiv -q_1Y^2 \pmod{q}$ and $(-q_1/q) = -(q_1/q) = +1$. Likewise if the sign is $-$, then $(q_1/q) = +1$, whereas $(q/q_1) = -1$. Q.E.D.

Thus, among other things, only one sign permits a solution of (6).

6. Kronecker's Symbol Reappraised

The intimate connection between reciprocity and ideal factorization theorems had led Kronecker to define quadratic character in accord with the ideal-theoretic versions of the completion theorems. In Chapter II we saw Kronecker's symbol grow as a "step-by-step" expediency. Now, we start anew by letting d be a discriminant (i.e., d square-free and $d \equiv 1$

modulo 4 or $d/4 \not\equiv 1$ modulo 4, whereas $d/4$ is square free.) For any prime p, odd or even, we *define* (by Theorem 1 of Chapter IX)

$$(d/p) = \begin{cases} 1 \text{ if } (p) = p_1 p_2, \text{ in } R(\sqrt{d}), \\ -1 \text{ if } (p) \text{ does not factor}, \\ 0 \text{ if } (p) = p_1^2. \end{cases}$$

We can then define (d/n) by the multiplicative law $(d/a)(d/b) = (d/ab)$, using the unique factorization of n. We call $(d/-n) = (d/n)$. This symbol is consistent with that of Chapter II.

Indeed in some fields of higher degree than 2, the (seemingly forbidding) analogue of Theorem 1 of Chapter IX is easier than the analogue of the reciprocity theorems.

For later purposes in Chapter XIII, §3 Exercises 6 to 13 on ambiguous ideals and classes will indicate an even deeper role of concepts introduced here for quadratic reciprocity.

chapter XII

Quadratic forms and ideals

1. The Problem of Distinguishing between Conjugates

We note that at many vital junctures in ideal theory we had to return to the corresponding forms to obtain quantitative results. For instance, we did so in the theory of units (Chapter VI), Minkowski's theorem (Chapter VIII), the class number formula (Chapter X), and reciprocity (Chapter XI).

Actually, we were presenting to the reader a truly historical view of the subject, for many of these impressive results were deduced from the theory of forms (and they preceded ideal theory by at least 50 years). Yet ideal theory is conceptually simpler.

Our first problem is to return to the naïve approach of Chapter III, §§1 and 2, and derive an *exact* correspondence between ideals of $\mathfrak{D}_1$ and certain forms. This is not so easy as these portions of Chapter III may lead us to believe. We must engage in a seemingly lengthy preparation, because, acting in haste we might think that an ideal and its conjugate correspond to the same form. For example, using module bases for the ideals $\mathfrak{j} = [\alpha, \beta]$ and $\mathfrak{j}' = [\alpha', \beta']$ we obtain the same form each time

$$N(\alpha x + \beta y) = N(\alpha' x + \beta' y) = (\alpha x + \beta y)(\alpha' x + \beta' y)$$
$$= Ax^2 + Bxy + Cy^2.$$

If we could not distinguish $\mathfrak{j}$ from $\mathfrak{j}'$ then, since $\mathfrak{j}\mathfrak{j}' \sim (1)$, it would follow that $\mathbf{J}^2 = \mathbf{I}$ for $\mathbf{J}$ the class of $\mathfrak{j}$, contradicting class structure, generally.

The precautions needed are quite deep and lead to a re-examination of the notion of equivalence.

2. The Ordered Bases of an Ideal

We first consider the ideal $\mathfrak{j}$ in $\mathfrak{O}_1$ for $R(\sqrt{d})$ with a module basis

$$\mathfrak{j} = [\alpha, \beta] = \{x\alpha + y\beta\}, \tag{1}$$

i.e., the set of all $\alpha x + \beta y$, where α, β are fixed elements in $\mathfrak{O}_1$ and x and y are variable rational integers. We have in the past drawn no essential distinction between $[\alpha, \beta]$ and $[\beta, \alpha]$. We must now consider each two-element basis to be *ordered* by the condition that the ratio of $\Delta/\sqrt{d}$ be positive, or,

$$\frac{\Delta}{\sqrt{d}} = \frac{1}{\sqrt{d}}\begin{vmatrix} \alpha & \beta \\ \alpha' & \beta' \end{vmatrix} = \frac{\alpha\beta' - \beta\alpha'}{\sqrt{d}} > 0 \tag{2}$$

We saw in Chapter IV, §10, that $\pm\Delta/\sqrt{d} = N[\mathfrak{j}] > 0$. If $(\alpha\beta' - \beta\alpha')/\sqrt{d} < 0$, then the ordering $[\beta, \alpha]$ instead (with the substitution of α for β, and conversely) will produce a positive ratio in (2). Thus, instead of saying

$$|\alpha\beta' - \beta\alpha'| = N[\mathfrak{j}]\,|\sqrt{d}|$$

for $N(\mathfrak{j})$, the norm, we say, more strongly,

$$\alpha\beta' - \beta\alpha' = N[\mathfrak{j}]\sqrt{d}. \tag{3}$$

Thus we consider a change of basis. If the bases are ordered, then

$$[\alpha, \beta] = [\gamma, \delta] \tag{4}$$

if and only if

$$\begin{cases} \alpha = P\gamma + Q\delta, \\ \beta = R\gamma + S\delta, \end{cases} \tag{5}$$

where P, Q, R, S determine a strictly unimodular transformation, i.e.,

$$PS - QR = +1. \tag{6}$$

Comparing the result with that of Chapter IV, §7, we see that (6) has $+1$ (not ± 1). We need note only for proof that

$$\begin{vmatrix} \alpha & \beta \\ \alpha' & \beta' \end{vmatrix} = \begin{vmatrix} P & Q \\ R & S \end{vmatrix}\begin{vmatrix} \gamma & \delta \\ \gamma' & \delta' \end{vmatrix}. \tag{7}$$

THEOREM 1. Two ordered bases of an ideal are equivalent under a strictly unimodular transformation, and conversely.

From now on all bases are imagined to be ordered. Thus, from (1),

$$\mathfrak{j}' = [\beta', \alpha'] = \{y\alpha' + x\beta'\} \tag{8}$$

and the operation of taking conjugates has the (rational) counterpart of changing order of elements.

3. Strictly Equivalent Ideals

Once we consider an ordered basis, we must revise our whole notion of equivalence. For instance, if ρ is an integer in $\mathfrak{O}_1$, not zero,

$$\text{(1)} \qquad \begin{cases} \rho\mathfrak{j} = \rho[\alpha, \beta] = [\rho\alpha, \rho\beta] \text{ if } N(\rho) > 0 \\ \qquad\qquad\quad = [\rho\beta, \rho\alpha] \text{ if } N(\rho) < 0, \end{cases}$$

since we find by direct substitution into determinants, as in §2, (2),

$$(\rho\alpha)(\rho\beta)' - (\rho\beta)(\rho\alpha)' = N(\rho)(\alpha\beta' - \beta\alpha').$$

We therefore find it wise to use the following definitions: Two ideals $\mathfrak{j}$ and $\mathfrak{k}$ are said to be *strictly equivalent* if they are equivalent in such a manner that

$$\text{(2)} \qquad \rho\mathfrak{j} = \sigma\mathfrak{k},$$

where $N(\rho\sigma) > 0$. We write this $\mathfrak{j} \approx \mathfrak{k}$.

Now, there could be, at the same time, *other* equivalence relations for which $N(\rho^*\sigma^*) < 0$, which we ignore in favor of (2). (Incidentally the transitivity, symmetry, and reflexivity all carry over very easily from equivalence, as defined in Chapter VIII, §2).

THEOREM 2. Two ideals are strictly equivalent if they are equivalent, in the ordinary sense, in a field for which

(*a*) $D < 0$ or

(*b*) $D > 0$ while the fundamental unit has negative norm.

In the remaining case in which

(*c*) $D > 0$ while the fundamental unit has positive norm, if $\mathfrak{j} \sim \mathfrak{k}$ (ordinary equivalence), either $\mathfrak{j} \approx \mathfrak{k}$ or $\sqrt{D}\mathfrak{j} \approx \mathfrak{k}$, not both (in the sense of strict equivalence).

Proof. The statements in case (*a*) are obvious, since the norms are always positive.

In case $D > 0$ let η_1 be the fundamental unit. If $N(\eta_1) = -1$, it is easy to guarantee $\mathfrak{j} \approx \mathfrak{k}$ in statement (2) by using $\rho\eta_1\mathfrak{j} = \sigma\mathfrak{k}$ instead of (2), when $N(\rho\sigma) < 0$ (since $N(\rho\eta_1\sigma) > 0$). This takes care of statement (*b*).

To take care of statement (*c*), we note that if $\mathfrak{j} \sim \mathfrak{k}$, whereas $\mathfrak{j} \not\approx \mathfrak{k}$, it must follow that $\rho\mathfrak{j} = \sigma\mathfrak{k}$ and $N(\rho\sigma) < 0$. We then set $\sqrt{D}\mathfrak{k} = \mathfrak{k}^*$, and it follows that $\sqrt{D}\rho\mathfrak{j} = \sigma\mathfrak{k}^*$, whereas $N(\rho\sigma\sqrt{D}) = N(\rho\sigma)\, N(\sqrt{D}) = -DN(\rho\sigma) > 0$. Thus, if $\mathfrak{j} \sim \mathfrak{k}$ and $\mathfrak{j} \not\approx \mathfrak{k}$ then $\mathfrak{j} \approx \sqrt{D}\mathfrak{k}$.

We need now verify only that $(1) \not\approx (\sqrt{D})$, when $N(\eta_1) > 0$, to see $\mathfrak{k}^* \not\approx \mathfrak{k}$. For, if $(1) \approx (\sqrt{D})$, then $\rho(1) = \sigma(\sqrt{D})$, which means, for some unit η^*, $\rho\eta^* = \sigma\sqrt{D}$, whereas $N(\rho\sigma) > 0$. But $N(\rho\eta^*) = N(\sigma\sqrt{D})$, whence, $N(\eta^*)N(\rho) = -DN(\sigma)$. Hence $N(\eta^*) < 0$ if $(1) \approx (\sqrt{D})$. Q.E.D.

We defined $h_+(d)$ as the number of *strict* equivalence classes of $R(\sqrt{D})$ with discriminant d. We find on comparison with $h(d)$ the number of ordinary equivalence classes:

THEOREM 3

$$\begin{cases} h(d) = h_+(d), \text{ if } D < 0, \quad \text{or} \quad \text{if } D > 0 \quad \text{and} \quad N(\eta_1) < 0, \\ 2h(d) = h_+(d), \text{ if } D > 0 \quad \text{and} \quad N(\eta_1) > 0. \end{cases}$$

4. Equivalence Classes of Quadratic Forms

We now shift our attention to changes of basis in quadratic forms. Let us denote a quadratic form as follows:

$$Q(x, y) = Lx^2 + Mxy + Ny^2, \tag{1}$$

where L, M, N are called the first (or leading), second (or middle), and third (or last) coefficient, respectively. Here we assume that $D = M^2 - 4LN$ (the discriminant) is not a perfect square. It need not yet be squarefree. We consider the change of variables with integral coefficients p, q, r, s:

$$\begin{cases} x = px^* + qy^*, \\ y = rx^* + sy^*. \end{cases} \tag{2}$$

(determinant $ps - qr \neq 0$). We find

$$\begin{aligned} Q(x, y) &= L(px^* + qy^*)^2 + M(px^* + qy^*)(rx^* + sy^*) \\ &\qquad + N(rx^* + sy^*)^2 \\ &= L^*x^{*2} + M^*x^*y^* + N^*y^{*2} = Q^*(x^*, y^*), \end{aligned} \tag{3}$$

where

$$\begin{cases} L^* = Lp^2 + Mpr + Nr^2, \\ M^* = 2Lpq + M(ps + qr) + Nrs, \\ N^* = Lq^2 + Mqs + Ns^2. \end{cases} \tag{4}$$

It is easy to see, by dint of a lengthy calculation, that

$$D^* = M^{*2} - 4L^*N^* = (ps - qr)^2(M^2 - 4LN) \tag{5a}$$

$$D^* = (ps - qr)^2 D. \tag{5b}$$

Thus the discriminant remains unaltered exactly when $ps - qr = \pm 1$.

If forms $Q(x, y)$ and $Q^*(x^*, y^*)$ are related by (2) and (3) for some transformation with determinant $ps - qr = 1$, the forms are said to be *properly* equivalent and written $Q \approx Q^*$. If the forms have a relationship in which $ps - qr = -1$, then (whether or not they have a relationship where $ps - qr = +1$) they are called *improperly* equivalent and written $Q \sim Q^*$.

Unless otherwise stated, *equivalence* (of forms) means *proper equivalence*. An equivalence class of forms is denoted by symbols $\mathbf{Q}$, $\mathbf{Q}_1$, etc., analogously with ideal classes.

Now the transformations of type (2), for which $ps - qr = 1$, are transitive, symmetric, and reflexive from elementary properties of determinants. Thus properly equivalent forms form an "equivalence class" in the usual sense. So do the improperly equivalent forms, (although this latter class is of no direct use in the text).

Therefore, trivially,

$$Q(x, y) \sim Q(y, x) \sim Q(-x, y) \sim Q(y, -x) \sim Q(-x, -y)$$
$$\sim Q(x, -y) \sim Q(-y, x) \sim Q(-y, -x),$$
$$Q(x, y) \approx Q(y, -x) \approx Q(-y, x) \approx Q(-x, -y),$$

whereas a relationship of the type

$$Q(x, y) \approx Q(y, x), \text{ or } Q(x, y) \approx Q(x, -y) \text{ etc.,}$$
$$Q(x, y) \approx -Q(x, y), \quad \text{or even} \quad Q(x, y) \sim -Q(x, y), \text{ etc.,}$$

can easily to be nontrivial, if it occurs.

We define the *conjugate* of $Q(x, y)$ as

$$Q(x, -y) = Q'(x, y)$$

and the *negative* as $-Q(x, y)$.

One should note also that in (4)

$$L^* = Q(p, r). \tag{6}$$

Such an expression as (6), in which $(p, r) = 1$, is called a *proper* representation of L^* by the form $Q(x, y)$. Gauss first noted the following theorem in 1796.

THEOREM 4. Every proper representation of an integer L^* by a form leads to a properly equivalent form with leading coefficient equal to the integer L^*.

Proof. In (6) we need only find q and s such that $ps - qr = 1$ (which is possible if and only if $(p, r) = 1$). Q.E.D.

Thus the study of proper representations somehow leads to the study of equivalent forms.

THEOREM 5. If two forms $Q_1(x, y)$ and $Q_2(x, y)$ have the same discriminant D and represent the same prime p, then either $Q_1(x, y) \approx Q_2(x, y)$ or $[Q_1'(x, y) =]\ Q_1(x, -y) \approx Q_2(x, y)$.

Proof. By Theorem 4, we can find forms Q_1^* and Q_2^* such that

$$(7) \qquad Q_i \approx Q_i^*(x, y) = px^2 + b_ixy + c_iy^2, \qquad (i = 1, 2).$$

Then

$$(8) \qquad D = b_1^2 - 4pc_1 = b_2^2 - 4pc_2.$$

Taking p to be odd, $b_1^2 \equiv b_2^2 \pmod p$; thus $b_1 \equiv \pm b_2 \pmod p$ and b_1 and b_2 are both odd or both even. Therefore, $b_2 = \pm b_1 + 2hp$. If we write $Q_1^*(x + hy, \pm y) = Q_3^*(x, y)$, then $Q_3^*(x, y) = Q_2^*(x, y)$ identically. (The first and second coefficients match, whereas the third coefficients are determined by the discriminant.)

Exercise 3 (below) covers $p = 2$. Q.E.D.

EXERCISE 1. Prove (5*b*) by writing $4LQ(x, y) = N(\xi x + \zeta y)$, where $\xi = 2L$, $\zeta = M + \sqrt{D}$, and using transformation (2). Note $4^2DL^2 = (\xi\zeta' - \zeta\xi')^2$.
EXERCISE 2. Using (3) in Chapter VI, §9, show that the forms $x^2 - 2y^2$, $-x^2 + 2y^2$ are properly equivalent. Show that they are also properly equivalent to $y^2 - 2x^2$, $-y^2 + 2x^2$.
EXERCISE 3. With $p = 2$ in (8), show how to choose the sign so that $b_1 \equiv \pm b_2 \pmod 4$.
EXERCISE 4. Show that $Q(x, y) = x^2 - 3y^2$ is not properly or improperly equivalent to $-Q(x, y)$. Note (6).
EXERCISE 5. Show that equivalent forms provide representations of the same integers. Show that this statement also holds for proper representations.

5. The Correspondence Procedure

We now set up a precise correspondence between forms and ideals. We first introduce two definitions:

The ideal $\mathfrak{j}$ is called *primitive* when it is not divisible by any rational ideal except (1).

The quadratic form Q is called *primitive* when its coefficients are not all divisible by any rational integer except ± 1.

LEMMA 1. If $\mathfrak{j} = [\alpha, \beta]$ is an (ordered) ideal in the field $R(\sqrt{d})$ of discriminant d, the form

$$(1) \qquad Q(x, y) = N(\alpha x + \beta y)/N[\mathfrak{j}] = ax^2 + bxy + cy^2$$

has integral coefficients and is a primitive form of discriminant d. (*Note.* $N[\mathfrak{j}]$, as an *ideal* norm, is always positive, while $N(\alpha x + \beta y)$, as the product of conjugates, could be negative.)

Proof. We expand

$$(2) \qquad N(\alpha x + \beta y) = (\alpha x + \beta y)(\alpha' x + \beta' y) = Ax^2 + Bxy + Cy^2,$$

where the coefficients belong to $\mathfrak{j}\mathfrak{j}'$. We write

$$(3) \qquad \begin{cases} A = \alpha\alpha' & = aN[\mathfrak{j}], \\ B = \alpha\beta' + \alpha'\beta = bN[\mathfrak{j}], \\ C = \beta\beta' & = cN[\mathfrak{j}], \end{cases}$$

since $(N[\mathfrak{j}]) = \mathfrak{j}\mathfrak{j}'$ contains (hence divides) A, B, and C, the coefficients of $Q(x, y)$ are integers. Furthermore,

$$(4) \qquad \begin{aligned} b^2 - 4ac &= (B^2 - 4AC)/N^2[\mathfrak{j}] \\ &= (\alpha\beta' - \beta\alpha')^2/N^2[\mathfrak{j}] = d, \end{aligned}$$

where d is the field discriminant (see Chapter IV, §10).

Now the field discriminant has no square divisor except possibly 4, hence the only possible common divisor of a, b, c is 2, when $d = 4D$ and $D \not\equiv 1 \pmod 4$. To exclude this, note that $[\alpha, \beta] = (\alpha, \beta)$ by Chapter VII Theorem 5, §4. If we write $(2) = \mathfrak{t}^2$ (as must happen when $4 \mid d$ and the Kronecker symbol $(d/2) = 0$, then

$$\begin{aligned} (\alpha) &= \mathfrak{t}^u\mathfrak{a}, & A &= N(\alpha) = 2^u \cdot \text{odd number}, \\ (\beta) &= \mathfrak{t}^v\mathfrak{b}, & C &= N(\beta) = 2^v \cdot \text{odd number}, \\ \mathfrak{j} &= \mathfrak{t}^w\mathfrak{j}_1, & N(\mathfrak{j}) &= 2^w \cdot \text{odd number}, \end{aligned}$$

where $w = \min(u, v)$ and ideals $\mathfrak{a}$, $\mathfrak{b}$, $\mathfrak{j}_1$ are prime to $\mathfrak{t}$ by Theorem 15 of Chapter VII, §9. Thus by (3) a or c is odd. Q.E.D.

The form defined by (1) is said to *belong to* the ideal $\mathfrak{j}$ with basis $[\alpha, \beta]$, written

$$(5) \qquad \begin{cases} Q = Q[\alpha, \beta] = Q(\mathfrak{j}), \\ \mathfrak{j} = [\alpha, \beta] \to Q \qquad \text{"}\mathfrak{j} \text{ leads to } Q.\text{"} \end{cases}$$

LEMMA 2. Suppose we are given a quadratic form, not necessarily primitive, which we write as

$$(6) \qquad \begin{cases} Q(x, y) = Ax^2 + Bxy + Cy^2 \\ \qquad\qquad = t(ax^2 + bxy + cy^2), \end{cases}$$

where $\pm t$ is the greatest common divisor of A, B, C. We let $t > 0$ if $B^2 - 4AC > 0$, but if $B^2 - 4AC < 0$ we choose t so that $a > 0$. We call $d = b^2 - 4ac$, and we suppose that d is the field discriminant for $R(\sqrt{d})$, i.e.,

$d \equiv 1 \pmod 4$ for d square-free or $d \equiv 0 \pmod 4$ for $d/4$, a square-free integer $\not\equiv 1 \pmod 4$. Then the ideal

$$(7) \qquad \mathfrak{j} = [\alpha, \beta] = \begin{cases} [a, (b - \sqrt{d})/2], & a > 0, \quad \text{any } d, \\ [a, (b - \sqrt{d})/2]\sqrt{d}, & a < 0, \quad d > 0, \end{cases}$$

is integral and has an ordered basis; $\mathfrak{j}$ is primitive when $a > 0$, whereas $\mathfrak{j}/\sqrt{d}$ is primitive when $a < 0$. (Note that the lemma excludes the case in which $a < 0$, $d < 0$.)

Proof. First we notice that if $d \equiv 1 \pmod 4$ the basis elements of $\mathfrak{j}$ are integers, since b is odd, and if $d \equiv 0 \pmod 4$ the same is true, since b is even. Next, using (2) in §2, we see that $\Delta/\sqrt{d} > 0$. In fact, assuming $d > 0$ (which is the difficult case), we find that $\Delta = a\sqrt{d}$, if $a > 0$, and $\Delta = a(-d)\sqrt{d}$, if $a < 0$; hence the basis is ordered.

To see that $\mathfrak{j}$ (or $\mathfrak{j}/\sqrt{d}$) is primitive, we note that otherwise a rational integer $u > 1$ exists which divides a and $\beta = (b - \sqrt{d})/2$. This integer then divides β' so that u divides $\beta - \beta' = -\sqrt{d}$ or u^2 divides d, which, by assumption, limits u to the value 2, when $d \equiv 0 \pmod 4$. But, even then, if $d \equiv 0 \pmod 4$ b as well as a must be even. Thus, in order that 2 divide β, $\sqrt{d}/2$ must be divisible by 2 (see Chapter III, §7) but 4 does not divide $\sqrt{d}$, by nature of the field discriminant; ($d/4$ is square-free).

Q.E.D.

If Q is a quadratic form with coefficients A, B, C satisfying the discriminant properties of Lemma 2, we write

$$(8) \qquad \begin{cases} [\alpha, \beta] = \mathfrak{j} = \mathfrak{j}(A, B, C) = \mathfrak{j}(Q), \\ \qquad Q \to \mathfrak{j} \text{ ``}Q \text{ leads to } \mathfrak{j}\text{,''} \end{cases}$$

where $\mathfrak{j}$ is the ideal determined by the form Q, with basis as shown in (7).

To summarize, let us first *start with a primitive form* $Q = ax^2 + bxy + cy^2$ whose discriminant $d - b^2 - 4ac$ is a field discriminant and for which $a > 0$. We construct the ideal $\mathfrak{j}(a, b, c)$ with ordered basis according to Lemma 2 (above). Starting with this ideal $\mathfrak{j}$, with its ordered basis, we reconstruct a quadratic form $Q^*(\alpha, \beta)$ according to Lemma 1. Then, merely *reversing the steps*, we see $Q^* = Q$. This is the easy part.

Let us next *start with a primitive ideal* with ordered basis $\mathfrak{j} = [\alpha, \beta]$. We construct from it the quadratic form $Q[\alpha, \beta]$ according to Lemma 1,

$$Q\,[\alpha, \beta] = \frac{N(\alpha)x^2 + (\alpha\beta' + \beta\alpha')xy + N(\beta)y^2}{N[\mathfrak{j}]}.$$

Starting with the form $Q[\alpha, \beta]$, we obtain the (generally different) ideal $\mathfrak{j}^*$ with ordered basis by Lemma 2:

$$(9) \quad \mathfrak{j}^* = [\alpha^*, \beta^*] = \begin{cases} \left[\dfrac{N(\alpha)}{N[\mathfrak{j}]}, \dfrac{\alpha\beta' + \beta\alpha'}{2N[\mathfrak{j}]} - \dfrac{\sqrt{d}}{2}\right], & \text{if } d < 0 \text{ or } d > 0, N(\alpha) > 0, \\ \left[\dfrac{N(\alpha)}{N[\mathfrak{j}]}, \dfrac{\alpha\beta' + \beta\alpha'}{2N[\mathfrak{j}]} - \dfrac{\sqrt{d}}{2}\right]\sqrt{d}, & \text{if } d > 0, N(\alpha) < 0. \end{cases}$$

Recall, of course, that $N[\mathfrak{j}] > 0$, whereas $N(\alpha)$ can be < 0 (when $d > 0$).

We can easily verify that

$$(10) \qquad N[\mathfrak{j}][\alpha^*, \beta^*] = [\alpha, \beta]\gamma,$$

where
$$\gamma = \begin{cases} \alpha' \text{ if } N(\alpha) > 0 \text{ and } d > 0, \text{ or if } d < 0 \\ \alpha'\sqrt{d} \text{ if } N(\alpha) < 0 \quad \text{and} \quad d > 0. \end{cases}$$

Thus $N(\gamma) > 0$ always. Actually, (10) is correct, basis element by basis element. It is easy to see $N[\mathfrak{j}]\alpha^* = \alpha\gamma$, whereas the statement about the second basis element follows from the ordering of the basis of $\mathfrak{j}$, namely,

$$\alpha\beta' - \beta\alpha' = \sqrt{d}\, N[\mathfrak{j}].$$

We see by transposition that

$$\frac{\alpha\beta' + \beta\alpha'}{2} - \frac{\sqrt{d}\, N[\mathfrak{j}]}{2} = \beta\alpha';$$

whence follows $N[\mathfrak{j}]\beta^* = \beta\gamma$.

Thus, if we have

$$(11) \qquad Q \to \mathfrak{j} \to Q^*,$$

then $Q = Q^*$ (the *same* form) (if we exclude $d < 0$, $a < 0$). If we set up

$$(12) \qquad \mathfrak{j} \to Q \to \mathfrak{j}^*,$$

the best we can expect is an *equivalence* (rather than an equality):

$$(13) \qquad \mathfrak{j}^* \approx \mathfrak{j}.$$

EXERCISE 6. Show that if a quadratic form is primitive all forms in their proper (or improper) equivalence classes are primitive.

EXERCISE 7. Start with $Q = -x^2 + 2xy - 7y^2$ and construct the corresponding $\mathfrak{j}$; from it construct $Q^*(= -Q)$ [see (11)].

EXERCISE 8. Start with $\mathfrak{j} = [\sqrt{-6}, 2]$ and construct the corresponding Q; from it construct $\mathfrak{j}^*$. How does $\mathfrak{j}^*$ compare with $\mathfrak{j}$?

EXERCISE 9. Follow Exercise 7 using $Q = -26x^2 + 2xy + 3y^2$ to form $Q^*(= Q)$.

EXERCISE 10. Follow Exercise 8, using $\mathfrak{j} = [1 + \sqrt{79}, 3]$ and once more compare $\mathfrak{j}^*$ with $\mathfrak{j}$.

(In these closely related exercises, observe the sign of the xy term and the radical very carefully.)

EXERCISE 11. Let $\mathfrak{j}'$ be the conjugate of $\mathfrak{j}$. Then show that if $\mathfrak{j} = [\alpha, \beta]$ then $\mathfrak{j}' = [\beta', \alpha']$, keeping an ordered basis, and, if $Q(\mathfrak{j}) = ax^2 + bxy + cy^2$, then show $Q(\mathfrak{j}') = cx^2 + bxy + ay^2 \approx Q'(\mathfrak{j}) = ax^2 - bxy + cy^2$.

EXERCISE 12. If $\mathfrak{j}_1 = \lambda\mathfrak{j}_2$ for $N(\lambda) > 0$, show that $Q(\mathfrak{j}_1) = Q(\mathfrak{j}_2)$ for some set of variables x, y; whereas, if $\mathfrak{j}_1 = \lambda\mathfrak{j}_2$ for $N(\lambda) < 0$, then $Q(\mathfrak{j}_1) \approx -Q'(\mathfrak{j}_2)$.

EXERCISE 13. Show that $Q_1 = x^2 + xy + ty^2 \approx Q_2 = x^2 - xy + ty^2 = Q_1'$ by a change of variables. Show directly that they lead to the same ideal. (Assume $1 - 4t \equiv d$, a field discriminant.)

6. The Correspondence Theorem

We see that the relation (12) of §5 forces us to construct a correspondence on a broader level.

THEOREM 6. If two forms with discriminant equal to a field discriminant, Q_1 and Q_2, satisfy

$$Q_1 \approx Q_2 \tag{1}$$

and $Q_1 \to \mathfrak{j}_1$, $Q_2 \to \mathfrak{j}_2$ (by Lemma 2, §5), then

$$\mathfrak{j}_1 \approx \mathfrak{j}_2. \tag{2}$$

Conversely, if two ideals satisfy (2), then the forms Q_1 and Q_2 found by Lemma 1, §5, designated by $\mathfrak{j}_1 \to Q_1$ and $\mathfrak{j}_2 \to Q_2$, have discriminants equal to a field discriminant and satisfy relation (1).

Proof. First of all, when $d < 0$, we can limit ourselves to forms $Q_i(x, y) = a_i x^2 + b_i xy + c_i y^2$, $a_i > 0$, without weakening the theorem (for any of its applications).

To prove the first part, let the properly equivalent forms $Q_1(x, y)$ and $Q_2(X, Y)$ be *regenerated in turn* by $\mathfrak{j}(Q_1)$ and $\mathfrak{j}(Q_2)$, according to the chain in (11), §5. We wish to show $\mathfrak{j}(Q_1) \approx \mathfrak{j}(Q_2)$.

Let $\mathfrak{j}(Q_1) = \mathfrak{j}_1 = [\alpha_1, \beta_1]$, $\mathfrak{j}(Q_2) = \mathfrak{j}_2 = [\alpha_2, \beta_2]$,

$$\begin{cases} Q_1(x, y) = (\alpha_1 x + \beta_1 y)(\alpha_1' x + \beta_1' y)/N[\mathfrak{j}_1], \\ Q_2(X, Y) = (\alpha_2 X + \beta_2 Y)(\alpha_2' X + \beta_2' Y)/N[\mathfrak{j}_2], \end{cases} \tag{3}$$

where for an integral transformation

$$\begin{aligned} X &= px + qy, \\ Y &= rx + sy, \end{aligned} \qquad ps - qr = 1. \tag{4}$$

But $Q_1(x, y) = Q_2(X, Y)$ "numerically" and identically under transformation (4). The locus of points where $Q_1(x, y) = 0$ must transform into the locus where $Q_2(X, Y) = 0$. Thus either

$$\frac{x}{y} = -\frac{\beta_1}{\alpha_1} \quad \text{and} \quad \frac{X}{Y} = -\frac{\beta_2}{\alpha_2} \tag{5}$$

must coincide under (4) or

$$\frac{x}{y} = -\frac{\beta_1'}{\alpha_1'} \quad \text{and} \quad \frac{X}{Y} = -\frac{\beta_2}{\alpha_2} \tag{6}$$

must coincide under (4).

First try (5). Then

$$-\frac{\beta_2}{\alpha_2} = \frac{X}{Y} = \frac{px + qy}{rx + sy} = \frac{p(x/y) + q}{r(x/y) + s} = \frac{p(-\beta_1/\alpha_1) + q}{r(-\beta_1/\alpha_1) + s} = \frac{-p\beta_1 + q\alpha_1}{-r\beta_1 + s\alpha_1}. \tag{7}$$

Hence, from the two "outside" terms,

$$\frac{\alpha_2}{s\alpha_1 - r\beta_1} = \frac{\beta_2}{-q\alpha_1 + p\beta_1} = \frac{\lambda}{\mu} \text{ (say)}$$

$$\begin{cases} \mu\alpha_2 = (-r\beta_1 + s\alpha_1)\lambda, \\ \mu\beta_2 = (p\beta_1 - q\alpha_1)\lambda. \end{cases} \tag{8}$$

We can assume, as we have often done before, that $N(\mu) > 0$. When $N(\mu) < 0$ (necessarily), $d > 0$, whence μ can be replaced by $\mu\sqrt{d}$ of norm $-dN(\mu)$. Next

$$\mu\mathfrak{j}_2 = [\mu\alpha_2, \mu\beta_2] = [s(\lambda\alpha_1) - r(\lambda\beta_1), -q(\lambda\alpha_1) + p(\lambda\beta_1)]. \tag{9}$$

But, since $ps - qr = +1$,

$$\mu\mathfrak{j}_2 = [\lambda\alpha_1, \lambda\beta_1], \tag{10}$$

and since both $\mathfrak{j}_1 = [\alpha_1, \beta_1]$ and $\mathfrak{j}_2 = [\alpha_2, \beta_2]$ are ordered, as well as $\mu\mathfrak{j}_2$, then $N(\lambda) > 0$ by (1), §3, and,

$$\mu\mathfrak{j}_2 = \lambda\mathfrak{j}_1 \quad \text{or} \quad \mathfrak{j}_1 \approx \mathfrak{j}_2. \tag{11}$$

Next we see alternative (6) is intrinsically impossible. Retracing our steps from (6) we obtain, instead of (8), the following:

$$\begin{cases} \mu\alpha_2 = (-r\beta_1' + s\alpha_1')\lambda, \\ \mu\beta_2 = (p\beta_1' - q\alpha_1')\lambda. \end{cases} \tag{12}$$

Again, restricting $N(\mu) > 0$ and going to the analogue of (10),

$$\mu \mathfrak{j}_2 = [\mu\alpha_2, \mu\beta_2] = [\lambda\alpha_1', \lambda\beta_1']. \tag{13}$$

Now $[\alpha_1', \beta_1']$ is not correctly ordered, but $[\lambda\alpha_1', \lambda\beta_1']$ is so ordered by the unimodular property (4). Thus, $N(\lambda) < 0$. Now from (4)

$$Q_2(1, 0) = Q_1(s, -r)$$

or

$$\frac{N(\alpha_2)}{N[\mathfrak{j}_2]} = \frac{N(\alpha_1 s - \beta_1 r)}{N[\mathfrak{j}_1]}. \tag{14}$$

Since $N[\mathfrak{j}] > 0$ (intrinsically), $N(\alpha_2)$ and $N(-r\beta_1' + s\alpha_1')$ agree in sign and, from (12), so do $N(\mu)$ and $N(\lambda)$, yielding a contradiction to alternative (6).

To prove that the "converse" portion of Theorem 6 is easier. First let us verify that if $\mathfrak{j} = [\alpha_1, \beta_1] = [\alpha_2, \beta_2]$ for two ordered bases, $Q[\alpha_1, \beta_1] \approx Q[\alpha_2, \beta_2]$ (formed under Lemma 1, §5). For then

$$\begin{cases} \alpha_2 = p\alpha_1 + q\beta_1, \qquad ps - qr = 1, \\ \beta_2 = r\alpha_1 + s\beta_1. \end{cases} \tag{15}$$

Thus $\alpha_2 x + \beta_2 y = \alpha_1(px + ry) + \beta_1(qx + sy)$ and from §5 (1)

$$Q_2(x, y) = Q_1(px + ry, qx + sy) \quad \text{or} \quad Q_1[\alpha_1, \beta_1] \approx Q_2[\alpha_2, \beta_2].$$

Finally, let $\mathfrak{j}_1 \approx \mathfrak{j}_2$ or $\rho\mathfrak{j}_1 = \sigma\mathfrak{j}_2$, where $N(\rho\sigma) > 0$. As usual, make $N(\rho) > 0$, $N(\sigma) > 0$ (by the $\sqrt{d}$ factor, if necessary). Then, since the basis of $\mathfrak{j}$ affects Q only within an equivalence class, we verify that

$$\rho[\alpha_1, \beta_1] = \sigma[\alpha_2, \beta_2], \tag{16}$$

$$Q[\alpha_1, \beta_1] = Q[\alpha_2, \beta_2], \tag{17}$$

operating on both sides of relation (16) according to Lemma 1, §5.

Q.E.D.

One remark might be in order here: the proof best indicates, in the rejection of alternative (6), how the ordered basis distinguishes the factors of $Q(x, y)$. An ordered basis therefore distinguishes an ideal from its conjugate. To maintain an ordered basis, we must use only strict or proper equivalence classes; hence the severity of this chapter!

An equivalent form of Theorem 6 is the following:

THEOREM 7. Under correspondence between Q_1 and $\mathfrak{j}_1$, either by Lemma 1 or 2 of §5, and under correspondence between Q_2 and $\mathfrak{j}_2$, either by Lemma 1 or 2 of §5 (not necessarily the same each time), these two equivalences imply one another;

$$\mathfrak{j}_1 \approx \mathfrak{j}_2, \tag{18}$$

$$Q_1 \approx Q_2 \tag{19}$$

(as long as $\mathfrak{j}(Q)$ is constructed only for forms Q which have discriminant d equal to a field discriminant and which have a positive leading coefficient if $d < 0$).

THEOREM 8. If Q is a form with discriminant d equal to a positive field discriminant d, then

$$Q(x, y) \approx -Q'(x, y) \text{ if and only if } N(\eta_1) = -1$$

for η_1 the fundamental unit in $R(\sqrt{d})$.

Proof. Suppose $N(\eta_1) = -1$. Let $\mathfrak{j} = [\alpha, \beta]$ lead to Q:

$$\mathfrak{j} \to Q(x, y) = N(\alpha x + \beta y)/N[\mathfrak{j}].$$

Now $\mathfrak{j} = \eta_1\mathfrak{j} = [\eta_1\alpha, -\eta_1\beta]$, remembering the order, and, since $N(\mathfrak{j}) = N(\eta_1\mathfrak{j})$,

$$N(\eta_1\alpha x - \eta_1\beta y)/N[\eta_1\mathfrak{j}] = -Q(x, -y).$$

Conversely, suppose $Q \approx -Q'$. Then, if $\mathfrak{j} \to Q$, as before,

$$\mathfrak{j}\sqrt{d} = [\sqrt{d}\alpha, -\sqrt{d}\beta], \quad N(\sqrt{d}\alpha x - \sqrt{d}\beta y)/N[\sqrt{d}\mathfrak{j}] = -Q(x, -y).$$

Then $\sqrt{d}\mathfrak{j} \approx \mathfrak{j}$ or $(\sqrt{d}) \approx (1)$, whence $N(\eta_1) = -1$ (compare the proof of Theorem 2). Q.E.D.

7. Complete Set of Classes of Quadratic Forms

To see how classes of quadratic forms come from ideal classes, let us take three typical cases, using the ideal classes[1] in Table III (appendix):

Case I. $d = -23$.

$$\mathfrak{j}_1 = (1) = [1, (1 - \sqrt{-23})/2]; \quad Q_1 = x^2 + xy + 6y^2$$
$$\mathfrak{j}_2 = 2_1 = [2, (1 - \sqrt{-23})/2]; \quad Q_2 = 2x^2 + xy + 3y^2$$
$$\mathfrak{j}_2' \approx \mathfrak{j}_3 = 2_2 = [-2, (1 + \sqrt{-23})/2]; \quad Q_3 = 2x^2 - xy + 3y^2 \approx Q_2'$$

Case II. $d = 4 \cdot 58, \quad \eta_1 = 99 + 13\sqrt{58}, \quad N(\eta_1) = -1.$

$$\mathfrak{j}_1 = (1) = [1, -\sqrt{58}]; \quad Q_1 = x^2 - 58y^2$$
$$\mathfrak{j}_2 = 2_1 = [2, -\sqrt{58}]; \quad Q_2 = 2x^2 - 29y^2$$

[1] The ideals in Table III (appendix) are written as modules in accordance with Exercise 4 of Chapter VIII, §1. The reader must sometimes order the basis, by adjusting the sign of the radical (for example).

Case III. $d = 4 \cdot 79, \quad \eta_1 = 80 + 9\sqrt{79}, \qquad N(\eta_1) = +1.$

$$\mathfrak{j}_1 = (1) = [1, -\sqrt{79}];$$
$$Q_1 = x^2 - 79y^2$$
$$\mathfrak{j}_2 = 3_1 = [3, 1 - \sqrt{79}];$$
$$Q_2 = 3x^2 + 2xy - 26y^2$$
$$\mathfrak{j}_2' \approx \mathfrak{j}_3 = 3_2 = [-3, 1 + \sqrt{79}];$$
$$Q_3 = 3x^2 - 2xy - 26y^2 \approx Q_2'$$
$$\mathfrak{j}_1^* = \sqrt{79}\,\mathfrak{j}_1 = [79, -\sqrt{79}];$$
$$Q_1^* = 79x^2 - y^2$$
$$\mathfrak{j}_2^* = \sqrt{79}\,\mathfrak{j}_2 = [79 - \sqrt{79}, -3\sqrt{79}];$$
$$Q_2^* = 26x^2 - 2xy - 3y^2$$
$$(\mathfrak{j}_2^*)' \approx \mathfrak{j}_3^* = \sqrt{79}\,\mathfrak{j}_3 = [79 + \sqrt{79}, -3\sqrt{79}];$$
$$Q_3^* = 26x^2 + 2xy - 3y^2 \approx (Q_2^*)'$$

(*Note.* Since $Q(x, y) \approx Q(y, -x)$, we can write $Q_1^* \approx -Q_1$, $Q_2^* \approx -Q_3$, $Q_3^* \approx -Q_2$ in the last case.)

In any case, Q_1, the form belonging to $\mathfrak{j}_1 = (1)$, is called *principal.* It takes the form $x^2 - Dy^2$ or $x^2 + xy - (D - 1)y^2/4$.

THEOREM 9. Let $Q(x, y)$ be a quadratic form whose discriminant is a field discriminant d. Let $Q(x, y)$ properly represent m. Then if a prime p divides m, $(d/p) = 0$ or 1.

Proof. To simplify matters, note that Q must be primitive and, for convenience, if $d < 0$, the leading coefficient can be made positive (by changing all signs if necessary). Now consider $\mathfrak{j}(Q)$, the ideal belonging to Q. It is equivalent to an ideal $\mathfrak{a}$ for which $(\mathfrak{a}, p) = (1)$ (by Theorem 7, Chapter VIII, §2), whence $p \nmid N[\mathfrak{a}]$. We can assume that the equivalence is strict (replacing $\mathfrak{a}$ by $\mathfrak{a}\sqrt{d}$ if necessary). Then $\mathfrak{a} \to Q^*$, a form equivalent to Q (by Theorem 6 above).

Now Q^* will also represent m properly (by Exercise 5, §4), or $Q^*(x_1, y_1) = m$, where $(x_1, y_1) = 1$. Then, writing $\mathfrak{a}$ in basis form as $[\alpha, \beta]$, we see by (1) in §5

$$m = Q^*(x_1, y_1) = N(\alpha x_1 + \beta y_1)/N[\mathfrak{a}].$$

Thus, writing $\lambda = \alpha x_1 + \beta y_1$, we see $p \mid m \mid N(\lambda)$.

If $p \nmid \lambda$, then, by Theorem 23, Chapter VII, §10, (p, λ) is a prime ideal divisor of p distinct from p, completing the proof (see Theorem 1, Chapter IX, §1). If, however, $p \mid \lambda$, then $p \mid \lambda' = \alpha' x_1 + \beta' y_1$, $p \mid (\beta'\lambda - \beta\lambda') = x_1\Delta$ where $\Delta = (\alpha\beta' - \beta\alpha')$, $p \mid (-\alpha\lambda + \alpha'\lambda') = y_1\Delta$, and $(p) \mid (x_1\Delta, y_1\Delta) = (\Delta)$. But, taking norms, $p^2 \mid \Delta^2 = N[\mathfrak{a}]d$, by Chapter IV, §10. In this case, $p \mid d$ and $(d/p) = 0$. Q.E.D.

COROLLARY. Let d be a field-discriminant and p be an odd prime for which $(d/p) = 1$ or 0. Then a quadratic form Q of discriminant d exists for which $Q(x, y) = p$ is solvable, and, if $d > 0$, a form of discriminant d exists for which $Q(x, y) = -p$.

Proof. Let $\mathfrak{p} \mid p$ where $N[\mathfrak{p}] = p$. Then, writing $\mathfrak{p} = [p, \pi]$ by the corollary to Theorem 23, Chapter VII, §10, we have

$$Q(x, y) = N(px + \pi y)/N[\mathfrak{p}];$$

hence $p = Q(1, 0)$. To complete the proof, note that $-Q(x, y)$ is also a form of discriminant d. Regardless of whether $-Q \approx Q$, it follows that $-Q(1, 0) = p$, trivially. Q.E.D.

EXERCISE 14. Find the complete set of forms for $D = -15, 15, 26, 34$, using Table III (appendix).

EXERCISE 15. Find the proper transformation of basis $x = aX + bY$, $y = cX + dY$ for which $Q(x, y) = -Q(X, -Y)$ where $Q(x, y) = x^2 - 10y^2$, $\eta_1 = 3 + \sqrt{10}$, by justifying

$$\eta_1[1, -\sqrt{10}] = [\eta_1, \eta_1\sqrt{10}] = [3 + \sqrt{10}, 10 + 3\sqrt{10}],$$
$$(x - \sqrt{10}y) = (3 + \sqrt{10})X + (10 + 3\sqrt{10})Y,$$
$$x = 3X + 10Y, \qquad y = -X - 3Y.$$

EXERCISE 16. Assuming $N(\eta_1) = -1$, find explicitly the transformation for which $Q \approx -Q'$ for Q the principal form for a given d:

$$(x, y) = \begin{cases} x^2 + xy - [(d-1)/4]y^2, & d \equiv 1 \pmod 4, \quad \eta_1 = (a + b\sqrt{d})/2, \\ x^2 - (d/4)y^2, & d \equiv 0 \pmod 4, \quad \eta_1 = a + b(\sqrt{d}/2). \end{cases}$$

EXERCISE 17. Find the smallest field discriminant (positive and negative) and the forms for which some $Q(x, y) \not\approx Q'(x, y)$. *Hint.* $\mathfrak{j} \not\approx \mathfrak{j}'$ means $\mathfrak{j}^2 \not\approx \mathfrak{j}\mathfrak{j}' \approx (1)$.

EXERCISE 18. If $\mathfrak{a} \to Q$, then, for some (x, y), $Q(x, y) = N[\mathfrak{a}]$. *Hint.* Note Q takes the values $N(\alpha)/N[\mathfrak{a}]$, where $\alpha \in \mathfrak{a}$.

8. Some Typical Representation Problems

These three sketchy examples will illustrate many possible situations. The reader will have many details to provide in any case.

Problem I. Solve $2x^2 + xy + 3y^2 = 78$.
If $\alpha = 2x + [(1 - \sqrt{-23})/2]y$, then $\alpha \in 2_1$, and by Lemma 1, §5

$$\begin{cases} N(\alpha) = N[2_1] \cdot 78 = 2^2 \cdot 3 \cdot 13, \\ 2_1 \mid (\alpha). \end{cases} \tag{1}$$

According to the class structure, $h = 3$ and $2_1 \approx 3_1^{-1} \approx 13_1^{-1} \approx 2_2^{-1} \approx 3_2 \approx 13_2$,

$$2_1 = [2, (1 - \sqrt{-23})/2],$$
$$3_1 = [3, (1 - \sqrt{-23})/2],$$
$$13_1 = [13, (9 - \sqrt{-23})/2].$$

(We just verify $2_1 3_1 = ((1 - \sqrt{-23})/2), 2_1 13_1 = ((9 - \sqrt{-23})/2)$.) Hence the norm in (1) is satisfied by

$$(2) \quad \begin{cases} (\alpha) = 2_1 2_2 3_1 13_2 = (8 - 2\sqrt{-23}), \\ (\alpha) = 2_1^2 13_1 3_1 = ((7 + 5\sqrt{-23})/2), \\ (\alpha) = 2_1 2_2 3_2 13_1 = (8 + 2\sqrt{-23}). \end{cases}$$

What we did was to consider all ideal factorizations of $\alpha \in 2_1$ whose norm is consistent with (1) knowing that the ideal structure serves as a limiting factor on the number of possible combinations producing a principal ideal. Note that with *smaller* class numbers like 1 there are *more* plausible combinations (2) yielding the norm. (This remark was essentially the basis of our computation of class number in Chapter X.)

By comparison with $\alpha = 2x + [(1 - \sqrt{-23})/2]y$, system (2) yields these solutions of Problem I:

$$(3) \quad \begin{cases} (x, y) = \pm(3, 4), \\ (x, y) = \pm(3, -5), \\ (x, y) = \pm(5, -4). \end{cases}$$

The $\pm$ sign factor arises from the unit, i.e., $(\pm\alpha) = (\alpha)$.

Problem II. Let us discuss primes: $p = (x^2 + 3y^2)/4 = Q(x, y)/4$.

We write $p = N(\pi)$ in $R(\sqrt{-3})$ a field of class number 1. Thus either $p = 3$, or $(-3/p) = 1$ and $p \equiv 1 \pmod 3$. But the associates of π are also solutions; e.g., there are six units ± 1 and $\pm(1 \pm \sqrt{-3})/2$ yielding the associates

$$\frac{x + y\sqrt{-3}}{2}\,\frac{1 \pm \sqrt{-3}}{2} = \frac{[(x \mp 3y)/2] \pm [(x \pm y)/2]\sqrt{-3}}{2}.$$

Therefore, there are at least *three* representations of p by $Q(x, y)/4$, ignoring trivial changes in sign, such as $Q(x, y) = Q(x, -y) = Q(-x, -y) = Q(-x, y)$.

Problem III. Solve $x^2 - 10y^2 = +10$.

We note that if

$$\alpha = x - \sqrt{10}y \in [1, -\sqrt{10}] = (1).$$

then we must solve

(4) $$\begin{cases} N(\alpha) = 10 = 2 \cdot 5, \\ \alpha \in (1). \end{cases}$$

Now $2_1 = 2_2 = [2, -\sqrt{10}]$, $5_1 = 5_2 = [5, -\sqrt{10}]$; $(h = 2)$. Hence $(\alpha) = 2_1 5_1 = (\sqrt{10}\,\xi)$, where $N(\xi) = \pm 1$. We easily see that if $\alpha = \sqrt{10}\xi$ then only $N(\xi) = -1$ is acceptable; thus, since $\eta_1 = 3 + \sqrt{10}$ is the fundamental unit, we set $\xi = \eta_1^{2k+1}$

$$\alpha = \pm\sqrt{10}(3 + \sqrt{10})(19 + 6\sqrt{10})^k, \qquad k = 0, \pm 1, \pm 2$$

(where $(3 + \sqrt{10})^2 = 19 + 6\sqrt{10}$ is the generator of all units of norm $+1$). If we set

$$\alpha = \pm(x_k + \sqrt{10}y_k),$$

we obtain $x_0 = 10$, $y_0 = 3$. We can form all other solutions by the recursion formula

$$(x_{k+1} + \sqrt{10}y_{k+1}) = (x_k + \sqrt{10}y_k)(19 + 6\sqrt{10})$$

or if

(5) $$\begin{cases} x_{k+1} = 19x_k + 60y_k, \qquad k = 0, \pm 1, \pm 2, \cdots, \\ y_{k+1} = 6x_k + 19y_k, \\ (x_0, y_0) = (10, 3), \end{cases}$$

then $\pm(x_k, y_k)$ generates the most general solution to the equation

(6) $$x_k^2 - 10y_k^2 = 10.$$

EXERCISE 19. Solve $2x^2 + 2xy + 3y^2 = 21$ and $= 42$ by ideal factorization.

EXERCISE 20. Show that for a properly chosen unit in Problem II we can have $y \equiv 0 \pmod 3$ so that

$$p = Q^*/4 = (X^2 + 27Y^2)/4 \qquad \text{if } p \equiv 1 \pmod 3$$

in precisely one way, ignoring changes of sign. Verify this by finding *three* different representations of $p = 13$ and $p = 31$ by $Q(x, y)/4$ and one by $Q^*(X, Y)/4$ (ignoring changes of sign).

EXERCISE 21. Referring to Exercise 20, can we always represent $p \equiv 1 \pmod 3$ by $Q(x, y)$ (instead of $Q(x, y)/4$)? What about $Q^*(X, Y)$?

EXERCISE 22. Repeat the process for $p = x^2 + y^2$ and show how the multiplicity of solutions is due to signs. If we set $5p = x^2 + y^2$, are the solutions trivially equivalent?

EXERCISE 23. Solve $x^2 - 82y^2 = 18$. (Note that the paucity of solutions is related to the high class number 4.) Compare $x^2 - 7y^2 = 18$.

EXERCISE 24. For fixed positive square-free D, under what conditions is the representation $8p = x^2 + Dy^2$ unique for p prime, assuming that such a representation is possible? (Consider whether D is $\equiv -1$ or $\not\equiv -1 \pmod 8$.)

chapter XIII

**Compositions, orders, and genera

1. Composition of Forms

Nowadays, we can recognize that ideal theory is a natural tool for the solution of diophantic quadratic equations, as in Chapter XII, §8. Yet the intricate factorizations used here were accomplished by Gauss in the following fashion (1796), seventy-five years before ideal theory.

THEOREM 1. Let two ideals $\mathfrak{j}_1$ and $\mathfrak{j}_2$ in $\mathfrak{O}_1$ correspond to quadratic forms $Q_1(x, y)$ and $Q_2(x, y)$ (with discriminant equal to a field discriminant) by Lemmas 1 or 2 in Chapter XII. Let us define

$$\mathfrak{j}_3 = \mathfrak{j}_1 \cdot \mathfrak{j}_2 \text{ (ideal multiplication)}. \tag{1}$$

Then, for a form Q_3 with the same discriminant,

$$Q_3(x_3, y_3) = Q_1(x_1, y_1)\, Q_2(x_2, y_2) \text{ (ordinary multiplication)}, \tag{2}$$

where $Q_3 = Q(\mathfrak{j}_3)$ and the new variables are defined by special bilinear expressions in integral coefficients A_i and B_i:

$$\begin{cases} x_3 = A_1x_1x_2 + A_2x_1y_2 + A_3x_2y_1 + A_4x_2y_2, \\ y_3 = B_1x_1x_2 + B_2x_1y_2 + B_3x_2y_1 + B_4x_2y_2. \end{cases} \tag{3}$$

As we saw in the Introductory Survey, this theorem originated in the special case of Fermat which involved $Q(x, y) = x^2 + y^2$. Gauss proved the general theorem by a manipulative *tour de force* that probably has few equals in any branch of mathematics. He actually substituted and worked with coefficients (whereas now we can benefit by ideal theory). For convenience, we make a restriction which Gauss did not make, limiting our attention to forms arising from ideals in $\mathfrak{O}_1$, or equivalently, to forms with discriminant equal to the discriminant of a quadratic field.[1]

We note first that the Theorem 1 permits us to use any convenient representative for the proper form-class and (strict) ideal-class in question. Then we make use of these four lemmas:

LEMMA 1. Numbers exist properly represented by an arbitrary quadratic form arising from an ideal in $\mathfrak{O}_1$, which are relatively prime to any preassigned integer. (For proof see Exercise 1, below.)

LEMMA 2. Every proper equivalence class of quadratic forms arising from ideals in $\mathfrak{O}_1$ contains a form $Q(x, y)$ whose first coefficient a is positive and relatively prime to some preassigned integer N.

Proof. The relative primeness follows from Lemma 1, using Theorem 4 in Chapter XII, §4. Our only concern is that a be positive in $Q(x, y) = ax^2 + bxy + cy^2$. But, if $d < 0$, there is no difficulty, since then each $Q(\mathfrak{j})$ necessarily has $a > 0$. If $d > 0$, however, we can note that if $x_0 = bNt + 1$ and $y_0 = -2aNt$ then $Q(x_0, y_0) = a(1 - dt^2N^2)$, which is positive if $a < 0$. Furthermore, since $(a, N) = 1$, likewise $(Q(x_0, y_0), N) = 1$, and x_0 and y_0 can be assumed relatively prime, (for example, if $t = 2ab$). We finally reapply Theorem 4 of Chapter XII, §4. Q.E.D.

LEMMA 3. Every two classes of primitive quadratic forms arising from ideals in $\mathfrak{O}_1$ in a given field have forms with the same middle coefficient:

$$(4)\qquad \begin{cases} Q_1(x, y) = a_1x^2 + bxy + c_1y^2, \\ Q_2(x, y) = a_2x^2 + bxy + c_2y^2, \end{cases}$$

and the further conditions that $(a_1, a_2) = 1$, $a_1 > 0$, $a_2 > 0$.

Proof. We start with representatives of two classes:

$$(5)\qquad \begin{cases} Q_1(x, y) = A_1x^2 + B_1xy + C_1y^2, \\ Q_2(x, y) = A_2x^2 + B_2xy + C_2y^2, \end{cases}$$

Now we can imagine $Q_1(x, y)$ given first with $A_1 > 0$, so that A_2 could be chosen positive and prime to A_1 by Lemma 2. Then we use two strictly

[1] We continue to exclude from consideration the negative definite quadratic forms (which cannot be generated by ideals).

unimodular transformations from (x, y) to (X_i, Y_i),

$$(6) \qquad \begin{cases} x = X_1 + h_1 Y_1, \\ y = \qquad\quad Y_1, \end{cases} \qquad \begin{cases} x = X_2 + h_2 Y_2, \\ y = \qquad\quad Y_2, \end{cases}$$

substituting the first in $Q_1(x, y)$ and the second in $Q_2(x, y)$.

$$(7) \quad \begin{cases} Q_1{}^*(X_1, Y_1) = Q_1(x, y) = A_1X_1{}^2 + (B_1 + 2A_1h_1)X_1Y_1 + \cdots, \\ Q_2{}^*(X_2, Y_2) = Q_2(x, y) = A_2X_1{}^2 + (B_2 + 2A_2h_2)X_2Y_2 + \cdots. \end{cases}$$

Now since

$$(8) \qquad d = B_1{}^2 - 4A_1C_1 = B_2{}^2 - 4A_2C_2,$$

$B_1 \equiv B_2 \pmod 2$, and we have only to choose h_1 and h_2 such that

$$(9) \qquad A_1h_1 - A_2h_2 = (B_2 - B_1)/2$$

in order to make the X_iY_i coefficients equal in (7). This is easy, since $(A_1, A_2) = 1$. In new notation we have (4). Q.E.D.

LEMMA 4. Every two classes of primitive quadratic forms arising from ideals in $\mathfrak{O}_1$ in a given field have representatives of this type:

$$(10) \quad \begin{cases} Q_1(x, y) = a_1x^2 + bxy + a_2c_0y^2, & a_1 > 0, \\ Q_2(x, y) = a_2x^2 + bxy + a_1c_0y^2, & a_2 > 0, \qquad (a_1, a_2) = 1 \end{cases}$$

Proof. Since the discriminants are equal, then, [keeping the notation of (4)], we see

$$b^2 - 4a_1c_1 = b^2 - 4a_2c_2 = d$$

and $a_1c_1 = a_2c_2$. Since $(a_1, a_2) = 1$, a_1 divides c_2 and a_2 divides c_1. This is shown in new notation in (10), and incidentally $d = b^2 - 4a_1 a_2c_0$. Q.E.D.

Proof of Theorem 1. We can now display a suitable system of type (3). We note that the forms in (10) are generated by

$$(11) \qquad \begin{cases} \mathfrak{j}_1 = [a_1, \lambda], \\ \mathfrak{j}_2 = [a_2, \lambda], \end{cases}$$

where, (restricting ourselves to the case $a > 0$ in (7), Chapter XII, §5)

$$(12) \qquad \lambda = (b - \sqrt{d})/2.$$

Here λ satisfies the equation

$$(13) \qquad \lambda^2 = b\lambda - a_1a_2c_0.$$

Now we note that $\mathfrak{j}_1$ and $\mathfrak{j}_2$ are the aggregate of

$$\text{(14)} \qquad \begin{cases} \alpha_1 = a_1x_1 + \lambda y_1, & \alpha_1 \in \mathfrak{j}_1, \\ \alpha_2 = a_2x_2 + \lambda y_2, & \alpha_2 \in \mathfrak{j}_2, \end{cases}$$

respectively, where x_i, y_i are arbitrary rational integers. Thus

$$\text{(15)} \qquad \begin{cases} \alpha_1\alpha_2 = a_1a_2x_1x_2 + a_1\lambda x_1y_2 + a_2\lambda x_2y_1 + y_1y_2\lambda^2, \\ \alpha_1\alpha_2 = a_1a_2(x_1x_2 - c_0y_1y_2) + \lambda(a_1x_1y_2 + a_2x_2y_1 + by_1y_2). \end{cases}$$

Otherwise expressed,

$$\text{(16)} \qquad \alpha_1\alpha_2 = a_1a_2x_3 + \lambda y_3,$$

where, in the manner of (3),

$$\text{(17)} \qquad \begin{cases} x_3 = x_1x_2 - c_0y_1y_2, \\ y_3 = a_1x_1y_2 + a_2x_2y_1 + by_1y_2. \end{cases}$$

Finally we infer from (16) that

$$\text{(18)} \qquad \mathfrak{j}_1\mathfrak{j}_2 = [a_1a_2, \lambda].$$

First we see that by the property of products $\mathfrak{j}_1\mathfrak{j}_2$ contains a_1a_2 and $a_1\lambda$, $a_2\lambda$, hence λ (since $(a_1, a_2) = 1$). Thus $\mathfrak{j}_1\mathfrak{j}_2 \supseteq [a_1a_2, \lambda]$. But the index of the module $[a_1a_2, \lambda]$ in $\mathfrak{D}_1 = [1, \lambda]$ is clearly a_1a_2 ($= N[\mathfrak{j}_1]\, N[\mathfrak{j}_2]$ which is the index of $\mathfrak{j}_1\mathfrak{j}_2$ in $\mathfrak{D}_1$) by the index definition of norm in Chapter IV, §§8, 10. Thus, since $\mathfrak{j}_1\mathfrak{j}_2$ and its subset $[a_1a_2, \lambda]$ have the same index in $\mathfrak{D}_1$, they must be the same by Lemma 8 of Chapter IV, §8. This completes the proof of Theorem 1. Incidentally, if $Q(\mathfrak{j}_1\mathfrak{j}_2) = Q_3$, then

$$\text{(19)} \qquad Q_3(x, y) = a_1a_2x^2 + bxy + c_0y^2. \qquad \text{Q.E.D}$$

Indeed, in the terminology of (10) and (17)

$$\text{(20)} \qquad Q_1(x_1, y_1)Q_2(x_2, y_2) = Q_3(x_3, y_3),$$

and we define the symbolic product for *composition of classes*

$$\text{(21)} \qquad \mathbf{Q}_1\mathbf{Q}_2 = \mathbf{Q}_3,$$

using the same laws of multiplication as those for the corresponding ideal product (1). We therefore can state that if some form in $\mathbf{Q}_1$ represents m_1 and some form in $\mathbf{Q}_2$ represents m_2 then a form in $\mathbf{Q}_1\mathbf{Q}_2$ exists which represents m_1m_2.

We note that the general element of ideal $\mathfrak{j}_1\mathfrak{j}_2$ is not the right-hand member of (16) but the sum of a finite set of such terms in keeping with the definition of ideal product.

EXERCISE 1. Prove Lemma 1 from Theorem 7 of Chapter VIII, §2. (Compare Theorem 9, Chapter XII, §7.)

EXERCISE 2. Work out the composition theorem for $R(\sqrt{-5})$ as done in the Introductory Survey. Do likewise for $R(\sqrt{34})$.

2. Orders, Ideals, and Forms

For this section we abandon a restriction which was accepted beginning with Chapter VIII, namely, that the ideals are those in the ring of *all* integers of $R(\sqrt{d})$ denoted by $\mathfrak{O}_1$. Using the old notation $R(\sqrt{d})$ for d, the field discriminant, we take the ring of integers congruent to a rational integer (mod f) called an *order* (compare Chapter IV, §10):

$$\mathfrak{O}_f = [1, f(d - \sqrt{d})/2], \tag{1a}$$

where $f \geq 1$ is now a fixed positive integer. For convenience we abbreviate

$$\omega_f = f(d - \sqrt{d})/2 = f\omega_1. \tag{1b}$$

Our purpose is to consider ideals in $\mathfrak{O}_f$. Considering $\bar{\mathfrak{j}}$ the ideal as a module in $\mathfrak{O}_f$, we can reduce it to the canonical form by using the techniques in §7, Chapter IV:

$$\bar{\mathfrak{j}} = [a, b + c\omega_f], \qquad c > 0, \qquad a > b \geq 0. \tag{2}$$

The norm $\bar{N}[\bar{\mathfrak{j}}]$ is ac, the index of $\bar{\mathfrak{j}}$ in $\mathfrak{O}_f$, as before. The discriminant of $\mathfrak{O}_f$ is seen to be

$$\Delta_f^2 = \begin{vmatrix} 1 & \omega_f \\ 1 & \omega_f' \end{vmatrix}^2 = f^2 d = d_f. \tag{3}$$

The basis, as written in (2), incidentally, is *ordered.*

We next examine Chapter VII, §§6 to 8, and learn (to our delight) that the ideals $\bar{\mathfrak{j}}$ for which

$$(\bar{N}[\bar{\mathfrak{j}}], f) = 1 \tag{4}$$

have a unique factorization. It is not hard to see, obeying the last restriction above, that we can reconstruct the finiteness of ideal classes, the basic factorization law, and relevant techniques such as equivalence classes and strict classes. Note that in $\mathfrak{O}_f$ the equivalence relation $\bar{\mathfrak{a}}\lambda = \bar{\mathfrak{b}}\mu$ requires that $\bar{\mathfrak{a}}$, $\bar{\mathfrak{b}}$, λ, μ *all* be in $\mathfrak{O}_f$. For strict equivalence we require $N(\lambda\mu) > 0$.

We shall dwell in some detail on the class number-formula. We call $h(f^2d)$ the class number of the ideal classes in the order $\mathfrak{O}_f$ and η_f the fundamental unit, chosen as before, >1. Clearly $\eta_f = \eta_1{}^u$ for some integer $u(> 0)$, since η_f is also a unit of $\mathfrak{O}_1$. (See Lemma 9, Chapter VI, §5).

We likewise define $h_+(f^2d)$ as the number of strict ideal classes.

In the cases in which $f > 1$ no complex roots of unity except ± 1 occur, as we can see by recalling the only two cases, $d = -3$ and $d = -4$, in which there are complex roots of unity (6 and 4, respectively) when $f = 1$.

THEOREM 2. The class number of the order $\mathfrak{O}_f$ for $f > 1$, is given by

$$(5) \qquad \begin{cases} h(f^2d) = h(d)\psi_d(f)/u, \\ \psi_d(f) \;= f \prod_{q \mid f} (1 - (d/q)/q). \end{cases}$$

Here $f = \Pi q_i^{a_i}$, and u is the "unit index," i.e., when $d > 0$, $u = \log \eta_f / \log \eta_1$, whereas, when $d < 0$, $u = 1$, except for $d = -4$, $u = 2$; and for $d = -3$, $u = 3$.

Proof. The proof of this theorem might be regarded as an opportunity for the reader to review Chapter X with greater maturity. The details of the proof are the same, step by step. First note $\bar{N}[(\alpha)] = |N(\alpha)|$.

We let $\bar{\mathfrak{a}}_1, \bar{\mathfrak{a}}_2, \cdots$ be $h(f^2d)$ ideals in $\mathfrak{O}_f$ belonging to the different classes, with $(\bar{N}(\bar{\mathfrak{a}}_t), f) = 1$, of course. Then, as in Chapter X, §3, we let

$$(6) \qquad F_f(T) = \begin{cases} \text{number of ideals } \mathfrak{a} \text{ in } \mathfrak{O}_f, \text{ for which} \\ 0 < \bar{N}(\bar{\mathfrak{a}}) \leq T, \\ (\bar{N}(\bar{\mathfrak{a}}), f) = 1. \end{cases}$$

$$(7) \qquad G(\bar{\mathfrak{a}}, T) = \begin{cases} \text{number of principal ideals } (\alpha) \leq \bar{\mathfrak{a}}, \text{ where} \\ 0 < |N(\alpha)| \leq T, \\ (N(\alpha), f) = 1. \end{cases}$$

Then, combining several steps of §§3 and 4 in Chapter X, we see that

$$\frac{F_f(T)}{T} = \sum_{t=1}^{h(f^2d)} \frac{G(\bar{\mathfrak{a}}_t, T \,|\, \bar{N}(\bar{\mathfrak{a}}_t)|)}{T},$$

$$(8) \qquad \frac{F_f(T)}{T} = h(f^2d)\kappa_f + (\text{error which} \to 0 \text{ as } T \to \infty),$$

where, analogous with Chapter X,

$$(9a) \qquad \kappa_f = \frac{2\pi}{2\sqrt{|f^2d|}} \lambda_f, \text{ if } d < 0,$$

$$(9b) \qquad \kappa_f = \frac{2\,|\log \eta_f|}{\sqrt{f^2d}} \lambda_f, \text{ if } d > 0,$$

but with a new factor

$$(10) \qquad \lambda_f = \prod_{q \mid f} (1 - 1/q).$$

Now our attention must focus on (10), the other components of (8) being fairly straightforward. The factor λ_f (<1) arises because for (7) we have to write a somewhat more restrictive set of conditions (for $\mathfrak{a}$ equal to each representative $\mathfrak{a}_t$), as compared with Chapter X, §3:

$$(11)\qquad \begin{cases} \bar{\mathfrak{a}} = [\alpha_1, \alpha_2], \\ \alpha = \alpha_1 x + \alpha_2 y, \\ N(\alpha) = ax^2 + bxy + cy^2, \qquad (b^2 - 4ac = \bar{N}(\bar{\mathfrak{a}})^2 f^2 d), \end{cases}$$

excluding cases in which $(N(\alpha), f) \neq 1$. Since $(\bar{N}(\bar{\mathfrak{a}}), f) = 1$, we can choose $N(\alpha_1) = a$ as relatively prime to f (for each $\bar{\mathfrak{a}} = \bar{\mathfrak{a}}_t$). The problem is to consider the effect on the density of algebraic integers in $\mathfrak{O}_f$ when we make the specific restriction imposed in (7)

$$(N(\alpha), f) \neq 1.$$

We note first that if $q \mid f$ and if $N(\alpha) \equiv 0 \pmod q$ for q odd, $(q \nmid a)$,

$$(12)\qquad 2aN(\alpha) = (2ax + by)^2 - df^2 N(\bar{\mathfrak{a}})^2 y^2$$

exactly when $2ax + by \equiv 0 \pmod q$. We therefore see that for any residue class of y modulo q, there are $q - 1$ admissible classes of x. Even when $q = 2$ (and $2 \mid f$) a is odd and b is even, since $b^2 - 4ac = df^2\bar{N}^2(\bar{\mathfrak{a}})$. Thus

$$(13a)\qquad ax^2 + bxy + cy^2 \equiv x^2 \pmod 2,$$

or

$$(13b)\qquad ax^2 + bxy + cy^2 \equiv x^2 + y^2 \pmod 2.$$

Thus it is still true for $q = 2$ that for every y there is only one $(= q - 1)$ admissible class foı x for which x^2 or $x^2 + y^2$ is odd. By the Chinese remainder theorem, there are only $\Pi q(q - 1)$ admissible congruence classes modulo Πq in a total of $(\Pi q)^2$, providing a ratio of

$$(14)\qquad \lambda_f = \prod_{q \mid f} (q(q-1)) \Big/ \prod_{q \mid f} (q^2).$$

On the other hand, analogous with Chapter X, §7,

$$(15a)\qquad \lim_{T \to \infty} \frac{F_f(T)}{T} = \lim_{s \to 1} \zeta_f(s; d)(s - 1) = h(f^2 d)\kappa_f,$$

$$(15b)\qquad \zeta_f(s; d) = \sum_{\bar{\mathfrak{a}}} \frac{1}{\bar{N}(\bar{\mathfrak{a}})^s} = \prod_{\bar{\mathfrak{p}} \nmid f} \left(1 - \frac{1}{\bar{N}(\bar{\mathfrak{p}})^s}\right)^{-1}$$

for $\bar{\mathfrak{a}}$ an ideal in $\mathfrak{O}_f$ and $(\bar{N}[\bar{\mathfrak{a}}], f) = 1$. Here, very easily, all we need to do is to avoid factoring the p in $\mathfrak{O}_f$ that divide f, in using *unique factorization.* Thus $\zeta_f(s; d)$ is compared with $\zeta(s)$ (see Chapter X, §7):

$$\zeta_f(s; d) = \prod_{p \mid f} \left(1 - \frac{1}{p^s}\right) \prod_{p \nmid f} \left(1 - \frac{(d/p)}{p^s}\right)^{-1} \zeta(s). \tag{16}$$

Since only a finite number of factors is involved, we can see that as $s \to 1$ (using $\zeta(s; d)$, as factored in §§6 and 7, Chapter X),

$$\frac{\zeta_f(s; d)}{\zeta(s; d)} \to \lambda_f \prod_{q \mid f} \left(1 - \frac{(d/q)}{q}\right). \tag{17}$$

Finally, as in Chapter X, §8, designating $h(d)$ by h_1 and the κ of Chapter X, §§3 and 4, by κ_1,

$$\lim_{s \to 1} \zeta_f(s; d)(s - 1) = h_1 \kappa_1 \lambda_f \prod_{q \mid f} \left(1 - \frac{(d/q)}{q}\right). \tag{18}$$

We now obtain the final result by canceling common factors from (18) compared with (15*a*). Q.E.D.

We can supplement Theorem 2 with the information that $h(f^2d) \geq h(d)$, but a proof is far from obvious (see Exercise 7, below). There is more to be seen, however. First of all, by Exercise 4 (below), in the case in which $d < 0$, $J = h(f^2d)/h(d)$ is an integer, and this integer is >1, generally. The real case is more difficult, but a method of matching ideals in $\mathfrak{O}$ and $\mathfrak{O}_f$ (as is performed in Exercise 7) will show that J, again, is an integer ≥ 1. This implies that the minimum $u > 0$ for which $\eta^u \in \mathfrak{O}_f$ (or $\eta^u = \eta_f$) divides $\psi_d(f)$, which is by no means obvious. Less obvious indeed is the unsettled question when $h(df^2) = h(d)$. For instance, when $d = 28$, $\eta_1 = 8 + 3\sqrt{7}$, we can calculate: $h(28f^2) = h(28) = 1$ when $f = 2, 7, 14, \cdots$. The last result means that $(8 + 3\sqrt{7})^v \not\equiv m \pmod{14}$ when $0 < v < \psi_{28}(14) = 14$ for any rational integer m. (It can be checked that $(8 + 3\sqrt{7})^{14} \equiv 1$ modulo 14.)

Dirichlet showed in 1856 that for certain d there exists an infinite number of f for which $h(f^2d) = h(d)$. It is not known if such an f exists for each d. (Compare Exercise 6, below.)

There is no difficulty setting up a correspondence between the $h_+(df^2)$ ideals and forms in $\mathfrak{O}_f$, the primitive forms having the form $Q(x, y) = ax^2 + bxy + cy^2$, where $(a, b, c) = 1$ and $b^2 - 4ac = f^2d$. The problem $Q(x, y) = m$ can be solved by ideal factorization or composition when $(m, f) = 1$. It is not our purpose to give the details here.

The study of orders does not, by itself, lead to particularly vital problems. The role of orders is still deeply entrenched in the theory by the old problem of finding solutions to $Q(x, y) = m$ for an arbitrary form. The following theorem even strengthens the importance of ideals in orders by making ideals "as general" as modules in the *quadratic* case.

THEOREM 3. Any module $[\alpha_1, \alpha_2]$ in $\mathfrak{O}_1$ corresponds to at least one ideal $\bar{\mathfrak{j}} = [\beta_1, \beta_2]$ in some order $\mathfrak{O}_f$ in the sense that $[\alpha_1, \alpha_2]\lambda_1 = \lambda_2[\beta_1, \beta_2]$ for some integers λ_1 and λ_2 in $\mathfrak{O}_1$.

Proof. We can appeal to the method of Theorem 6 in Chapter XII. We assume that the module $[\alpha_1, \alpha_2]$ is ordered. We write a quadratic form:

$$N(\alpha_1 x + \alpha_2 y) = t(ax^2 + bxy + cy^2), \quad a > 0 \tag{19}$$

where t is chosen so that $(a, b, c) = 1$. (From §5 of Chapter XII, e.g., if $\mathfrak{a} = [\alpha_1, \alpha_2]$ is actually an ideal of $\mathfrak{O}_1$, then t would be $N[\mathfrak{a}]$.) But, by the same process, extended to the case of an order, we can regard $ax^2 + bxy + cy^2$ as the form generated by the ideal in $\mathfrak{O}_f$:

$$[a, (b - \sqrt{d_f})/2],$$

where $d_f = b^2 - 4ac = f^2 d$ (d is the field discriminant). This ideal is our $\bar{\mathfrak{j}}$, by a comparison with the method in §6 of Chapter XII, (the details are left to the reader). Q.E.D.

The construction of the ideal classes in $\mathfrak{O}_f$ is not necessarily dull. We could, of course, factor all $p \leq f\sqrt{d}$ required by Minkowski's theorem (Chapter VIII) and form the classes.

Relatively recently, however, Weber (1897) and Fueter (1903) showed how to obtain ideal classes in $\mathfrak{O}_f$ from a new type of modified equivalence class in $\mathfrak{O}_1$, designated by "$\sim$ (mod f)." For two ideals $\mathfrak{a}, \mathfrak{b}$ in $\mathfrak{O}_1$, which are relatively prime to f, we write

$$\mathfrak{a} \sim \mathfrak{b} \pmod{f}, \tag{20}$$

when

$$\lambda\mathfrak{a} = \mu\mathfrak{b} \quad \text{and} \quad \lambda \equiv \mu r \pmod{f} \tag{21}$$

for a rational integer r and integers λ, μ of $\mathfrak{O}_1$, which are all relatively prime to f. Thus far the definition remains entirely in $\mathfrak{O}_1$.

We then take, $\mathfrak{a}_t$, the representatives of the different equivalence classes (mod f), and call $\bar{\mathfrak{a}}_t$ the aggregate of elements common to $\mathfrak{a}_t$ and $\mathfrak{O}_f$, e.g., $\bar{\mathfrak{a}}_t = \mathfrak{a}_t \cap \mathfrak{O}_f$. The $\bar{\mathfrak{a}}_t$ can be shown to be ideals in $\mathfrak{O}_f$, representing all its ideal classes. The *strict* equivalence $\mathfrak{a} \approx \mathfrak{b}$ in $\mathfrak{O}_1$ fits in, semantically at least, by the designation "$\mathfrak{a} \sim \mathfrak{b}$ (mod ∞)," and, if $N(\lambda\mu) > 0$ in (21), we would say "$\mathfrak{a} \sim \mathfrak{b}$ (mod $f\,\infty$)," like Hasse's congruence in Chapter II, §6.

EXERCISE 3. Show $h(-5^2 \cdot 3) = 2$. Find the *two* representative ideal classes and forms for $\mathfrak{O}_5$ by starting with these modules and using Theorem 3 to construct ideals:

$$[1, 5\rho], [5, \rho], [5, 1 + \rho], [5, 2 + \rho], [5, 3 + \rho], [5, 4 + \rho],$$

where $\rho = (-1 + \sqrt{-3}/2)$. (Note the three-to-one correspondence of modules and ideals.)

EXERCISE 4. Verify the earlier statement that when $d < 0$, $h(f^2d)/h(d)$ is an integer. It is 1 only when $d = -4, f = 2$; $d = -3, f = 2$ or 3, and $d \equiv -1 \pmod 8$, $f = 2$.

EXERCISE 5. Verify that $h(4 \cdot 28) = h(28)$. Find $h(9 \cdot 28)$.

EXERCISE 6. Verify that if $(1 + \sqrt{2})^m = A_m + B_m\sqrt{2}$ and if $m = 2^s n$ for n odd, then $2^s \parallel B_m$. (Use induction on s.) Hence show $h(2^{2k} \cdot 8) = 1$. Likewise, show $h(5^{2t} \cdot 5) = 1$ and $h(2^{2k} \cdot 5) = 2$ for $k \geq 3$.

EXERCISE 7. Show that if $\mathfrak{a} = [a, b + c\omega_1]$ and $(ac, f) = 1$ then $\bar{\mathfrak{a}} = [a, bf + cf\omega_1] = [a, B + c\omega_f]$, where $B = bf \pmod a$. Show that if $\bar{\mathfrak{a}} = \bar{\mathfrak{b}}$ then $\mathfrak{a} = \mathfrak{b}$, and that if $\bar{\mathfrak{a}} \sim \bar{\mathfrak{b}}$ in $\mathfrak{O}_f$ then $\mathfrak{a} \sim \mathfrak{b}$ (in $\mathfrak{O}_1$). Show that $\bar{\mathfrak{j}}_1\bar{\mathfrak{j}}_2 = \overline{\mathfrak{j}_1\mathfrak{j}_2}$, using (11) of §1 (above), and that if $\mathfrak{c} = \mathfrak{a}\mathfrak{b}$ then $\overline{\alpha\mathfrak{a}}\,\overline{\beta\mathfrak{b}} = \overline{\alpha\beta\mathfrak{c}}$ for some α and β in $\mathfrak{O}_1$ relatively prime to f.

EXERCISE 8. Show that for $\alpha \in \mathfrak{O}_f$, $(N(\alpha), f) = 1$ then $(\alpha) \sim (1)$ in $\mathfrak{O}_f$ if and only if $(\alpha) \sim (1) \pmod f$. From this show the two definitions of equivalence can be identified.

EXERCISE 9. Referring to Exercise 3, Chapter IX, §1, show that $\Phi[(f)] = \phi(f)\psi_d(f)$. Also show that as U and r vary, $r\eta_1{}^U$ takes on $u\phi(f)$ values modulo f. Thus show that (5) also gives the correct class number for definition (21).

EXERCISE 10. Derive a formula for $h_+(f^2d)/h(f^2d)$ analogous with Theorem 3, Chapter XII, §3.

3. Genus Theory of Forms

As a final topic[1] we consider the basic problem of whether congruence classes for a prime p determine its representability by a quadratic form in the manner of Fermat's famous result that for odd p

$$(1) \qquad p = x^2 + y^2 \text{ exactly when } p \equiv 1 \pmod 4.$$

Ideal theory would be enormously helpful. For instance, if $h_+(d) = 1$, only one form enters, and we easily see for $(p, 2d) = 1$,

$$(2) \qquad p = Q(x, y) = \begin{cases} x^2 + xy - \dfrac{d-1}{4}y^2 & (\text{if } d \equiv 1 \pmod 4) \\ \text{or} & \\ x^2 - (d/4)y^2 & (\text{if } d \equiv 0 \pmod 4) \end{cases}$$

[1] In the remainder of this chapter we return to the usual convention that d is a field discriminant (or $f = 1$). Thus §2, where $f > 1$, is not necessary for any other part of the book.

if and only if $(d/p) = 1$. If $d > 0$ and $N(\eta_1) = +1$, then $h_+(d) = 2h(d) = 2$, and there is some question whether $Q(x, y)$ or $-Q(x, y)$ should represent p. A fairly simple result still holds (see Exercise 17 below). When $h(d) > 1$, the matter is not so trivial; we find that we must first divide the forms into *genera* (plural of *genus*) in considering which forms represent a prime. The composition and ideal multiplication theorems make it suffice to build on prime factors p, except for sign considerations.

As a preliminary step let us consider a field discriminant d. There are several cases according to whether the square-free kernel D is odd and $D \equiv \pm 1 \pmod 4$ or even and $D/2 \equiv \pm 1 \pmod 4$:

$$\left.\begin{aligned} &(3a) && d = q_1 q_2 \cdots q_r \equiv 1 \pmod 4, \\ &(3b) && d = 4q_2 q_3 \cdots q_r \equiv 12 \pmod{16}, \\ &(3c) && d = 8q_2 q_3 \cdots q_r \equiv 8 \pmod{32}, \\ &(3d) && d = 8q_2 q_3 \cdots q_r \equiv 24 \pmod{32}, \end{aligned}\right\} \quad d > 0.$$

and

$$\left.\begin{aligned} &(4a) && d = -q_1 q_2 \cdots q_r \equiv 1 \pmod 4, \\ &(4b) && d = -4q_2 q_3 \cdots q_r \equiv 12 \pmod{16}, \\ &(4c) && d = -8q_2 q_3 \cdots q_r \equiv 8 \pmod{32}, \\ &(4d) && d = -8q_2 q_3 \cdots q_r \equiv 24 \pmod{32}. \end{aligned}\right\} \quad d < 0.$$

Here q_i are different positive odd primes and r is the total number of different prime divisors of d.

For each d we consider certain *Jacobi* characters as functions of m (relatively prime to $2d$), noting that when $d < 0$ we choose only positive forms, $(m > 0)$:

$$(5) \qquad \chi_i(m) = \left(\frac{m}{q_i}\right) \begin{cases} i = 1, 2, \cdots, r, & d \equiv 1 \pmod 4, \\ i = \ 2, \cdots, r, & d \not\equiv 1 \pmod 4. \end{cases}$$

Now the law of quadratic reciprocity tells us, easily, that in cases $(3a)$ and $(4a)$

$$(6) \qquad \prod_{i=1}^{r} \chi_i(m) = \prod_{1}^{r} \left(\frac{m}{q_i}\right) = \prod_{1}^{r} \left(\frac{q_i^*}{m}\right) = \left(\frac{d}{m}\right),$$

using $q_i^* = q_i(-1/q_i) \equiv 1 \pmod 4$ and recalling that d *must* have the sign of Πq_i^*. In the remaining cases the matter is more detailed, but we *define* $\chi_1(m)$ by the relation

$$(7) \qquad \chi_1(m) \cdot \prod_{i=2}^{r} \chi_i(m) = (d/m), \qquad (m > 0 \text{ for } d < 0).$$

We thus have defined $\chi_1(m)$ as a multiplicative function of m on the basis of similar properties of $(d/m) = (d/|m|)$ and $\chi_i(m)$, $i \geq 2$. In fact we verify by substitution (see Exercise 11, below):

LEMMA 5. The remaining character in (7) is given by

$$\chi_1(m) = \begin{cases} (-1/|m|) \operatorname{sgn} m & \text{in cases (3b) and (4b),} \\ (2/m) & \text{in cases (3c) and (4c),} \\ (-2/|m|) \operatorname{sgn} m & \text{in cases (3d) and (4d).} \end{cases} \tag{8}$$

Here, sgn m means "the sign of m" ($+1$ when $m > 0$ and -1 when $m < 0$). When $d < 0$, it is understood that $m > 0$ only.

We now begin the process of classifying quadratic forms.

LEMMA 6. All integers m represented by a quadratic form of field discriminant d and relatively prime to $2d$ have the same-value of $\chi_i(m)$ for each i and satisfy $(d/m) = 1$.

Proof. Let m_1 and m_2 be represented by the form

$$Q(x_j, y_j) = ax_j^2 + bx_jy_j + cy_j^2 = m_j, \qquad (j = 1, 2), \tag{9}$$

and suppose that $(a, 2d) = 1$, by choice of the form keeping the same proper equivalence class. Then, with $d = b^2 - 4ac$, we easily have

$$4am_j = (2ax_j + by_j)^2 - dy_j^2 \tag{10}$$

Then if χ_i is of the form (m/q) where $q \mid d$ and q is odd,

$$4am_j \equiv \text{perfect square (mod } q)$$

and $\chi_i(am_j) = 1$; thus

$$\chi_i(m_1) = \chi_i(m_2) = \chi_i(a) \tag{11}$$

Furthermore, if χ_1 is not of the form (m/q_1), then (11) holds for $i = 1$ by the product formula (7), once we verify that $(d/m_j) = 1$.

To see this last statement, let $Q(x, y) = m$ be a representation in which $(x, y) = 1$. Then all prime factors, p, of m, by Theorem 9 of Chapter XII, satisfy $(d/p) = 1$ (as $p \nmid d$). If (x, y) becomes (gx, gy), then a factor g^2 is contributed to m, not affecting the residue symbols. Q.E.D.

We therefore define for every form the *generic characters* $\chi_i(Q)$ *of the form* to be these values of $\chi_i(m)$ where $(m, 2d) = 1$ and Q represents m:

$$\chi_1(Q), \chi_2(Q), \cdots, \chi_r(Q) \tag{12}$$

The characters of course apply to the class of *properly equivalent forms* as they represent the same numbers. For the same reason the conjugates $Q(x, y)$ and $Q(x, -y)$ have the same generic character even if they are not properly equivalent.

We next consider the forms Q for which the $\chi_i(Q)$ are equal to some preassigned array of signs $e_i = +1$ or -1,

$$(13) \qquad e_1, e_2, \cdots, e_r, \text{ subject to } \prod_{i=1}^{r} e_i = 1.$$

There are 2^{r-1} possible arrays, and the set of forms corresponding to each array is called a *genus of forms*. The forms for which all $e_i = +1$ are naturally called the *principal genus of forms. Each genus is also a collection of proper equivalence classes*. The genera are multiplied by composition of classes, in the notation of (20) and (21) of §1. Thus, if $Q_1 \in \mathbf{Q}_1$, $Q_2 \in \mathbf{Q}_2$, and $Q_3 \in \mathbf{Q}_1\mathbf{Q}_2$,

$$(14) \qquad \chi_i(Q_1)\chi_i(Q_2) = \chi_i(Q_3), \qquad 1 \leq i \leq r,$$

since ordinary multiplication is involved in composition. The classes $\mathbf{Q}$ belonging to the principal genus accordingly form a *subgroup* of the class group. Each genus constitutes a coset of the principal genus (see Exercise 9, Chapter I, §6).

LEMMA 7. There is an odd prime p (indeed infinitely many) for which the generic characters $\chi_i(p) = e_i$ for a preassigned array of signs e_i whose product is 1.

Proof. Consider $\chi_i(p) = (p/q_i) = e_i$ for $i = 2, \cdots, r$. Since the q_i are odd primes, we choose p congruent to a residue or nonresidue modulo q_i, as the case may require, whereas $\chi_1(p) = e_1$ is a condition on p modulo an odd q_1 or modulo 4 or 8. The Chinese remainder theorem determines p from several independent arithmetic progressions, whereas $(d/p) = 1$ automatically from (7). Dirichlet's theorem (see Chapter X, §1), yields the result. Q.E.D.

There is therefore at least one class of forms in each genus by the corollary to Theorem 9, Chapter XII, §7. We then use the coset property to see that there is an equal number of classes in each genus, and we have proved the following result:

THEOREM 4. If we consider $h_+(d)$ proper equivalence classes of forms with discriminant d equal to a field discriminant, then they can be subdivided equally into 2^{r-1} genera of $h_+(d)/2^{r-1}$ forms which form a subgroup of the proper equivalence class group under composition.

This theorem was proved by Gauss in 1801. Since Dirichlet's theorem was not available then, another proof had to be given (in fact with no aid from infinitesimal analysis). Gauss further showed the famous *duplication* (meaning "squaring") theorem:

THEOREM 5. The principal genus contains precisely those forms that are squares of some form under composition.

Proof. We shall sketch a proof of this very important result. First we make a transition to *ideal theory* and *nonstrict equivalence classes.* We consider four different types of field of discriminant d with r distinct prime factors and fundamental unit η_1. (We let A, B, M be general symbols for rational integers.)

Type 1: $d < 0$.
Type 2: $d > 0, N(\eta_1) = -1$. Here $d = a^2 + b^2$ and $(4M - 1) \nmid d$.
Type 3: $d > 0, N(\eta_1) = +1$ and $d = a^2 + b^2$. Here $(4M - 1) \nmid d$.
Type 4: $d > 0, N(\eta_1) = +1$ and $d \neq A^2 + B^2$. Here $q_r = (4s - 1) \mid d$

(by convention the *last* prime factor of d is taken $\equiv -1 \bmod 4$).

Now consider the following table which summarizes past results on ambiguous ideals:

TABLE 1

I	II	III	IV	V	VI	VII	VIII
Field Type	Principal Ambiguous Ideals	Independent Ambiguous Ideals	Ambiguous Classes	Relation of Form Q to Negatives (conj.)	Form Genera	Form Genera without Negatives	t
1	2	2^{r-1}	2^{r-1}	No negatives	2^{r-1}	2^{r-1}	$r-1$
2	2	2^{r-1}	2^{r-1}	Equivalent, hence same genus	2^{r-1}	2^{r-1}	$r-1$
3	4	2^{r-2}	2^{r-1}	Inequivalent but same genus	2^{r-1}	2^{r-1}	$r-1$
4	4	2^{r-2}	2^{r-2}	Inequivalent and $\chi_r(Q) = -\chi_r(-Q)$	2^{r-1}	2^{r-2}	$r-2$

The first four columns embody Exercises 8 to 11 and 13 of Chapter XI, §3, if we note that the r different prime factors of d produce r ramified (ambiguous) prime ideals and 2^r possible ambiguous ideals by selection of subsets. Column II expresses Exercise 13, whereas Column III expresses the number of ambiguous ideals *independent to within principal nonunit ideal factors* (Column II divided into 2^r). Column IV reflects the presence of the special class for Type 3 (see Exercise 8). As in Chapter XI, our ideal classes here are *nonstrict.*

Looking at Column V we see an application of Theorem 8, Chapter XII, §6, together with the fact that the generic characters of fields of Type 3 satisfy $\chi_i(-m) = \chi_i(m)$ for all i (whereas "taking the conjugate" preserves genus). To cope with the sign relation involved in fields of Type 4, we cut down the number of proper equivalence classes **Q** of forms to onehalf by considering only the forms $Q \in \mathbf{Q}$ with a definite choice of sign e:

$$eQ(x, y) \tag{15}$$

where $e = \chi_r(Q(x, y)) = \chi_r(\mathbf{Q}) = \chi_r(m)$. Here m is, as before, any integer represented by $Q(x, y)$ and such that $(m, 2d) = 1$. Of course $\chi_r(eQ(x, y)) = 1$, thus the *product of the remaining* $\chi_1(eQ) \cdots \chi_{r-1}(eQ)$ *is always* 1.

We then obtain 2^t genera for the forms (see Columns VII and VIII) that are determined by the *independent* characters

$$\chi_1(eQ), \chi_2(eQ), \cdots, \chi_t(eQ) \tag{16a}$$

for $h_+(d)$ classes of forms in fields of Types 1 to 3 and $h_+(d)/2$ classes in fields of Type 4 according to (15). Note that $-Q$ and Q lead to the *same* eQ, since $\chi_r(-1) = -1$.

We can now make the transition to *ideals* by considering *only nonstrict equivalence classes*. In fields of Type 3 (or 4) this amounts to using only one of the equivalence classes of $\mathfrak{a}$ and $\mathfrak{a}\sqrt{d}$, since they both lead to the same genus (or the same eQ). We call the 2^t genera thus determined the *ideal genera* as distinguished from the 2^{r-1} *form genera* originally introduced.

Table III (appendix) lists the values of *all* $(t + 1)$ character symbols for the *ideal* genera in order of the indicated prime factors of d (except that for Type 4, $q_r \equiv -1 \pmod 4$ is listed first in the factors of d, then omitted in the list of character symbols).

We use Exercise 18, Chapter XII, §7, to write the generic character of an ideal class directly in terms of an ideal $\mathfrak{a}$ it contains:

$$\chi_1(eN[\mathfrak{a}]), \cdots, \chi_t(eN[\mathfrak{a}]), \qquad (N[\mathfrak{a}], 2d) = 1, \tag{16b}$$

where $e = 1$ except for fields of Type 4 where $e = \chi_r(N[\mathfrak{a}])$. These t characters are independent in value.

Let us now visualize the (nonstrict) equivalence class group for ideals. Using Kronecker's decomposition theorem, we set

$$\begin{aligned} \mathbf{G} &= \mathbf{Z}(2^{s_1}) \times \cdots \times \mathbf{Z}(2^{s_t}) \times \mathbf{Z}(r_1) \times \cdots \times \mathbf{Z}(r_u), \\ h(d) &= 2^{s_1 + \cdots + s_t} r_1 \cdots r_t \geq 2^t, \end{aligned} \tag{17a}$$

where the indicated groups are cyclic with generators (say) $\mathbf{g}_1, \cdots, \mathbf{g}_t$, $\mathbf{g}_1', \cdots, \mathbf{g}_u'$ and the values of $s_i > 0$, whereas $r_i > 1$ are odd. Here the value of t must agree with Column VIII of the table, since the ambiguous classes (whose square is the identity) are 2^t in number:

$$\mathbf{g} = \mathbf{g}_1^{m_1} \cdots \mathbf{g}_t^{m_t}, \qquad m_i = 0 \quad \text{or} \quad 2^{s_i}/2. \tag{17b}$$

We can verify that precisely $h(d)/2^t$ elements are perfect squares. Let

$$\mathbf{g} = \mathbf{g}_1^{v_1} \cdots \mathbf{g}_t^{v_t} \mathbf{g}_1'^{w_1} \cdots \mathbf{g}_1'^{w_u} \tag{17c}$$

where we can solve for x_i and x_i' in

$$v_i \equiv 2x_i \pmod{2^{s_i}} \quad \text{and} \quad w_i \equiv 2x_i' \pmod{r_i}. \tag{17d}$$

Then the v_i are each restricted to half the values modulo 2^{s_i}, but the w_i are not restricted modulo r_i (odd). On the other hand, all perfect squares are in the principal genus by composition formula (14), but the principal genus numbers as many classes as squares, by Theorem 4 (modified to accommodate ideal genera for Type 4). Hence Theorem 5 follows. Q.E.D.

We can pursue Theorem 5 one step further: the square of a form necessarily represents a perfect square (prime to $2d$), since composition simulates ordinary multiplication. A form representing a perfect square (prime to $2d$) is necessarily in the principal genus, which is determined by the odd primes in d. Hence Gauss noted the following:

THEOREM 6. A quadratic form whose discriminant d equals the discriminant of a field represents a square prime to $2d$ if and only if the form is in the principal genus.

COROLLARY. The quadratic form in the foregoing theorem represents a perfect square (relatively prime to $2d$) if and only if it represents a quadratic residue (relatively prime to $2d$) modulo every odd prime divisor of d. Negative definite forms remain excluded.

The corollary is due in one form to Legendre (1785), and it can be proved in a fairly elementary manner. Its proof is still of sufficient depth to deserve a special analysis in the Concluding Survey. Gauss reversed the procedure to prove Theorem 4 from Theorem 6. (Legendre and Gauss, in fact, considered forms with no restriction on discriminant.)

We can now return to the motivating question of this section, namely, when do congruence properties of a prime p determine its representability by a quadratic form? Now we see a partial answer in that this is always the case when $h(d) = 2^t$ or when there is *exactly one ideal class in each genus*. For this reason *ideal* genera were used (although a corresponding theory of ambiguous forms was thereby ignored). Gauss, indeed, felt that the search for fields of class number unity was less meaningful than the search for fields of one class per (ideal) genus.

The easy result that 2^t divides $h(d)$ helps us to understand to some extent why certain fields with very composite d *must* have large class numbers, but an adequate understanding of the reason for *odd* cyclic groups in the structure of the class group is still, generally speaking, on the outer frontiers[1] of number theory.

[1] The nature of the odd cyclic structures is even more mysterious in regard to the occurrence of repeated prime powers in the class group. The smallest known instance in complex fields is $d = -3299$ where $\mathbf{Z}(3) \times \mathbf{Z}(3)$ occurs in the class group (Gauss). For real fields the smallest known instance is $d = 62{,}501$ where the same factors occur (Pall).

EXERCISE 11. Complete Lemma 5 by verifying (7) in each case for positive and negative m.

EXERCISE 12. Write out the characters for $D = -3, -5, -62, -78$; $D = 3, 5, 6, 34, 65, 82$.

EXERCISE 13. Verify the genera in Table III (appendix) for $D = -62$ (taking the class structure for granted). Show that exactly one of the forms listed satisfies the representation:

$$\left.\begin{array}{r} x^2 + 62y^2 \\ 7x^2 + 2xy + 9y^2 \\ 2x^2 + 31y^2 \end{array}\right\} = p \text{ if } (2/p) = (p/31) = 1$$

$$\left.\begin{array}{r} 3x^2 + 2xy + 21y^2 \\ 11x^2 + 4xy + 6y^2 \end{array}\right\} = p \text{ if } (2/p) = (p/31) = -1,$$

where $(p, 62) = 1$. Verify this numerically for two values of $p(>11)$ in each genus. (Note that there is no need to put a $\pm$ sign on the middle term and that the uniqueness of the form representing p comes from Theorem 5, Chapter XII, §4).

EXERCISE 14. Construct the genus classifications analogous to Exercise 13, when $D = 10$, $D = 65$, and verify each genus for at least one prime.

EXERCISE 15. Construct the genus classifications for $D = 14$, $D = 42$, and determine which sign $\chi_r(Q)$ has to be selected for each form.

EXERCISE 16. Do likewise for $D = 34$. Are both $+p$ and $-p$ representable are by each form?

EXERCISE 17. Referring to fields of Type 4, explain the choice of sign in (2) if $h_+(d) = 2$, $h(d) = 1$.

EXERCISE 18. Using techniques analogous to Chapter II, §7, show that there are exactly 2^t real characters in the (nonstrict) ideal class group [see (17a)] and show that these characters must be those formed by taking products $\chi_a(Q)\chi_b(Q) \cdots \chi_m(Q)$ where $a, b, \cdots, m$ is a subset of the indices $1, 2, \cdots, t$.

EXERCISE 19. Define the real characters for Exercise 18 when $D = -65$, $D = 34$.

EXERCISE 20. Prove the lemma in Chapter X, §12.

EXERCISE 21. Show that the corollary to Theorem 6 would not hold true if negative definite forms were permitted, by considering a form that represents -1, in connection with (4b, d).

EXERCISE 22. Consider $\mathbf{G}_+$ the (proper) equivalence class group for forms. Show that, for fields of Type 1 or 2, $\mathbf{G}$ and $\mathbf{G}_+$ have the same decomposition, whereas, for Type 3, one factor (say) $\mathbf{Z}(2^{s_1})$ is replaced by $\mathbf{Z}(2^{s_1+1})$ in $\mathbf{G}_+$, and, for Type 4, $\mathbf{G}_+$ has an extra factor $\mathbf{Z}(2)$ as compared with $\mathbf{G}$.

4. Hilbert's Description of Genera

Hilbert devised a remarkably general quadratic character symbol to avoid the inelegant specialization of cases required in the last section (for example, in Lemma 5, §3). The symbol proved to be more easily generalized to fields of higher degree.

Let n and m be rational integers, $m \neq$ perfect square, $n \neq 0$; let q be any prime. Hilbert defined

$$\left(\frac{n, m}{q}\right) = +1 \text{ or } -1, \tag{1}$$

depending on whether or not it is true that for *each* power q^s (s an integer ≥ 1) we can solve

$$n \equiv N(\xi_s) \quad \text{modulo } q^s \tag{2}$$

for some integer ξ_s in $R(\sqrt{m})$. (When m is a nonzero perfect square, the symbol is taken as 1 for completeness.) There are many rules of calculation for which we refer to more advanced treatises. We state only three and without proof:

$$\left(\frac{n_1, m}{q}\right)\left(\frac{n_2, m}{q}\right) = \left(\frac{n_1 n_2, m}{q}\right), \tag{3}$$

$$\left(\frac{n, m}{q}\right) = 1 \text{ if } q \nmid 2mn, \tag{4}$$

$$\left(\frac{n, m}{q}\right) = \left(\frac{n}{q}\right) \text{ if } \begin{cases} q \mid m, \\ q \nmid n. \end{cases} \quad q \text{ odd}, \tag{5}$$

These rules state (respectively) that the symbol is multiplicative in n, that it is "interesting" for only a finite set of primes (divisors of $2mn$), and that it is really a quadratic residue symbol. (The last equation (5) is essentially the matter discussed in Exercises 14 and 15 in Chapter I).

In this terminology we can finally identify the genera. We let $q_1, q_2, \cdots, q_r$ be the prime divisors of d, as before. Then the array of signs [corresponding to (12), §3] identifying a genus of forms, one of which represents the value m (prime to $2d$), is

$$\left(\frac{m, d}{q_1}\right), \cdots, \left(\frac{m, d}{q_r}\right). \tag{6}$$

If we consider the genus of an ideal $\mathfrak{a}$ we can take $m = N[\mathfrak{a}]$, if $(N[\mathfrak{a}], 2d) = 1$ (by Exercise 18, Chapter XII, §7). The independent characters [corresponding to (16*b*), §3] are

$$\left(\frac{eN[\mathfrak{a}], d}{q_1}\right), \cdots, \left(\frac{eN[\mathfrak{a}], d}{q_t}\right) \tag{7}$$

where $e = 1$, except for fields of Type 4 in which $q_r \equiv -1 \pmod 4$ and $e = \left(\frac{N[\mathfrak{a}], d}{q_r}\right)$. Here q_1 can be taken as 2 for some of the cases in Lemma 5, §3.

Moreover, it would follow that the product of the indicated signs in (6) is 1. Hilbert found, however, that to achieve the height of perfection a new symbol must be introduced

$$\left(\frac{n, m}{\infty}\right) = \pm 1, \tag{8}$$

depending on whether or not n can agree with $N(\alpha)$ in sign (mod ∞) for an α in $R(\sqrt{m})$. (The only case in which the value is -1 is easily $n < 0$ and $m < 0$.) We now have the unconditional statement

$$\prod_{q \mid 2mn} \left(\frac{n, m}{q}\right) = \left(\frac{n, m}{\infty}\right); \tag{9}$$

or, still more elegantly,

$$\prod_{q} \left(\frac{n, m}{q}\right) = 1, \qquad q = \text{all primes and } \infty. \tag{10}$$

The reader will notice that in some way (10) provides a connection between quadratic reciprocity and unique factorization. The fact that the product equals 1 is a manifestation of quadratic reciprocity, and the fact that the only interesting q are divisors of $2mn$ is a manifestation of unique factorization, if we compare Lemma 5, §3.

*CONCLUDING SURVEY

The reader may very well ask were this subject leads, and he should be prepared to receive a variety of answers.

Some of the new directions are so strongly algebraic, combinatorial, or analytic as to be lacking in direct appeal to the main tradition. We shall select three topics, which we believe have such appeal and which are closely related although seemingly different in origin.

We shall combine legitimate deductions and rash conclusions, freely intermingled for quick reading. The more serious student, of course, will refer to advanced treatises for details.

The new directions, inevitably, involve algebraic numbers of degree n. An *algebraic* (rational) *number* of degree n is defined as a root of the irreducible equation with rational coefficients a_t:

$$(1) \qquad \theta^n + a_1\theta^{n-1} + \cdots + a_{n-1}\theta + a_n = 0.$$

An *algebraic integer* is one whose defining equation has only rational integers, as coefficients a_t. A field $R(\theta)$ is defined as the set of values resulting from rational operations with θ, as in Chapter III. Sometimes it is convenient to consider some $R(\theta_1, \theta_2)$, a field generated by two elements, but, indeed, any field can be generated in a variety of ways. Algebraic integers again form a ring. We speak of fields including one another, $R(\theta_1) \supseteq R(\theta_2)$, in the usual way for sets.

The ideal theory is no harder than in the quadratic case, in principle. There is unique factorization into ideals but a finite class number which generally exceeds 1 (producing nonunique factorization into principal ideals). The laws of decomposition of Chapter IX are more complicated, but a rational prime breaks up into no more than n ideals (in a generally irregular fashion). The theory of quadratic *forms* is replaced by a much less attractive theory that makes us willingly confine all our attention to *ideals* in the ring of all algebraic integers of a given field.

CYCLOTOMIC FIELDS AND GAUSSIAN SUMS

Aside from the quadratic field, the most important is the one generated by a root of unity of index n, or the *cyclotomic* ("circle-cutting") field, generated by powers of

$$\theta = \exp 2\pi i/n = \cos 2\pi/n + i \sin 2\pi/n. \tag{2}$$

This root and its powers $\theta^t = \exp 2\pi it/n = \cos 2\pi t/n + i \sin 2\pi t/n$ are represented as the vectors from the origin to points of a regular polygon of n sides with center at origin and one vertex at $1 = \theta^n$. This follows from de Moivre's theorem in elementary fashion. The important feature is that the powers θ^t are determined by the exponent t (mod n).

When n is prime, the reduced equation defining θ is

$$\frac{\theta^n - 1}{\theta - 1} = \theta^{n-1} + \theta^{n-2} + \cdots + \theta + 1 = 0. \tag{3}$$

(The irreducibility of (3) is not obvious, but we always omit details for the sake of the survey.) This cyclotomic field first was seeen to be more basic when Gauss (1800) separated the exponents as residues or nonresidues (mod n), excluding 0, and wrote

$$\begin{cases} R = \sum \theta^t & t \text{ residue}, \\ N = \sum \theta^u & u \text{ nonresidue}. \end{cases} \tag{4}$$

Thus, when $n = 5$, $R = \theta + \theta^4$, $N = \theta^2 + \theta^3$, and, when $n = 7$, $R = \theta + \theta^2 + \theta^4$, $N = \theta^3 + \theta^5 + \theta^6$, etc. By (3), however,

$$R + N + 1 = 0. \tag{5}$$

Gauss next introduced the so-called *Gaussian sum*

$$S = \sum_{r=0}^{n-1} \theta^{r^2} = 1 + 2R \tag{6}$$

(as we can see by noticing that the squares, r^2, equal each residue twice). Another closely related expression, using Legendre's symbol (r/n) is

$$T = \sum_{r=0}^{n-1} \theta^r \left(\frac{r}{n}\right); \tag{7}$$

and, by (5) and (6),

$$R - N = T = S. \tag{8}$$

The expression (6) gives us some insight into S physically, since it suggests a superposition of n units with directions determined by r^2. If the superposition is "random," well-established laws for estimating "probable error" would suggest a total length $|S|$ equal to $\sqrt{n}$.

In fact, we can rigorously establish $|S| = \sqrt{n}$, for, taking complex conjugates, we see that

$$\bar{S} = \sum_{s=0}^{n-1} \theta^{-s^2}, \tag{9}$$

$$|S|^2 = S\bar{S} = \sum_{r,s=0}^{n-1} \theta^{r^2-s^2} = \sum_{r,s=0}^{n-1} \theta^{(r+s)(r-s)}. \tag{10}$$

If we let $r - s = q, r + s = 2s + q$, we find that q and s take on independently all values modulo n. So do q and $v = 2s + q$. Thus

$$|S|^2 = \sum_{q,v=0}^{n-1} \theta^{qv} = \sum_{q=0}^{n-1} 1 + \sum_{q=0}^{n-1} \sum_{v=1}^{n-1} (\theta^v)^q = n + 0, \tag{11}$$

since the powers $(\theta^v)^q$ take on all exponents from 0 to $(n - 1)$ when q varies (for $v = 1, 2, \cdots, n - 1$). Finally

$$|S| = \sqrt{n}. \tag{12}$$

Gauss showed also (with more delicacy) that

$$\begin{cases} S = \sqrt{n}, & n \text{ prime} \equiv 1 \pmod 4, \\ S = i\sqrt{n}, & n \text{ prime} \equiv -1 \pmod 4). \end{cases} \tag{13}$$

The transition from (12) to (13) involved an entire world of mathematics equivalent in depth to quadratic reciprocity. The result we need is only slightly stronger.

Letting d be the positive or negative *discriminant* of a quadratic field, we redefine θ and T by using the Kronecker symbol:

$$\begin{cases} \theta = \exp 2\pi i/d, \\ T = \displaystyle\sum_{r=0}^{|d|-1} \theta^r \left(\frac{d}{r}\right). \end{cases} \tag{14}$$

The startlingly simple result is

$$T = \sqrt{d}. \tag{15}$$

Dirichlet (1837) used the result (15) for an elegant computational purpose: in deriving the class number formula in Chapter X, §9, we were confronted with a decomposition into partial fractions. It generally takes the form

$$\frac{f_d(x)}{1 - x^{|d|}} = \sum_{r=1}^{|d|-1} \frac{f_d(\theta^{-r})}{1 - \theta^r x} \frac{1}{|d|}. \tag{16}$$

Here $f_d(\theta^r)$ can be shown as, essentially, a Gaussian sum of type (14). Then, following out the calculations of Chapter X generally, we obtain the class numbers

$$h = \begin{cases} \dfrac{-\sum_{r=1}^{d-1} \left(\dfrac{d}{r}\right) \log \left|\sin \dfrac{\pi r}{d}\right|}{2 \log \eta_1}, & \text{if } d > 0, \\[2ex] -\dfrac{w}{2|d|} \sum_{r=1}^{|d|-1} \left(\dfrac{d}{r}\right) r, & \text{if } d < 0, \end{cases} \tag{17}$$

in the terminology of Chapter X. The reader can convince himself, somewhat, by testing with small d.

As elegant as an "actual class-number formula" may be, Dirichlet's successors have read an even more significant meaning into formula (15). The formula shows an *imbeddability* result for the quadratic field of discriminant d: *the cyclotomic field generated by* θ, *of index* $|d|$, *contains* $R(\sqrt{d})$. *Incidentally, no lower index will suffice.* Now we recall that d is the modulus that determines whether a prime p splits (by the residue class of p modulo $|d|$). This is more than coincidence; it is the basis of a completely independent proof of the imbeddability theorem!

Can such properties be established for other fields (which need not be contained in any cyclotomic field)? The answer is still incomplete.

CLASS FIELDS

We digress to make the historical observation that Dedekind's ideal theory was the *third* major attempt to cope with nonunique factorization. The first was, of course, Gauss's composition of forms (1800), and the second was a relatively neglected explanation of Kummer (1854) that unique factorization can be achieved by "actual" ideal numbers. Looking back at Kummer's work (with the wisdom acquired from Dedekind), we would say that if the class group has order (class number) h then for any ideal $\mathfrak{j}$, $\mathfrak{j}^h = (\alpha)$, a principal ideal. Thus $\mathfrak{j}$ can be replaced by $\sqrt[h]{\alpha}$, the "actual" value of the ideal $\mathfrak{j}$.

To illustrate, take the field $R(\sqrt{-5})$ where $h = 2$. We recall from the genus theory that for a given rational prime p

(*a*) if $(-20/p) = -1$, p does not factor in $R(\sqrt{-5})$;

(*b*) if $(5/p) = (-1/p) = 1$ (or $p = 5$), p factors into two principal ideals and $p = x^2 + 5y^2$;

(*c*) if $(5/p) = (-1/p) = -1$ (or $p = 2$), p factors into two nonprincipal ideals and $p = 2x^2 + 2xy + 3y^2$.

Let us consider the nonprincipal factors of 2. Since $(2, 1 + \sqrt{-5})^2 = (2)$, then by introducing $\sqrt{2}$ into $R(\sqrt{-5})$ we have an ideal number for $(2, 1 + \sqrt{-5})$. The ideal $(2, 1 + \sqrt{-5})$ can be described as precisely those numbers in $R(\sqrt{-5})$ that are divisible by $\sqrt{2}$ in the sense that the quotient is an algebraic integer, as in (1). For example,

$$(18a)\qquad \begin{cases} \dfrac{2}{\sqrt{2}} = \sqrt{2} = \beta, & \beta^2 - 2 = 0, \\[2ex] \dfrac{1 + \sqrt{-5}}{\sqrt{2}} = \gamma, & (\gamma^2 + 2)^2 + 5 = 0. \end{cases}$$

We are now in possession of a field $R(\sqrt{2}, \sqrt{-5})$ generated by *adjoining* $\sqrt{2}$ to $R(\sqrt{-5})$. It could also be regarded as a *relative-quadratic* field or a quadratic field over $R(\sqrt{-5})$. For instance, the algebraic integer γ in (18*a*) could be regarded as a root of the *quadratic* equation with coefficients in $R(\sqrt{-5})$:

$$(18b)\qquad \gamma^2 = (1 + \sqrt{-5})^2/2.$$

Not only $2_1 = (2, 1 + \sqrt{-5})$, but *all* nonprincipal prime ideals $\mathfrak{r}$ become principal in $R(\sqrt{-5}, \sqrt{2})$. This is true because there is only one non-principal class, 2_1. Thus $2_1\mathfrak{r}$ is principal. For example,

$$2_1 3_1 = 2_1(3, 1 + 2\sqrt{-5}) = (1 - \sqrt{-5}),$$

$$2_1 7_1 = 2_1(7, 1 + 2\sqrt{-5}) = (3 - \sqrt{-5}).$$

If $2_1 = (\sqrt{2})$ in $R(\sqrt{-5}, \sqrt{2})$, it is clear that $2_1 \sim 3_1 \sim 7_1 \sim 1$. Yet trouble lies ahead, since $R(\sqrt{-5}, \sqrt{2})$ has some nonprincipal ideals, in fact, its class number is 2.

Now, have we really simplified factorization theory by making non-principal prime ideals become principal? On one hand, 3_1 and 2_1 are happily both *prime* and principal in $R(\sqrt{-5}, \sqrt{2})$, as can be shown, but, on the other hand, although 7_1 becomes principal, it is unhappily capable of further factorization, indeed into two *nonprincipal* ideals. (We recall that since $R(\sqrt{-5}, \sqrt{2})$ has algebraic numbers of degree 4, the rational primes could have four ideal factors.) The easiest way of explaining this fact is to note that, since $(8/7) = +1$, 7 factors, (indeed as $(3 + \sqrt{2})(3 - \sqrt{2})$), in $R(\sqrt{2})$. These are "irreconcilable" with the two factors of 7 in $R(\sqrt{-5})$, since $3 \pm \sqrt{2}$ does not divide all elements of 7_1 (for instance, $1 + 2\sqrt{-5}$) in the sense that the quotient does not satisfy an algebraic equation (1) in rational integral coefficients.

We must note, incidentally, that since $(-2) = (+2)$, as ideals, $R(\sqrt{-2}, \sqrt{-5})$ would illustrate the same thing as $R(\sqrt{2}, \sqrt{-5})$. Dirichlet, some time earlier, had noticed that $R(\sqrt{-5}, \sqrt{-1})$ has class-number unity. This field contains $\sqrt{5} = -\sqrt{-5}\sqrt{-1}$. Thus, to factor a prime in $R(\sqrt{-5}, \sqrt{-1})$, we can begin with any of the three fields $R(\sqrt{-5})$, $R(\sqrt{-1})$, or $R(\sqrt{5})$ and consider the effect of adjoining one other radical to it.

Hilbert later noticed certain further remarkable properties that created the designation that $R(\sqrt{-5}, \sqrt{-1})$ is the *class field* of $R(\sqrt{-5})$.

(a) *The nonprincipal prime ideals in* $R(\sqrt{-5})$ *become principal prime ideals in* $R(\sqrt{-5}, \sqrt{-1})$ (*without splitting*).

For example, let q be a rational prime in the second genus:

$$q = 2a^2 + 2ab + 3b^2 \tag{19}$$

$$\begin{cases} 2q = (2a + b)^2 + 5b^2 \\ 4q = (2a + b + \sqrt{5}b)^2 + (2a + b - \sqrt{5}b)^2. \end{cases} \tag{20}$$

Thus, finally,

$$\begin{aligned} q = &\left\{\frac{2a + b + \sqrt{5}b}{2} + \sqrt{-1}\,\frac{2a + b - \sqrt{5}b}{2}\right\} \\ &\times \left\{\frac{2a + b + \sqrt{5}b}{2} - \sqrt{-1}\,\frac{2a + b - \sqrt{5}b}{2}\right\} \end{aligned} \tag{21}$$

and q has the two indicated principal ideals in $R(\sqrt{-5}, \sqrt{-1})$, although not in $R(\sqrt{-5})$. The reader may wonder if q has further factors, but the answer is negative, since the genus $\{(5/q) = (-1/q) = -1\}$ makes q unfactorable in $R(\sqrt{5})$, hence factorable into at most two factors (as shown) when $\sqrt{-1}$ is adjoined to make $R(\sqrt{5}, \sqrt{-1}) = R(\sqrt{-5}, \sqrt{-1})$.

(*b*) *The principal prime ideals in* $R(\sqrt{-5})$ *split in* $R(\sqrt{-5}, \sqrt{-1})$ *into prime ideals.*

We note that the principal prime ideals in $R(\sqrt{-5})$ are either unfactorable rationals, p, such that $(-20/p) = -1$, or divisors of p in the principal genus for which $(5/p) = (-1/p) = 1$. In either case, $(5/p)$ or $(-1/p)$ is $+1$. Thus the prime p splits in $R(\sqrt{5})$ or $R(\sqrt{-1})$. To reconcile this factorization with the behavior in $R(\sqrt{-5})$, we must assume a further split in $R(\sqrt{5}, \sqrt{-1})$. For example, take $29 = (3 + 2\sqrt{-5})(3 - 2\sqrt{-5}) = (5 + 2\sqrt{-1})(5 - 2\sqrt{-1})$. We could show the four ideals generated by the pair of elements: $((3 \pm 2\sqrt{-5}), (5 \pm 2\sqrt{-1}))$ are the four factors of 29 in $R(\sqrt{-5}, \sqrt{-1})$. In fact, they happen to be *principal.*

$$29 = \left(2 + \sqrt{-1}\left(\frac{1+\sqrt{5}}{2}\right)\right)\left(2 - \sqrt{-1}\left(\frac{1+\sqrt{5}}{2}\right)\right)$$
$$\times\left(2 + \sqrt{-1}\left(\frac{1-\sqrt{5}}{2}\right)\right)\left(2 - \sqrt{-1}\left(\frac{1-\sqrt{5}}{2}\right)\right).$$

Now this formulation is based on $R(\sqrt{-5})$, a quadratic field of class number 2. The definition can be extended to other quadratic fields and to nonquadratic fields. We find a peculiar occurrence in that the value of the theory depends on selecting definitions that make possible interesting theorems!

To give an example of the power of semantics, consider one further interpretation. We start this time with the rational field R and consider all "principal" ideals to be only those (x) for which $(d/x) = 1$ where d is the discriminant of a fixed quadratic field. The ideals for which $(d/x) = -1$ are called "nonprincipal," whereas the ideals for which $(d/x) = 0$ are ignored in the designation. We can say that every ideal (x) has a "principal" square since $(d/x)^2 = (d/x^2) = 1$. Thus the "class number" is 2. The quadratic field $R(\sqrt{d})$ then provides a "class field" for R, for we note the analogous properties of primes p:

(a) If $(d/p) = -1$, (p) does not split in $R(\sqrt{d})$.

(b) If $(d/p) = 1$, (p) splits in $R(\sqrt{d})$.

This interpretation is more than a trick. It was developed in detail by Takagi (1920). We must, however, abandon this line of speculation with the remark that *determining the extent to which these interpretations of class fields can be stretched is the principal unsolved problem* (rather than a lack of proofs to well-defined conjectures).

GLOBAL AND LOCAL VIEWPOINTS

We begin with a conventional problem of deciding whether or not the equation

$$(22) \qquad ax^2 + by^2 + cz^2 = 0$$

in rational integral (nonzero) coefficients a, b, c has a rational integral solution x, y, z not all zero.

As a preliminary matter, we can make changes of variables so that a, b, and c are square-free (by absorbing a square factor in x^2, y^2, or z^2), and we can easily arrange that a, b, c be relatively prime in pairs (by a similar change of variables). Now Legendre's theorem (1785) states that (22), subject to

$$(23) \qquad \begin{cases} a, b, c \text{ square-free}, \\ (a, b) = (b, c) = (c, a) = 1, \end{cases}$$

has a solution if and only if

$$(24a) \qquad a, b, c \text{ are not all of the same sign},$$

$$(24b) \qquad ar_1^2 + b \equiv 0 \pmod{c} \text{ solvable for } r_1,$$

$$(24c) \qquad br_2^2 + c \equiv 0 \pmod{a} \text{ solvable for } r_2,$$

$$(24d) \qquad cr_3^2 + a \equiv 0 \pmod{b} \text{ solvable for } r_3.$$

The conditions are clearly necessary for solvability of (22) indeed, $r_1 \equiv x/y \pmod{c}$, etc. (once the solution triple has been relieved of trivial common factors).

The sufficiency is not easy. It is accomplished by the *method of descent*. We define the *index* of (22) as

$$(25) \qquad I = \max(|a|, |b|, |c|) \times \min(|a|, |b|, |c|).$$

We shall show that if the theorem is true for all indices $I < I_0$ it is true for an equation of index I_0. We see by inspection that the theorem is true when $I = 2$. Listing the equations (and avoiding trivial repetitions), we find

$$(26) \qquad \begin{cases} I = 1, & x^2 + y^2 + z^2 = 0; & \text{unsolvable}, \\ I = 1, & x^2 + y^2 - z^2 = 0; & (x, y, z) = (1, 0, 1), \\ I = 2, & x^2 + y^2 + 2z^2 = 0; & \text{unsolvable}, \\ I = 2, & x^2 + y^2 - 2z^2 = 0; & (x, y, z) = (1, 1, 1), \\ I = 2, & x^2 - y^2 + 2z^2 = 0; & (x, y, z) = (1, 1, 0). \end{cases}$$

Here the conditions (24*b*, *c*, *d*) are trivial, but condition (24*a*) does the job.

The proof is basically a transformation to an equation of lower index. Let I_0 (≥ 3) be the index of (22) and let

$$|a| \leq |b| < |c| \tag{27}$$

(the equality holding only if $|a| = |b| = 1$). Then

$$I_0 = |ac| > |ab|. \tag{28}$$

We solve (24*b*) by reducing r_1 modulo c so that it is $< c/2$ in absolute value. Then the new integer q can be defined:

$$q = (ar_1^2 + b)/c, \tag{29}$$

$$|q| < (|a|\, c^2/4 + |c|)/|c| = |ac|/4 + 1 < |ac| = I_0. \tag{30}$$

Now the transformation

$$\begin{cases} x = bX + r_1 Y, \\ y = r_1 aX - Y, \\ z = qZ \end{cases} \tag{31}$$

changes (22) to

$$abX^2 + Y^2 + qZ^2 = 0. \tag{32}$$

If I' is the index of (32),

$$I' = \max(|ab|, |q|) < I_0. \tag{33}$$

If the coefficients ab and q have a common factor (>1), (32) bears further reduction but I' can be shown to decrease or remain unchanged. This completes the descent.

Now our problem is to prove that if the conditions (24*a*, *b*, *c*, *d*) are valid for (23), they must also be valid for (32) (or whatever its reduced form might become if $(ab, q) > 1$). This is not easy and is indeed rather manipulative.

Condition (24*a*) on the sign of a, b, and c is easy, however, if we rephrase it as

$$a\xi^2 + b\eta^2 + c\zeta^2 = 0 \begin{cases} \text{solvable for real } (\xi, \eta, \zeta) \\ \text{not all zero.} \end{cases} \tag{34}$$

Then it is clear that the linear transformation (31) transfers the real solution from (22) to (32). [The reader will find it easy but more annoying to prove the transfer of (24*a*) directly, i.e., to show from (29) that ab and q, in (32), are not both positive!]

Can we similarly transfer conditions (24 *b*, *c*, *d*) from (22) to (32)? The transfer was originally made by brute force!

A more elegant procedure emerged in the work of Hensel (1884) on p-adic integers. *A p-adic integer* is defined for a given prime p as the infinite formal power series with integral coefficients:

$$\xi = x_0 + px_1 + p^2x_2 + \cdots + p^nx_n + \cdots \tag{35}$$

The series may terminate, representing an ordinary integer, or it may be infinite. For example, in Chapter I, §8, we saw how to construct a series which formally satisfies (for odd p)

$$X^2 \equiv A \pmod{p^n} \text{ if } (A/p) = 1 \tag{36}$$

by taking the terms of (35) up to p^nx_n. The whole series would be a p-adic representation of $\sqrt{A}$.

By the Chinese remainder theorem, condition (24b), for instance, states that for all odd p that divide c

$$ar^2 + b \equiv 0 \pmod{p} \text{ is solvable.} \tag{37}$$

(Here $p = 2$ is trivial.) Now an equivalent form of (37) is

$$a\xi^2 + b = 0, \text{ solvable for } \xi \text{ a } p\text{-adic integer.} \tag{38}$$

Finally it is fully equivalent to write (when $p \mid c$)

$$a\xi^2 + b\eta^2 + c\zeta^2 = 0 \begin{cases} \text{solvable for } \xi, \eta, \zeta \\ p\text{-adic integers not all } 0. \end{cases} \tag{39}$$

Here condition (39) has the advantage that a, b, c need not be square-free or relatively prime in pairs (for the elimination of square-divisors and common factors would leave a p-adic solution alone). Furthermore, a solution in p-adic integers is completely transferable by transformations (31) (using elementary manipulation of power series). The only "catch" is that the primes p that divide abc in (22) are not necessarily those that divide abq in (32). Thus we must have one additional minor proof to show that (22) is readily solvable in p-adic integers when $p \nmid abc$, in other words, for all p.

Now Legendre's theorem states:

The equation (22) *is solvable in rational integers x, y, z if it is solvable in p-adic integers for all odd p and at the same time for real x, y, z.*

We note that the sign condition (24) can be interpreted as the existence of a p-adic solution for $p = \infty$.

We also note that there is an absence of concern for $p = 2$. Actually, this is an alternate (traditional) form of Legendre's theorem, under conditions (23), in which the sign condition (24) is replaced by

$$ax^2 + by^2 + cz^2 \equiv 0 \pmod{8} \begin{cases} \text{solvable for } x, y, z, \\ \text{not all even.} \end{cases} \tag{40a}$$

In the case of system (26) the reader can verify that condition (40*a*) picks out the unsolvable cases as surely as does (24*a*). (We recall that if x is odd, $x^2 \equiv 1 \bmod 8$). More generally, if we dropped conditions (23), we should have to replace (40*a*) by

$$(40b) \quad a\xi^2 + b\eta^2 + c\zeta^2 = 0 \begin{cases} \text{solvable for } p\text{-adic integers } (p = 2 \text{ incl.}), \\ \xi, \eta, \zeta \text{ not all zero.} \end{cases}$$

In modern parlance, the p-adic conditions for $p = 2$, p odd, and $p = \infty$ are "local" solutions to (22) in the "neighborhood" of a prime p. The solution in rational integers is a "global" solution. The global solution is (easily) a local solution for each prime, but these solutions also lead to a global solution (nontrivially) by Legendre's method of descent.

At this juncture number theory was strongly influenced by Riemann's theory of complex functions (1852) [through a famous sequel of Dedekind and Weber (1880)]. According to Riemann's approach, a complex function is completely determined by knowing the power series around the *singularities*. For instance, consider the equation in which the unknown x is a *polynomial* (not an integer),

$$(41) \qquad x^2 = (t - \alpha_1)^{m_1}(t - \alpha_2)^{m_2} \cdots (t - \alpha_k)^{m_k}, \qquad (m_1 \geq 0).$$

Here the α_i are (different) real or complex numbers and m_i are integers. We ask if a polynomial $x = \phi(t)$ exists to satisfy (41). The answer is trivial, since (41) is solvable if and only if all m_i are *even* integers. The "evenness of m_i" can be interpreted as saying that in the neighborhood of any α_i a Taylor series exists for x such that

$$(42a) \qquad x = \beta_0 + \beta_1(t - \alpha_i) + \beta_2(t - \alpha_i)^2 + \cdots.$$

If m_i is odd, we should have, on the contrary,

$$(42b) \qquad x = (t - \alpha_i)^{1/2}[\beta_0 + \beta_1(t - \alpha_i) + \beta_2(t - \alpha_i)^2 + \cdots].$$

The $t = \alpha_i$ are similar to the p which divide abc (and $p = 2$) of (22). In the neighborhood of the other values of t it is easy to show that a power series always exists as for the odd $p \nmid abc$.

The "$p = \infty$" case is taken care of by the "order of magnitude":

$$(42c) \qquad x \sim t^{(m_1+m_2+\cdots m_k)/2} \qquad \text{as } t \to \infty.$$

Thus we say x has a Taylor series at $t = \infty$ when Σm_i is even. It is now clear that if we know about $t = \infty$ we can afford to be "ignorant" about, say, $t = \alpha_1$ (in the knowledge that m_1 is even if the same is true about each m_i $(i > 1)$, and Σm_i). This is like ignoring $p = 2$ when we know about $p = \infty$ and odd primes.

At any rate, the "local" solutions (42*a*), (42*c*) determine a so-called "global" solution: $x = (t - \alpha_1)^{m_1/2} \cdots (t - \alpha_k)^{m_k/2}$, which in effect looks very different! The reader should ponder the analogy carefully.

Returning to number theory, the connection involving $p = 2$, p odd, and $p = \infty$ happens to be very much like the Hilbert condition

$$\prod \left(\frac{a, b}{p}\right) = 1. \tag{43}$$

The connection with the genus theory and quadratic reciprocity, however, is too involved to pursue at this juncture.

There are other equations like (22) for which local solvability determine global solvability, but, unfortunately, they are both few in number and also esthetically unattractive. Generally, one might well wonder if ideal theory creates a sufficiently large number of primes to give "enough" local solvability conditions for global solvability.

Just as the properties of the solution to a diophantine equation can be discussed locally, so can the properties (*a*) and (*b*) of the class field be discussed locally by working "modulo p" with p-adic series. The harder part consists in showing when the local properties can be "built into" a global solution or class field, as the case may be. This new development involved the complete rewriting of the foundations of algebraic number theory mostly by Artin, Chevalley, and Hasse.

Thus the axiomatic formulation of number theory is still fluid after some 2500 years.

Bibliography and comments

Some Classics Prior to 1900

Making no effort at completeness and limiting ourselves to "books," we start with

Euclid, *Elements* (books VII, VIII, and IX), circa 300 B.C. (edition of T. L. Heath, Cambridge 1908).

Euclid, typically, did not use formal arithmetic, his "numbers" being geometric lengths. Archimedes, circa 260 B.C. (edition of Heath 1897), and Diophantus, circa 275 A.D. (edition of Heath 1910), tended more towards numerical work.

Number theory revived in the seventeenth century, along with infinitesimal analysis, but the earliest major works were Euler's *Algebra* (1770) and Legendre's *Essai de la théorie des nombres* (1798). Soon thereafter came

C. F. Gauss, *Disquisitiones Arithmeticus*, 1801 (German edition by H. Maser, Springer, Berlin, 1889).

Now the algebraic structures were most plainly visible. Stronger analytic methods and more perfected algebraic techniques appeared in the famous "progress report" which gave a glimpse of Kummer's Theory:

H. J. S. Smith, *Report on the Theory of Numbers*, (British Association, 1865).

Many books, including two noteworthy tabulations of power residues, began to appear:

C. G. J. Jacobi, *Canon Arithmeticus* (1839),

and of factorizations of algebraic numbers:

C. G. Reuschle, *Tafeln komplexer Primzahlen* (1875).

Ideal theory (of Dedekind) appeared as the famous "Supplement XI" (second edition, 1871) appended to

P. G. L. Dirichlet and R. Dedekind, *Vorlesungen über Zahlentheorie* (fourth edition, Braunschweig, 1894),

and was followed by works consciously integrating number theory and algebra on the elementary and higher levels:

J. A. Serret, *Cours d'algébre supérieure* (first edition 1875, seventh edition, Paris 1928, vol. II).

H. Weber, *Lehrbuch der Algebra* (Braunschweig, 1896, vol. II).

At the same time, one of the major "separatist" tendencies became apparent in

H. Minkowski, *Geometrie der Zahlen* (Leipzig, 1896).

The century initiated by Gauss was appropriately terminated by the great work which made algebraic number theory an end in itself rather than a tool for handling the rationals:

D. Hilbert, *Die Theorie der algebraischen Zahlkörper*, (1894/1895 Jahresbuch der deutschen Mathematiker Vereinigung, issued 1897).

Some Recent Books (after 1900)

No attempt is made at critical comments because of the complexity of such a task.

A. A. Albert, *Modern Higher Algebra*, University of Chicago Press, Chicago, 1937.

E. Artin, *Theory of Algebraic Numbers* (lecture notes by G. Würges), Gottingen, 1956.

P. Bachmann (a), *Das Fermatproblem*, Walter de Gruyter, Berlin, 1919.

P. Bachmann (b), *Grundlehren der neueren Zahlentheorie*, Walter de Gruyter, Berlin-Leipzig, 1921.

G. Birkhoff and S. MacLane, *Survey of Modern Algebra*, Macmillan, New York, 1953.

J. W. S. Cassels, *An Introduction to the Geometry of Numbers*, Springer, Berlin, 1959.

C. Chevalley, *Class Field Theory*, Nagoya University, Nagoya, 1953–1954.

H. Davenport, *The Higher Arithmetic*, Hutchinson House, London, 1952.

L. E. Dickson (a), *History of the Theory of Numbers* III, Carnegie Institution, Washington, 1923.

L. E. Dickson (b), *Introduction to the Theory of Numbers*, University of Chicago Press, Chicago, 1929.

M. Eichler, *Quadratische Formen und Orthogonale Gruppen*, Springer, Berlin, 1952.

R. Fricke, *Lehrbuch der Algebra* III, Vieweg, Braunschweig, 1928.

R. Füeter, *Synthetische Zahlentheorie*, Walter de Gruyter, Berlin, 1950.

G. H. Hardy and E. M. Wright, *An Introduction to the Theory of Numbers*, Clarendon, Oxford, 1938.

H. Hasse (a), *Zahlentheorie*, Akademie-Verlag, Berlin, 1949.

H. Hasse (b), *Vorlesungen über Zahlentheorie*, Springer, Berlin, 1950.

M. A. Heaslet (see J. V. Uspensky).

E. Hecke, *Vorlesungen über die Theorie der algebraischen Zahlen*, Akademische Verlag, Leipzig, 1923.

L. Holzer, *Zahlentheorie*, Teubner, Leipzig, 1958.

B. W. Jones (a), *The Arithmetic Theory of Quadratic Forms*, Wiley, New York, 1950.

B. W. Jones (b), *The Theory of Numbers*, Rinehart, New York, 1955.

E. Landau (a), *Handbuch der Lehre von der Verteilung der Primzahlen*, II, Teubner, Leipzig, 1909.

E. Landau (b), *Vorlesungen über Zahlentheorie* (1927), reprinted, Chelsea, New York, 1946–1947.

W. J. LeVeque, *Topics in Number Theory* II, Addison-Wesley, New York, 1956.
S. MacLane (see Birkhoff).
L. J. Mordell, *Three Lectures on Fermat's Last Theorem*, University Press, Cambridge, 1921.
I. Niven and H. S. Zuckerman, *Introduction to the Theory of Numbers*, Wiley, New York, 1960.
O. Ore (a) *Les Corps algébriques et la théorie des ideaux*, Gauthier Villars, Paris, 1934.
O. Ore (b), *Number Theory and its History*, McGraw-Hill, New York, 1949.
G. Pall, *Integral Quadratic Forms* (in preparation).
O. F. G. Schilling, *The Theory of Valuations*, American Mathematical Society, New York, 1950.
C. L. Siegel (a), *Analytic Number Theory* (lecture notes by B. Friedman), New York University, 1945.
C. L. Siegel (b), *Geometry of Numbers* (lecture notes by B. Friedman), New York University, 1946.
J. Sommer, *Introduction a la théorie des nombres algébriques*, translation from German (1907), Hermann, Paris, 1911.
B. M. Stewart, *Theory of Numbers*, Macmillan, New York, 1952.
J. V. Uspensky and M. A. Heaslet, *Elementary Number Theory*, McGraw-Hill, New York, 1939.
H. S. Vandiver and G. E. Wahlin, "Algebraic Numbers II," *Bull. nat. Research Coun.*, Wash., 1928.
J. M. Vinogradov, *Elements of Number Theory* (translation from Russian, 1952), Dover, New York, 1954.
B. L. van der Waerden, *Modern Algebra*, vol. I, II (translation from German, 1931), Ungar, New York, 1949–1950.
G. E. Wahlin (see H. S. Vandiver).
G. L. Watson, *Integral Quadratic Forms*, Cambridge University Press, Cambridge, 1960.
H. Weyl, *Algebraic Theory of Numbers*, Princeton University Press, Princeton, 1940.
E. M. Wright (see G. H. Hardy).
H. S. Zuckerman (see I. Niven).

Special References by Chapter

Introductory Survey: Ore (b); Dickson (a); Mordell; Bachmann (a).
Chapter I: Niven and Zuckerman; Hasse (b); Dickson (b); Birkhoff and MacLane.
Chapter II: Hasse (b); Dirichlet and Dedekind.
Chapter III: Sommer; Hasse (b).
Chapter IV: Cassels; Minkowski; Ore (a).
Chapter V: §2, Van der Waerden, II, Chapter XV; §3, Birkhoff-MacLane; §5, Hecke; §6, Cassels; §7, Dickson (b); §8, Siegel (b), *Lecture* XIII.
Chapter VI: Sommer; §§7 and 8, Hardy and Wright (pp. 211–217); E. S. Barnes and H. P. F. Swinnerton-Dyer (The Inhomogeneous Minima of Binary Quadratic Forms I), *Acta. Math. Stockh.*, **87,** 1952, pp. 259–323); §9, Dickson (a); §10, Fricke, Hilbert.
Chapter VII: Dirichlet and Dedekind; Weyl; Hilbert; Hecke; Sommer; §11, E. C. Dade, O. Taussky, and H. Zassenhaus (On the Semigroup of Ideal Classes in an Order of an Algebraic Number Field), *Bull. Amer. Math. Soc.*, **67,** (1961, pp. 305–308).
Chapter VIII: Minkowski; Hasse (a), p. 437, ff.

Chapter IX: Sommer; Hasse (b); §6, D. H. Lehmer (On Imaginary Quadratic Fields whose Class Number Is Unity), *Bull. Amer. math. Soc.*, **39,** (1933, p. 360); §8, G. Rabinowitsch (Eindeutigkeit der Zerlegung in Primzahlfaktoren in quadratischen Zahlkörpern), *J. reine angew. Math.*, **142,** (1913, pp. 153–164).

Chapter X: Dirichlet and Dedekind; Hasse (b); §10, Siegel (a); §11, Landau (a), p. 451, §12, A. Selberg (An Elementary Proof of the Prime Number Theorem for Arithmetic Progressions), *Canad. J. Math.*, **2** (1950, pp. 66–78); H. Weber (Beweis des Satzes dass jede eigentliche primitive quadratische Form unendlich viele Primzahlen darstellen fähig ist), *Math. Ann.*, **20** (1882, pp. 301–329).

Chapter XI: Sommer; Hasse (b); Bachmann (b); Uspensky and Heaslet.

Chapter XII: Hecke; Fricke; Davenport; Pall.

Chapter XIII: §1, Hecke; §2, Fueter; Fricke, p. 255; Dickson (a), p. 104; P. G. L. Dirichlet (Une Propriété des formes quadratiques à déterminant positive), *J. Math. pures appl.*, II, **1,** (1856, pp. 76–79); §3, Sommer; G. Pall (Note on Irregular Determinants), *J. Lond. math. Soc.*, **11** (1936, pp. 34–35); §4, Hilbert; Hasse (a), p. 65; Jones (a).

Concluding Survey: Hilbert; Artin; Hasse (a); Weyl; Chevalley.

appendix

Table I Minimum prime divisors of numbers not divisible by 2, 3, or 5 from 1 to 18,000

The material in this table has been extracted from *Table de diviseurs pour tous les nombres des* 1*e*, 2*e*, *et* 3*e* million, *ou plus exactement*, *depuis la* 3,036,000, *avec les nombres premiers qui s'y trouvent*, by J. C. Burckhardt, Paris, 1817.

The reader might accustom himself to usage of the table by these observations: in Table I-A the extreme lower right-hand entry (*) states that 19 is the minimum prime divisor of 8797; and in Table I-B the extreme upper left-hand entry (†) states that 9001 is prime.

TABLE I

A: 1—8999

	00	03	06	09	12	15	18	21	24	27	30	33	36	39	42	45	48	51	54	57	60	63	66	69	72	75	78	81	84	87
01	—	7	—	17	—	19	—	11	7	37	—	—	13	47	—	7	—	—	11	—	17	—	7	67	19	13	29	—	31	7
07	—	—	—		17	11	13	7	29	—	31	—	—	—	7	—	11	—	—	13	—	7	—	—	—	—	37	11	7	—
11	—	—	13	—	7	—	—	—	—	—	—	7	23	—	—	13	17	19	7	—	—	—	11	—	—	7	73	—	13	31
13	—	—	—	11	—	17	7	—	19	—	23	—	—	7	11	—	—	—	—	29	7	59	17	31	—	11	13	7	47	—
17	—	—	—	7	—	37	23	29	—	11	7	31	—	—	—	—	—	7	—	—	11	—	13	—	7	—	—	—	19	23
19	—	11	—	—	23	7	17	13	41	—	—	—	7	—	—	—	61	—	—	7	13	71	—	11	—	73	7	23	—	—
23	—	17	7	13	—	—	—	11	—	7	—	—	—	—	41	—	7	47	11	59	19	—	37	7	31	—	—	—	—	11
29	—	7	17	—	—	11	31	—	7	—	13	—	19	—	—	7	11	23	61	17	—	—	7	13	—	—	—	11	—	7
31	—	—	—	7	—	—	—	—	11	—	7	—	—	—	—	23	—	7	—	11	37	13	19	29	7	17	41	47	—	—
37	—	—	7	—	—	29	11	—	—	7	—	47	—	31	19	13	7	11	—	—	—	—	—	7	—	—	17	79	11	—
41	—	11	—	—	17	23	7	—	—	—	—	13	11	7	—	19	47	53	—	—	7	17	29	11	13	—	—	7	23	—
43	—	7	—	23	11	—	19	—	7	13	17	—	—	—	—	7	29	37	—	—	—	—	7	53	—	19	11	17	—	7
47	—	—	—	—	29	7	—	19	—	41	11	—	7	—	31	—	37	—	13	7	—	11	17	—	—	—	7	—	—	—
49	7	—	11	13	—	—	43	7	31	—	—	17	41	11	7	—	13	19	—	—	23	7	61	—	11	—	47	29	7	13
53	—	—	—	—	7	—	17	—	11	—	43	7	13	59	—	29	23	—	7	11	—	—	—	17	—	7	—	31	79	—
59	—	—	—	7	—	—	11	17	—	31	7	—	—	37	—	47	43	7	53	13	73	—	—	—	7	—	29	41	11	19
61	—	19	—	31	13	7	—	—	23	11	—	—	7	17	—	—	—	13	43	7	11	—	—	—	53	—	7	—	—	—
67	—	—	23	—	7	—	—	11	—	—	—	7	19	—	17	—	31	—	7	73	—	—	59	—	13	7	—	—	—	11
71	—	7	11	—	31	—	—	13	7	17	37	—	—	11	—	7	—	—	—	29	13	23	7	—	11	67	17	—	43	7
73	—	—	—	7	19	11	—	41	—	47	7	—	—	29	—	17	11	7	13	23	—	—	—	19	7	—	—	11	37	31
77	7	13	—	—	—	19	—	7	—	—	17	11	—	41	7	23	—	31	—	53	59	7	11	—	19	—	—	13	7	67
79	—	—	7	11	—	—	—	—	37	7	—	31	13	23	11	19	7	—	—	—	—	—	—	7	29	11	—	—	61	—
83	—	—	—	—	—	—	7	37	13	11	—	17	29	7	—	—	19	71	—	—	7	13	41	—	—	—	—	7	17	—
89	—	—	13	23	—	7	—	11	19	—	—	—	7	—	—	13	—	—	11	7	—	—	—	29	37	—	7	19	13	11
91	7	17	—	—	—	37	31	7	47	—	11	—	—	13	7	—	67	29	17	—	—	7	—	—	23	—	13	—	7	59
97	—	—	17	—	—	—	7	13	11	—	19	43	—	7	—	—	59	—	23	11	7	—	37	—	—	71	53	7	29	* 19

TABLE I (continued)

B: 9001—17999

	90	93	96	99	1 02	05	08	11	14	17	20	23	26	29	32	35	38	41	44	47	50	53	56	59	62	65	68	71	74	77
01	† —	71	—	—	101	—	7	17	13	—	11	—	—	7	43	23	37	59	—	61	7	11	—	—	17	29	53	7	—	31
07	—	41	13	—	59	7	101	29	11	23	—	31	7	—	47	13	—	—	—	7	43	—	—	—	19	17	7	—	13	—
11	—	—	7	11	—	23	19	41	—	7	—	13	—	—	11	59	7	103	—	47	17	61	67	7	13	11	—	71	23	89
13	—	67	—	23	7	—	11	—	101	13	41	7	—	37	73	—	19	11	7	—	—	—	13	—	31	7	17	109	11	—
17	71	7	59	47	17	13	29	—	7	—	61	109	11	—	—	7	41	19	13	—	—	17	7	11	—	83	67	—	—	7
19	29	—	—	7	11	67	31	—	19	—	7	97	—	—	—	11	13	7	—	41	23	—	—	—	7	—	11	17	—	13
23	7	—	—	—	—	17	79	7	—	19	11	—	13	—	7	—	23	29	—	—	83	7	17	—	—	13	—	—	7	37
29	—	19	—	—	53	—	7	31	11	37	23	—	73	7	—	83	—	71	47	11	7	—	—	17	—	—	—	7	29	—
31	11	7	—	—	13	—	—	—	7	—	53	11	17	67	101	7	—	13	—	—	—	—	7	89	—	61	—	37	—	7
37	7	—	23	19	29	41	—	7	—	11	—	13	—	17	7	—	101	67	—	—	11	7	19	—	13	23	113	—	7	—
41	—	—	31	—	7	83	37	13	17	59	—	7	—	—	—	11	—	79	7	—	13	23	—	19	109	7	11	61	107	113
43	—	—	—	61	—	13	7	11	—	—	—	—	47	7	17	29	109	—	11	23	7	67	—	107	37	71	—	7	—	11
47	83	13	11	7	—	53	—	71	—	17	7	—	—	11	13	19	61	7	—	—	41	103	—	37	7	—	17	13	73	—
49	—	—	—	—	37	7	19	—	107	31	—	53	7	23	—	17	11	—	—	7	101	—	—	41	—	13	7	11	—	—
53	11	47	7	37	—	61	—	19	13	7	17	11	—	—	29	—	7	—	97	—	—	13	11	7	—	—	19	17	31	41
59	—	7	13	23	—	—	—	—	7	11	31	17	—	—	—	7	—	—	19		11	—	7	—	71	29	23	—	13	7
61	13	11	—	7	31	59	—	—	73	19	7	47	11	13	89	71	83	7	—	29	—	—	—	11	7	—	13	131	19	—
67	—	17	7	—	—	—	—	13	—	7	11	83	53	—	—	—	7	31	17	—	13	11	—	7	—	—	101	—	—	109
71	47	—	19	13	—	11	7	—	—	79	—	89	—	7	23	41	11	37	29	—	7	19	—	—	53	73	—	7	—	13
73	43	7	17	—	—	97	83	—	7	61	—	—	19	—	13	7	—	—	41	11	—	—	7	—	—	—	47	13	101	7
77	29	—	—	11	43	7	73	—	23	—	13	—	7	19	11	—	—	—	31	7	—	—	61	13	41	11	7	89	—	29
79	7	83	—	17	19	71	11	7	13	—	47	—	31	—	7	37	—	11	—	—	17	7	—	19	73	59	—	41	7	23
83	31	11	23	67	7	19	—	53	—	—	43	7	11	—	37	17	—	13	7	—	—	—	—	11	19	7	—	—	—	—
89	61	41	—	7	—	—	—	67	—	—	7	13	—	31	97	107	17	7	—	23	79	11	29	59	7	53	—	—	—	—
91	—	—	11	97	41	7	—	19	—	13	107	—	7	11	—	—	29	23	43	7	—	—	13	—	11	47	7	—	—	—
97	11	—	—	13	7	—	17	—	—	47	—	7	—	41	—	—	13	—	7	—	31	89	11	17	43	7	61	29	—	13

TABLE I (continued)

C: 1—8999

	01	04	07	10	13	16	19	22	25	28	31	34	37	40	43	46	49	52	55	58	61	64	67	70	73	76	79	82	85	88
01	—	—	—	7	—	—	—	31	41	—	7	19	—	—	11	43	13	7	—	—	—	37	—	—	7	11	—	59	—	13
03	—	13	19	17	—	7	11	—	—	—	29	41	7	—	13	—	—	11	—	7	17	19	—	47	67	—	7	13	11	—
07	—	11	7	19	—	—	—	—	23	7	13	—	11	—	59	17	7	41	—	—	31	43	19	7	—	—	—	29	47	—
09	—	—	—	—	7	—	23	47	13	53	—	7	—	19	31	11	—	—	7	37	41	13	—	43	—	7	11	—	67	23
13	—	7	23	—	13	—	—	—	7	29	11	—	47	—	19	7	17	13	37	—	—	11	7	—	71	23	41	43	—	7
19	7	—	—	—	—	—	19	7	11	—	—	13	—	—	7	31	—	17	—	11	29	7	—	—	13	19	—	—	7	—
21	11	—	7	—	—	—	17	—	—	7	—	11	61	—	29	—	7	23	—	—	—	—	11	7	—	—	89	—	—	—
27	—	7	—	13	—	—	41	17	7	11	53	23	—	—	—	7	13	—	—	—	11	—	7	—	17	29	—	19	—	7
31	—	—	17	—	11	7	—	23	—	19	31	47	7	29	61	11	—	—	—	7	—	59	53	79	—	13	7	—	19	—
33	7	—	—	—	31	23	—	7	17	—	13	—	—	37	7	41	—	—	11	19	—	7	—	13	—	17	—	—	7	11
37	—	19	11	17	7	—	13	—	43	—	—	7	37	11	—	—	—	—	7	13	17	41	—	31	11	7	—	—	—	—
39	—	—	—	—	13	11	7	—	—	17	43	19	—	7	—	—	11	13	29	—	7	47	23	—	41	—	17	7	—	—
43	11	—	—	7	17	31	29	—	—	—	7	11	19	13	43	—	—	7	23	—	—	17	11	—	7	—	13	—	—	37
49	—	—	7	—	19	17	—	13	—	7	47	—	23	—	—	—	7	29	31	—	11	—	17	7	—	—	—	73	83	—
51	—	11	—	—	7	13	—	—	—	—	23	7	11	—	19	—	—	59	7	—	—	—	43	11	—	7	—	37	17	53
57	—	—	—	7	23	—	19	37	—	—	7	—	13	—	—	—	—	7	—	—	47	11	29	—	7	13	73	23	43	17
61	7	—	—	—	—	11	37	7	13	—	29	—	—	31	7	59	11	—	67	—	61	7	—	23	17	47	19	11	7	—
63	—	—	7	—	29	—	13	31	11	7	—	—	53	17	—	—	7	19	—	11	—	23	—	7	37	79	—	—	—	—
67	—	—	13	11	—	—	7	—	17	47	—	—	—	7	11	13	—	23	19	—	7	29	67	37	53	11	31	7	13	—
69	13	7	—	—	37	—	11	—	7	19	—	—	—	13	17	7	—	11	—	—	31	—	7	—	—	—	13	—	11	7
73	—	11	—	29	—	7	—	—	31	13	19	23	7	—	—	—	—	—	—	7	—	—	13	11	73	—	7	—	—	19
79	—	—	19	13	7	23	—	43	—	—	11	7	—	—	29	—	13	—	7	—	37	11	—	—	47	7	79	17	23	13
81	—	13	11	23	—	41	7	—	29	43	—	59	19	7	13	31	17	—	—	—	7	—	—	73	11	—	23	7	—	83
87	11	—	—	—	19	7	—	—	13	—	—	11	7	61	41	43	—	17	37	7	23	13	11	19	83	—	7	—	31	—
91	—	—	7	—	13	19	11	29	—	7	—	—	17	—	—	—	7	11	—	43	41	—	—	7	19	—	61	—	11	17
93	—	17	13	—	7	—	—	—	—	11	31	7	—	—	23	13	—	67	7	71	11	43	—	41	—	7	—	—	13	—
97	—	7	—	—	11	—	—	—	7	—	23	13	—	17	—	7	19	—	29	—	—	73	7	47	13	43	11	—	—	7
99	—	—	17	7	—	—	—	11	23	13	7	—	29	—	53	37	—	7	11	17	—	67	13	31	7	—	19	43	—	11

TABLE I (continued)

D: 9001—17999

	91	94	97	1 00	03	06	09	12	15	18	21	24	27	30	33	36	39	42	45	48	51	54	57	60	63	66	69	72	75	78
01	19	7	89	73	—	—	11	23	7	—	—	—	13	—	47	7	—	11	17	19	—	—	7	—	—	13	—	103	11	7
03	—	—	31	7	—	23	—	17	—	11	7	79	—	—	53	61	—	7	—	113	11	73	41	13	7	—	—	—	23	19
07	7	23	17	—	11	—	13	7	37	—	—	19	97	—	7	11	—	—	89	13	—	7	113	—	23	—	11	—	7	—
09	—	97	7	—	13	103	—	11	17	7	—	—	71	—	—	31	7	13	11	59	29	19	23	7	47	17	37	—	—	11
13	13	—	11	17	—	—	7	—	29	—	—	—	—	7	—	—	—	61	23	—	7	—	19	67	11	37	13	7	83	47
19	11	—	—	43	17	7	61	13	—	53	—	11	7	47	19	—	31	59	—	7	13	17	11	83	—	—	7	67	—	103
21	7	—	—	11	—	13	67	7	41	—	17	—	—	29	7	53	—	—	13	—	—	7	79	37	19	11	—	17	7	71
27	—	11	71	37	23	—	7	103	—	—	67	17	11	7	—	—	19	41	73	—	7	—	—	11	29	13	—	7	17	—
31	23	—	37	7	—	—	17	11	13	—	7	31	29	83	—	43	—	7	11	—	—	13	—	17	7	—	—	—	47	11
33	—	—	—	79	—	7	13	47	19	—	11	—	7	—	67	—	—	43	—	7	37	11	—	—	—	—	7	19	89	17
37	—	—	7	—	—	11	—	17	83	7	53	—	47	—	—	13	7	23	—	37	—	43	—	7	17	127	—	11	13	—
39	13	—	—	—	7	—	—	—	11	—	61	7	—	13	—	23	53	29	7	11	—	—	—	43	—	7	13	—	—	—
43	41	7	—	11	—	29	31	—	7	13	—	23	—	—	11	7	73	—	—	—	19	—	7	61	59	11	—	43	53	7
49	7	11	—	13	79	23	—	7	—	17	—	59	11	—	7	—	13	—	—	31	—	7	—	11	—	—	17	47	7	13
51	—	13	7	19	11	—	47	—	—	7	29	—	41	31	13	11	7	—	—	—	109	—	19	7	83	—	11	13	—	—
57	—	7	11	89	—	—	—	—	7	71	—	—	—	11	19	7	17	53	—	83	23	13	7	—	11	—	31	—	97	7
61	—	—	43	—	13	7	97	—	11	29	—	17	7	37	31	19	23	13	—	7	—	—	—	—	—	—	7	41	17	53
63	7	—	13	29	43	—	19	7	31	—	—	11	—	—	7	13	—	17	—	89	59	7	11	—	—	19	—	61	7	—
67	89	—	—	—	7	—	11	19	43	—	23	7	17	73	—	79	—	11	7	—	29	—	—	—	13	7	19	31	11	17
69	53	17	—	—	—	47	7	59	23	11	43	37	113	7	29	—	61	19	17	—	7	31	13	—	—	79	71	7	—	107
73	—	—	29	7	11	13	—	—	71	31	7	—	53	17	43	11	89	7	13	107	—	—	—	—	7	—	11	23	—	61
79	67	—	7	—	97	59	—	—	—	7	19	—	13	11	17	—	7	109	61	—	43	23	31	7	11	13	—	37	—	19
81	—	19	—	17	7	11	79	29	37	109	13	7	—	103	—	—	11	—	7	23	17	113	43	13	—	7	—	11	—	—
87	—	53	—	7	13	—	—	—	—	—	7	—	19	23	11	—	71	7	29	—	—	17	—	—	7	11	—	59	43	31
91	7	—	—	—	—	—	29	7	67	11	73	—	—	13	7	—	17	31	—	—	11	7	—	—	37	—	13	—	7	—
93	29	11	7	—	19	17	—	23	—	7	89	13	11	—	59	—	7	—	—	53	—	—	17	7	13	—	—	—	73	29
97	17	—	97	23	37	19	7	11	—	—	—	—	67	7	—	—	—	17	11	—	7	—	—	—	19	59	23	7	—	11
99	—	7	41	—	—	13	17	—	7	73	11	29	—	—	—	7	—	79	13	47	—	11	7	17	23	—	89	—	—	7

TABLE I (continued)

E: 1—8999

	02	05	08	11	14	17	20	23	26	29	32	35	38	41	44	47	50	53	56	59	62	65	68	71	74	77	80	83	86	89
03	7	—	11	—	23	13	—	7	19	—	—	31	—	11	7	—	—	—	13	—	—	7	—	—	11	—	53	19	7	29
09	11	—	—	—	—	—	7	—	—	—	—	11	13	7	—	17	—	—	71	19	7	23	11	—	31	13	—	7	—	59
11	—	7	—	11	17	29	—	—	7	41	13	—	37	—	11	7	—	47	31	23	—	17	7	13	—	11	—	—	79	7
17	7	11	19	—	13	17	—	7	—	—	—	—	11	23	7	53	29	13	41	61	—	7	17	11	—	—	—	—	7	37
21	13	—	—	19	7	—	43	11	—	23	—	7	—	13	—	—	—	17	7	31	—	—	19	—	41	7	13	53	37	11
23	—	—	—	—	—	—	7	23	43	37	11	13	—	7	—	—	—	—	—	—	7	11	—	17	13	—	71	7	—	—
27	—	17	—	7	—	11	—	13	37	—	7	—	43	—	19	29	11	7	17	—	13	61	—	—	7	—	23	11	—	79
29	—	23	—	—	—	7	—	17	11	29	—	—	7	—	43	—	47	73	13	7	—	—	—	—	17	59	7	—	—	—
33	—	13	7	11	—	—	19	—	—	7	53	—	—	—	11	—	7	—	43	17	23	47	—	7	—	11	29	13	89	—
39	—	7	—	17	—	37	—	—	7	—	41	—	11	—	23	7	—	19	—	—	17	13	7	11	43	71	—	31	53	7
41	—	—	29	7	11	—	13	—	19	17	7	—	23	41	—	11	71	7	—	13	79	31	—	37	7	—	11	19	—	—
47	13	—	7	31		—	23	—	—	7	17	—	—	11	—	47	7	—	—	19	—	—	41	7	11	61	13	17	—	23
51	—	19	23	—	—	17	7	—	11	13	—	53	—	7	—	—	—	—	—	11	7	—	13	—		23	83	7	41	—
53	11	7	—	—	—	—	—	13	7	—	—	11	—	—	61	7	31	53	—	—	13	—	7	23	29	—	—	—	17	7
57	—	—	—	13	31	7	11	—	—	—	—	—	7	—	—	67	13	11	—	7	—	79	—	17	—	—	7	61	11	13
59	7	13	—	19	—	—	29	7	—	11	—	—	17	—	7	—	—	23	—	59	11	7	19	—	—	—	—	13	7	17
63	—	—	—	—	7	41	—	17	—	—	13	7	—	23	—	11	61	31	7	67	—	—	—	13	17	7	11	—	—	—
69	—	—	11	7	13	29	—	23	17	—	7	43	53	11	41	19	37	7	—	47	—	—	—	67	7	17	—	—	—	—
71	—	—	13	—	—	7	19	—	—	—	—	—	7	43	17	13	11	41	53	7	—	—	—	71	31	19	7	11	13	—
77	—	—	—	11	7	—	31	—	—	13	29	7	—	—	11	17	—	19	7	43	—	—	13	—	—	7	41	—	—	47
81	—	7	—	—	—	13	—	—	7	11	17	—	—	37	—	7	—	—	13	—	11	—	7	43	—	31	—	17	—	7
83	—	11	—	7	—	—	—	—	—	19	7	—	11	47	—	—	13	7	—	31	61	29	—	11	7	43	59	83	19	13
87	7	—	—	—	—	—	—	7	—	29	19	17	13	53	7	—	—	—	11	—	—	7	71	—	—	13	—	—	7	11
89	17	19	7	29	—	—	—	—	—	7	11	37	—	59	67	—	7	17	—	53	19	11	83	7	—	—	—	—	—	89
93	—	—	19	—	—	11	7	—	—	41	37	—	17	7	—	—	11	—	—	13	7	19	61	—	59	—	—	7	—	17
99	13	—	29	11	—	7	—	—	—	—	—	59	7	13	11	—	—	—	41	7	—	—	—	23	—	11	7	37	—	—

TABLE I (continued)

F: 9001—17999

	92	95	98	1 01	04	07	10	13	16	19	22	25	28	31	34	37	40	43	46	49	52	55	58	61	64	67	70	73	76	79
03	—	13	—	—	101	7	—	89	41	—	—	—	7	—	13	71	11	—	17	7	23	37	—	—	47	—	7	11	29	—
09	—	37	17	11	7	—	101	43	13	—	29	7	—	—	11	—	—	41	7	17	67	13	—	89	61	7	73	19	—	—
11	61	—	—	—	29	—	7	—	17	43	—	—	23	7	—	—	—	11	19	13	7	—	97	—	—	17	—	7	11	—
17	13	31	—	67	11	7	23	—	—	17	19	—	7	13	—	11	107	103	47	7	—	59	—	71	—	73	7	—	79	19
21	—	—	7	29	17	71	103	—	—	7	11	19	—	—	—	—	7	—	—	43	31	11	13	7	—	23	—	—	67	—
23	23	89	11	53	7	—	73	13	59	—	17	7	—	11	31	—	37	—	7	—	13	19	—	23	11	7	29	17	—	—
27	—	7	31	13	—	17	—	47	7	—	—	—	101	—	29	7	13	—	—	11	—	—	7	—	—	43	—	—	—	7
29	11	13	—	7	—	—	41	—	29	79	7	11	—	19	13	—	—	7	—	—	97	53	11	127	7	—	—	13	17	—
33	7	—	—	—	—	—	11	7	—	—	13	83	41	23	7	31	—	11	—	109	—	7	71	13	—	29	—	—	7	79
39	—	—	—	—	11	—	7	17	103	—	—	—	37	7	89	11	101	13	—	—	7	41	47	—	17	19	11	7	31	—
41	—	7	13	—	53	23	61	11	7	—	—	—	—	17	—	7	19	—	11	67	—	—	7	—	41	—	—	—	13	7
47	7	—	43	73	31	11	—	7	19	13	37	—	29	—	7	59	11	—	97	—	79	7	13	67	—	—	—	11	7	131
51	11	—	—	—	7	13	43	—	61	17	—	7	71	—	—	—	—	113	7	—	101	—	11	31	—	7	17	—	19	29
53	19	41	59	11	—	—	7	—	43	—	—	—	—	7	11	17	13	31	—	19	7	103	83	29	—	11	—	7	127	13
57	—	19	—	7	—	31	—	41	—	11	7	29	13	59	—	—	—	7	—	—	11	47	101	107	7	13	37	17	—	—
59	47	11	—	—	—	7	—	37	89	—	13	19	7	—	43	—	17	83	107	7	—	—	—	11	109	—	7	—	—	—
63	59	73	7	—	—	47	13	11	107	7	—	17	19	—	—	—	7	53	11	13	—	79	29	7	101	—	113	97	17	11
69	13	7	71	—	19	11	—	—	7	—	—	—	17	13	—	7	11	—	—	—	—	—	7	19	43	41	13	11	—	7
71	73	17	—	7	37	—	—	83	11	—	7	13	61	—	19	47	—	7	17	11	—	23	59	103	7	31	43	29	41	—
77	—	61	7	—	—	13	11	31	—	7	—	—	79	—	—	23	7	11	13	17	—	37	—	7	—	19	—	—	11	—
81	—	11	41	—	47	—	7	19	—	—	—	23	11	7	13	—	—	73	53	71	7	—	—	11	—	97	19	7	—	—
83	—	7	—	17	11	41	—	—	7	23	71	—	13	—	97	7	—	19	—	—	17	—	7	—	53	13	11	—	—	7
87	37	—	—	61	—	7	—	59	13	—	11	41	7	—	—	17	—	—	19	7	—	11	—	—	—	—	7	—	23	—
89	7	43	11	23	17	—	13	7	—	19	—	—	—	11	7	—	73	—	37	13	—	7	—	—	11	103	23	—	7	—
93	—	53	13	—	7	43	—	—	11	67	19	7	—	79	103	13	17	37	7	11	41	31	23	—	—	7	—	—	13	19
99	17	29	—	7	—	—	11	—	—	13	7	43	—	67	—	—	23	7	—	53	—	19	13	97	7	107	—	127	11	41

Table II Power residues for primes less than 100

The following table has been taken from *Canon Arithmeticus*, by C. G. J. Jacobi, 1839 (reprinted Berlin 1956). Here the base g is the least primitive root modulo p and $g^I \equiv N \pmod{p}$. The reader might check his reading of the tables by noting that for $p = 37$ the entry marked (*) means $2^{32} \equiv 7 \pmod{37}$, whereas the entry marked (†) means $2^3 \equiv 8 \pmod{37}$.

TABLE II

p = 3, g = 2

N											I										
	0	1	2									0	1	2							
0		2	1								0		2	1							

p = 5, g = 2

N											I										
	0	1	2	3	4							0	1	2	3	4					
0		2	4	3	1						0		4	1	3	2					

p = 7, g = 3

N											I										
	0	1	2	3	4	5	6					0	1	2	3	4	5	6			
0		3	2	6	4	5	1				0		6	2	1	4	5	3			

p = 11, g = 2

N											I										
	0	1	2	3	4	5	6	7	8	9		0	1	2	3	4	5	6	7	8	9
0		2	4	8	5	10	9	7	3	6	0		10	1	8	2	4	9	7	3	6
1	1										1	5									

p = 13, g = 2

N											I										
	0	1	2	3	4	5	6	7	8	9		0	1	2	3	4	5	6	7	8	9
0		2	4	8	3	6	12	11	9	5	0		12	1	4	2	9	5	11	3	8
1	10	7	1								1	10	7	6							

p = 17, g = 3

N											I										
	0	1	2	3	4	5	6	7	8	9		0	1	2	3	4	5	6	7	8	9
0		3	9	10	13	5	15	11	16	14	0		16	14	1	12	5	15	11	10	2
1	8	7	4	12	2	6	1				1	3	7	13	4	9	6	8			

p = 19, g = 2

N											I										
	0	1	2	3	4	5	6	7	8	9		0	1	2	3	4	5	6	7	8	9
0		2	4	8	16	13	7	14	9	18	0		18	1	13	2	16	14	6	3	8
1	17	15	11	3	6	12	5	10	1		1	17	12	15	5	7	11	4	10	9	

TABLE II (continued)

p = 23, g = 5

N	0	1	2	3	4	5	6	7	8	9
0		5	2	10	4	20	8	17	16	11
1	9	22	18	21	13	19	3	15	6	7
2	12	14	1							

I	0	1	2	3	4	5	6	7	8	9
0		22	2	16	4	1	18	19	6	10
1	3	9	20	14	21	17	8	7	12	15
2	5	13	11							

p = 29, g = 2

N	0	1	2	3	4	5	6	7	8	9
0		2	4	8	16	3	6	12	24	19
1	9	18	7	14	28	27	25	21	13	26
2	23	17	5	10	20	11	22	15	1	

I	0	1	2	3	4	5	6	7	8	9
0		28	1	5	2	22	6	12	3	10
1	23	25	7	18	13	27	4	21	11	9
2	24	17	26	20	8	16	19	15	14	

p = 31, g = 3

N	0	1	2	3	4	5	6	7	8	9
0		3	9	27	19	26	16	17	20	29
1	25	13	8	24	10	30	28	22	4	12
2	5	15	14	11	2	6	18	23	7	21
3	1									

I	0	1	2	3	4	5	6	7	8	9
0		30	24	1	18	20	25	28	12	2
1	14	23	19	11	22	21	6	7	26	4
2	8	29	17	27	13	10	5	3	16	9
3	15									

p = 37, g = 2

N	0	1	2	3	4	5	6	7	8	9
0		2	4	8	16	32	27	17	34	31
1	25	13	26	15	30	23	9	18	36	35
2	33	29	21	5	10	20	3	6	12	24
3	11	22	*7	14	28	19	1			

I	0	1	2	3	4	5	6	7	8	9
0		36	1	26	2	23	27	32	†3	16
1	24	30	28	11	33	13	4	7	17	35
2	25	22	31	15	29	10	12	6	34	21
3	14	9	5	20	8	19	18			

p = 41, g = 6

N	0	1	2	3	4	5	6	7	8	9
0		6	36	11	25	27	39	29	10	19
1	32	28	4	24	21	3	18	26	33	34
2	40	35	5	30	16	14	2	12	31	22
3	9	13	37	17	20	38	23	15	8	7
4	1									

I	0	1	2	3	4	5	6	7	8	9
0		40	26	15	12	22	1	39	38	30
1	8	3	27	31	25	37	24	33	16	9
2	34	14	29	36	13	4	17	5	11	7
3	23	28	10	18	19	21	2	32	35	6
4	20									

TABLE II (continued)

p = 43, g = 3

	N										I									
	0	1	2	3	4	5	6	7	8	9		0	1	2	3	4	5	6	7	8	9
0		3	9	27	38	28	41	37	25	32	0		42	27	1	12	25	28	35	39	2
1	10	30	4	12	36	22	23	26	35	19	1	10	30	13	32	20	26	24	38	29	19
2	14	42	40	34	16	5	15	2	6	18	2	37	36	15	16	40	8	17	3	5	41
3	11	33	13	39	31	7	21	20	17	8	3	11	34	9	31	23	18	14	7	4	33
4	24	29	1								4	22	6	21							

p = 47, g = 5

	N										I									
	0	1	2	3	4	5	6	7	8	9		0	1	2	3	4	5	6	7	8	9
0		5	25	31	14	23	21	11	8	40	0		46	18	20	36	1	38	32	8	40
1	12	13	18	43	27	41	17	38	2	10	1	19	7	10	11	4	21	26	16	12	45
2	3	15	28	46	42	22	16	33	24	26	2	37	6	25	5	28	2	29	14	22	35
3	36	39	7	35	34	29	4	20	6	30	3	39	3	44	27	34	33	30	42	17	31
4	9	45	37	44	32	19	1				4	9	15	24	13	43	41	23			

p = 53, g = 2

	N										I									
	0	1	2	3	4	5	6	7	8	9		0	1	2	3	4	5	6	7	8	9
0		2	4	8	16	32	11	22	44	35	0		52	1	17	2	47	18	14	3	34
1	17	34	15	30	7	14	28	3	6	12	1	48	6	19	24	15	12	4	10	35	37
2	24	48	43	33	13	26	52	51	49	45	2	49	31	7	39	20	42	25	51	16	46
3	37	21	42	31	9	18	36	19	38	23	3	13	33	5	23	11	9	36	30	38	41
4	46	39	25	50	47	41	29	5	10	20	4	50	45	32	22	8	29	40	44	21	28
5	40	27	1								5	43	27	26							

p = 59, g = 2, $p - 1 = 2 \cdot 29$

	N										I									
	0	1	2	3	4	5	6	7	8	9		0	1	2	3	4	5	6	7	8	9
0		2	4	8	16	32	5	10	20	40	0		58	1	50	2	6	51	18	3	42
1	21	42	25	50	41	23	46	33	7	14	1	7	25	52	45	19	56	4	40	43	38
2	28	56	53	47	35	11	22	44	29	58	2	8	10	26	15	53	12	46	34	20	28
3	57	55	51	43	27	54	49	39	19	38	3	57	49	5	17	41	24	44	55	39	37
4	17	34	9	18	36	13	26	52	45	31	4	9	14	11	33	27	48	16	23	54	36
5	3	6	12	24	48	37	15	30	1		5	13	32	47	22	35	31	21	30	29	

TABLE II (continued)

p = **61**, g = 2, p − 1 = 2² · 3 · 5

N

	0	1	2	3	4	5	6	7	8	9
0		2	4	8	16	32	3	6	12	24
1	48	35	9	18	36	11	22	44	27	54
2	47	33	5	10	20	40	19	38	15	30
3	60	59	57	53	45	29	58	55	49	37
4	13	26	52	43	25	50	39	17	34	7
5	14	28	56	51	41	21	42	23	46	31
6	1									

I

	0	1	2	3	4	5	6	7	8	9
0		60	1	6	2	22	7	49	3	12
1	23	15	8	40	50	28	4	47	13	26
2	24	55	16	57	9	44	41	18	51	35
3	29	59	5	21	48	11	14	39	27	46
4	25	54	56	43	17	34	58	20	10	38
5	45	53	42	33	19	37	52	32	36	31
6	30									

p = **67**, g = 2, p − 1 = 2 · 3 · 11

N

	0	1	2	3	4	5	6	7	8	9
0		2	4	8	16	32	64	61	55	43
1	19	38	9	18	36	5	10	20	40	13
2	26	52	37	7	14	28	56	45	23	46
3	25	50	33	66	65	63	59	51	35	3
4	6	12	24	48	29	58	49	31	62	57
5	47	27	54	41	15	30	60	53	39	11
6	22	44	21	42	17	34	1			

I

	0	1	2	3	4	5	6	7	8	9
0		66	1	39	2	15	40	23	3	12
1	16	59	41	19	24	54	4	64	13	10
2	17	62	60	28	42	30	20	51	25	44
3	55	47	5	32	65	38	14	22	11	58
4	18	53	63	9	61	27	29	50	43	46
5	31	37	21	57	52	8	26	49	45	36
6	56	7	48	35	6	34	33			

p = **71**, g = 7, p − 1 = 2 · 5 · 7

N

	0	1	2	3	4	5	6	7	8	9
0		7	49	59	58	51	2	14	27	47
1	45	31	4	28	54	23	19	62	8	56
2	37	46	38	53	16	41	3	21	5	35
3	32	11	6	42	10	70	64	22	12	13
4	20	69	57	44	24	26	40	67	43	17
5	48	52	9	63	15	34	25	33	18	55
6	30	68	50	66	36	39	60	65	29	61
7	1									

I

	0	1	2	3	4	5	6	7	8	9
0		70	6	26	12	28	32	1	18	52
1	34	31	38	39	7	54	24	49	58	16
2	40	27	37	15	44	56	45	8	13	68
3	60	11	30	57	55	29	64	20	22	65
4	46	25	33	48	43	10	21	9	50	2
5	62	5	51	23	14	59	19	42	4	3
6	66	69	17	53	36	67	63	47	61	41
7	35									

p = **73**, g = 5

N

	0	1	2	3	4	5	6	7	8	9
0		5	25	52	41	59	3	15	2	10
1	50	31	9	45	6	30	4	20	27	62
2	18	17	12	60	8	40	54	51	36	34
3	24	47	16	7	35	29	72	68	48	21
4	32	14	70	58	71	63	23	42	64	28
5	67	43	69	53	46	11	55	56	61	13
6	65	33	19	22	37	39	49	26	57	66
7	38	44	1							

I

	0	1	2	3	4	5	6	7	8	9
0		72	8	6	16	1	14	33	24	12
1	9	55	22	59	41	7	32	21	20	62
2	17	39	63	46	30	2	67	18	49	35
3	15	11	40	61	29	34	28	64	70	65
4	25	4	47	51	71	13	54	31	38	66
5	10	27	3	53	26	56	57	68	43	5
6	23	58	19	45	48	60	69	50	37	52
7	42	44	36							

TABLE II (continued)

p = **79**, g = 3

N											I										
	0	1	2	3	4	5	6	7	8	9		0	1	2	3	4	5	6	7	8	9
0		3	9	27	2	6	18	54	4	12	0		78	4	1	8	62	5	53	12	2
1	36	29	8	24	72	58	16	48	65	37	1	66	68	9	34	57	63	16	21	6	32
2	32	17	51	74	64	34	23	69	49	68	2	70	54	72	26	13	46	38	3	61	11
3	46	59	19	57	13	39	38	35	26	78	3	67	56	20	69	25	37	10	19	36	35
4	76	70	52	77	73	61	25	75	67	43	4	74	75	58	49	76	64	30	59	17	28
5	50	71	55	7	21	63	31	14	42	47	5	50	22	42	77	7	52	65	33	15	31
6	62	28	5	15	45	56	10	30	11	33	6	71	45	60	55	24	18	73	48	29	27
7	20	60	22	66	40	41	44	53	1		7	41	51	14	44	23	47	40	43	39	

p = **83**, g = 2

N											I										
	0	1	2	3	4	5	6	7	8	9		0	1	2	3	4	5	6	7	8	9
0		2	4	8	16	32	64	45	7	14	0		82	1	72	2	27	73	8	3	62
1	28	56	29	58	33	66	49	15	30	60	1	28	24	74	77	9	17	4	56	63	47
2	37	74	65	47	11	22	44	5	10	20	2	29	80	25	60	75	54	78	52	10	12
3	40	80	77	71	59	35	70	57	31	62	3	18	38	5	14	57	35	64	20	48	67
4	41	82	81	79	75	67	51	19	38	76	4	30	40	81	71	26	7	61	23	76	16
5	69	55	27	54	25	50	17	34	68	53	5	55	46	79	59	53	51	11	37	13	34
6	23	46	9	18	36	72	61	39	78	73	6	19	66	39	70	6	22	15	45	58	50
7	63	43	3	6	12	24	48	13	26	52	7	36	33	65	69	21	44	49	32	68	43
8	21	42	1								8	31	42	41							

p = **89**, g = 3

N											I										
	0	1	2	3	4	5	6	7	8	9		0	1	2	3	4	5	6	7	8	9
0		3	9	27	81	65	17	51	64	14	0		88	16	1	32	70	17	81	48	2
1	42	37	22	66	20	60	2	6	18	54	1	86	84	33	23	9	71	64	6	18	35
2	73	41	34	13	39	28	84	74	44	43	2	14	82	12	57	49	52	39	3	25	59
3	40	31	4	12	36	19	57	82	68	26	3	87	31	80	85	22	63	34	11	51	24
4	78	56	79	59	88	86	80	62	8	24	4	30	21	10	29	28	72	73	54	65	74
5	72	38	25	75	47	52	67	23	69	29	5	68	7	55	78	19	66	41	36	75	43
6	87	83	71	35	16	48	55	76	50	61	6	15	69	47	83	8	5	13	56	38	58
7	5	15	45	46	49	58	85	77	53	70	7	79	62	50	20	27	53	67	77	40	42
8	32	7	21	63	11	33	10	30	1		8	46	4	37	61	26	76	45	60	44	

p = **97**, g = 5

N											I										
	0	1	2	3	4	5	6	7	8	9		0	1	2	3	4	5	6	7	8	9
0		5	25	28	43	21	8	40	6	30	0		96	34	70	68	1	8	31	6	44
1	53	71	64	29	48	46	36	83	27	38	1	35	86	42	25	65	71	40	89	78	81
2	93	77	94	82	22	13	65	34	73	74	2	69	5	24	77	76	2	59	18	3	13
3	79	7	35	78	2	10	50	56	86	42	3	9	46	74	60	27	32	16	91	19	95
4	16	80	12	60	9	45	31	58	96	92	4	7	85	39	4	58	45	15	84	14	62
5	72	69	54	76	89	57	91	67	44	26	5	36	63	93	10	52	87	37	55	47	67
6	33	68	49	51	61	14	70	59	4	20	6	43	64	80	75	12	26	94	57	61	51
7	3	15	75	84	32	63	24	23	18	90	7	66	11	50	28	29	72	53	21	33	30
8	62	19	95	87	47	41	11	55	81	17	8	41	88	23	17	73	90	38	83	92	54
9	85	37	88	52	66	39	1				9	79	56	49	20	22	82	48			

Table III Class structures of quadratic fields of $\sqrt{m}$ for m less than 100

This two-part table has been taken from *Introduction à la théorie des nombres algébriques*, by J. Sommer, Paris, 1911. In the following tables d stands for the discriminant and the basis is $[1, \omega]$. (Here ω' denotes the conjugate.) The factors of the discriminant are listed in the order of the corresponding character symbols; except that when a *real* field has discriminant divisible by a prime $q \equiv -1 \pmod 4$ such a prime is listed first in d but omitted in the list of character symbols (see Chapter XIII, §3).

TABLE III

Part 1. Imaginary Fields

	ω	d	Ideal Structure		Genus Structure	
			Ideal‡	Class	Class	Character
-1*	$\sqrt{-1}$	-2^2	$(1)E$	I	I	$+$
-2	$\sqrt{-2}$	-2^3	$(1)E$	I	I	$+$
-3†	$\frac{1+\sqrt{-3}}{2}$	-3	$(1)E$	I	I	$+$
-5	$\sqrt{-5}$	$-2^2 \cdot 5$	(1) $(2, 1+\sqrt{-5})$	A^2 A	A^2 A	$+\ +$ $-\ -$
-6	$\sqrt{-6}$	$-2^3 \cdot 3$	(1) $(2, \sqrt{-6})$	A^2 A	A^2 A	$+\ +$ $-\ -$
-7	$\frac{1+\sqrt{-7}}{2}$	-7	$(1)E$	I	I	$+$
-10	$\sqrt{-10}$	$-2^3 \cdot 5$	(1) $(2, \sqrt{-10})$	A^2 A	A^2 A	$+\ +$ $-\ -$
-11	$\frac{1+\sqrt{-11}}{2}$	-11	$(1)E$	I	I	$+$
-13	$\sqrt{-13}$	$-2^2 \cdot 13$	(1) $(2, 1+\sqrt{-13})$	A^2 A	A^2 A	$+\ +$ $-\ -$
-14	$\sqrt{-14}$	$-2^3 \cdot 7$	(1) $(3, 1-\sqrt{-14})$ $(2, \sqrt{-14})$ $(3, 1+\sqrt{-14})$	J^4 J^3 J^2 J	$\left.\begin{matrix} J^4 \\ J^2 \end{matrix}\right\}$ $\left.\begin{matrix} J^3 \\ J \end{matrix}\right\}$	$+\ +$ $-\ -$
-15	$\frac{1+\sqrt{-15}}{2}$	$-3 \cdot 5$	(1) $(2, 1+\omega)$	A^2 A	A^2 A	$+\ +$ $-\ -$

* This field contains the units $\pm\sqrt{-1}$ in addition to ± 1.

† This field contains the units $\pm\omega$, $\pm\omega'$ in addition to ± 1.

‡ The Euclidean fields are designated by E; there is no other quadratic field beyond this table.

TABLE III, Part 1 (continued)

	ω	d	Ideal Structure		Genus Structure	
			Ideal‡	Class	Class	Character
-17	$\sqrt{-17}$	$-2^2 \cdot 17$	(1)	J^4	J^4	$+\ +$
			$(3, 1 - \sqrt{-17})$	J^3	J^2	
			$(2, 1 + \sqrt{-17})$	J^2	J^3	$-\ -$
			$(3, 1 + \sqrt{-17})$	J	J	
-19	$\frac{1 + \sqrt{-19}}{2}$	-19	(1)	I	I	$+$
-21	$\sqrt{-21}$	$-2^2 \cdot 3 \cdot 7$	(1)	$A^2A_1^2$	$A^2A_1^2$	$+\ +\ +$
			$(5, 3 + \sqrt{-21})$	AA_1	AA_1	$+\ -\ -$
			$(3, \sqrt{-21})$	A_1	A_1	$-\ +\ -$
			$(2, 1 + \sqrt{-21})$	A	A	$-\ -\ +$
-22	$\sqrt{-22}$	$-2^3 \cdot 11$	(1)	A^2	A^2	$+\ +$
			$(2, \sqrt{-22})$	A	A	$-\ -$
-23	$\frac{1 + \sqrt{-23}}{2}$	-23	(1)	J^3	J^3	$+$
			$(2, \omega')$	J^2	J^2	
			$(2, \omega)$	J	J	
-26	$\sqrt{-26}$	$-2^3 \cdot 13$	(1)	J^6	J^6	$+\ +$
			$(5, 2 - \sqrt{-26})$	J^5	J^4	
			$(3, 1 + \sqrt{-26})$	J^4	J^2	
			$(2, \sqrt{-26})$	J^3	J^5	$-\ -$
			$(3, 1 - \sqrt{-26})$	J^2	J^3	
			$(5, 2 + \sqrt{-26})$	J	J	
-29	$\sqrt{-29}$	$-2^2 \cdot 29$	(1)	J^6	J^6	$+\ +$
			$(3, 1 - \sqrt{-29})$	J^5	J^4	
			$(5, 1 - \sqrt{-29})$	J^4	J^2	
			$(2, 1 + \sqrt{-29})$	J^3	J^5	$-\ -$
			$(5, 1 + \sqrt{-29})$	J^2	J^3	
			$(3, 1 + \sqrt{-29})$	J	J	
-30	$\sqrt{-30}$	$-2^3 \cdot 3 \cdot 5$	(1)	$A^2A_1^2$	$A^2A_1^2$	$+\ +\ +$
			$(2, \sqrt{-30})$	AA_1	AA_1	$+\ -\ -$
			$(3, \sqrt{-30})$	A_1	A_1	$-\ +\ -$
			$(5, \sqrt{-30})$	A	A	$-\ -\ +$

TABLE III, Part 1 (continued)

	ω	d	Ideal Structure		Genus Structure	
			Ideal‡	Class	Class	Character
-31	$\frac{1+\sqrt{-31}}{2}$	-31	(1)	J^3	J^3	$+$
			$(2, 1+\omega')$	J^2	J^2	
			$(2, 1+\omega)$	J	J	
-33	$\sqrt{-33}$	$-2^2 \cdot 3 \cdot 11$	(1)	$A^2A_1^2$	$A^2A_1^2$	$+ + +$
			$(2, 1+\sqrt{-33})$	AA_1	AA_1	$+ - -$
			$(3, \sqrt{-33})$	A_1	A_1	$- + -$
			$(6, 3+\sqrt{-33})$	A	A	$- - +$
-34	$\sqrt{-34}$	$-2^3 \cdot 17$	(1)	J^4	J^4	$+ +$
			$(5, 1-\sqrt{-34})$	J^3	J^2	
			$(2, \sqrt{-34})$	J^2	J^3	$- -$
			$(5, 1+\sqrt{-34})$	J	J	
-35	$\frac{1+\sqrt{-35}}{2}$	$-5 \cdot 7$	(1)	A^2	A^2	$+ +$
			$(5, 2+\omega)$	A	A	$- -$
-37	$\sqrt{-37}$	$-2^2 \cdot 37$	(1)	A^2	A^2	$+ +$
			$(2, 1+\sqrt{-37})$	A	A	$- -$
-38	$\sqrt{-38}$	$-2^3 \cdot 19$	(1)	J^6	J^6	$+ +$
			$(3, 1-\sqrt{-38})$	J^5	J^4	
			$(7, 2+\sqrt{-38})$	J^4	J^2	
			$(2, \sqrt{-38})$	J^3	J^5	$- -$
			$(7, 2-\sqrt{-38})$	J^2	J^3	
			$(3, 1+\sqrt{-38})$	J	J	
-39	$\frac{1+\sqrt{-39}}{2}$	$-3 \cdot 13$	(1)	J^4	J^4	$+ +$
			$(2, \omega')$	J^3	J^2	
			$(3, 1+\omega)$	J^2	J^3	$- -$
			$(2, \omega)$	J	J	
-41	$\sqrt{-41}$	$-2^2 \cdot 41$	(1)	J^8	J^8	$+ +$
			$(3, 1-\sqrt{-41})$	J^7	J^6	
			$(5, 2-\sqrt{-41})$	J^6	J^4	
			$(7, 1-\sqrt{-41})$	J^5	J^2	
			$(2, 1+\sqrt{-41})$	J^4	J^7	$- -$
			$(7, 1+\sqrt{-41})$	J^3	J^5	
			$(5, 2+\sqrt{-41})$	J^2	J^3	
			$(3, 1+\sqrt{-41})$	J	J	

TABLE III, Part 1 (continued)

	ω	d	Ideal Structure: Ideal‡	Ideal Structure: Class	Genus Structure: Class	Genus Structure: Character
-42	$\sqrt{-42}$	$-2^3 \cdot 3 \cdot 7$	(1)	$A^2A_1^2$	$A^2A_1^2$	$+\ +\ +$
			$(7, \sqrt{-42})$	AA_1	AA_1	$+\ -\ -$
			$(3, \sqrt{-42})$	A_1	A_1	$-\ +\ -$
			$(2, \sqrt{-42})$	A	A	$-\ -\ +$
-43	$\frac{1+\sqrt{-43}}{2}$	-43	(1)	I	I	$+$
-46	$\sqrt{-46}$	$-2^3 \cdot 23$	(1)	J^4	J^4, J^2	$+\ +$
			$(5, 2-\sqrt{-46})$	J^3		
			$(2, \sqrt{-46})$	J^2	J^3, J	$-\ -$
			$(5, 2+\sqrt{-46})$	J		
-47	$\frac{1+\sqrt{-47}}{2}$	-47	(1)	J^5	J^5, J^4, J^3, J^2, J	$+$
			$(2, 1+\omega')$	J^4		
			$(3, \omega')$	J^3		
			$(3, \omega)$	J^2		
			$(2, 1+\omega)$	J		
-51	$\frac{1+\sqrt{-51}}{2}$	$-3 \cdot 17$	(1)	A^2	A^2	$+\ +$
			$(3, 1+\omega)$	A	A	$-\ -$
-53	$\sqrt{-53}$	$-2^2 \cdot 53$	(1)	J^6	J^6, J^4, J^2	$+\ +$
			$(3, 1-\sqrt{-53})$	J^5		
			$(9, 1-\sqrt{-53})$	J^4		
			$(2, 1+\sqrt{-53})$	J^3	J^5, J^3, J	$-\ -$
			$(9, 1+\sqrt{-53})$	J^2		
			$(3, 1+\sqrt{-53})$	J		
-55	$\frac{1+\sqrt{-55}}{2}$	$-5 \cdot 11$	(1)	J^4	J^4, J^2	$+\ +$
			$(2, \omega')$	J^3		
			$(5, 2+\omega)$	J^2	J^3, J	$-\ -$
			$(2, \omega)$	J		
-57	$\sqrt{-57}$	$-2^2 \cdot 3 \cdot 19$	(1)	$A^2A_1^2$	$A^2A_1^2$	$+\ +\ +$
			$(2, 1+\sqrt{-57})$	AA_1	AA_1	$+\ -\ -$
			$(3, \sqrt{-57})$	A_1	A_1	$-\ +\ -$
			$(6, 3+\sqrt{-57})$	A	A	$-\ -\ +$

TABLE III, Part 1 (continued)

	ω	d	Ideal Structure		Genus Structure	
			Ideal‡	Class	Class	Character
-58	$\sqrt{-58}$	$-2^3 \cdot 29$	(1)	A^2	A^2	$+\ +$
			$(2, \sqrt{-58})$	A	A	$-\ -$
-59	$\frac{1+\sqrt{-59}}{2}$	-59	(1)	J^3	J^3	$+$
			$(3, \omega')$	J^2	J^2	
			$(3, \omega)$	J	J	
-61	$\sqrt{-61}$	$-2^2 \cdot 61$	(1)	J^3	J^3	$+\ +$
			$(5, 2-\sqrt{-61})$	J^2	J^2	
			$(5, 2+\sqrt{-61})$	J	J	
			$(7, 3-\sqrt{-61})$	AJ^2	AJ^2	$-\ -$
			$(7, 3+\sqrt{-61})$	AJ	AJ	
			$(2, 1+\sqrt{-61})$	A	A	
-62	$\sqrt{-62}$	$-2^3 \cdot 31$	(1)	J^8	J^8	$+\ +$
			$(3, 1-\sqrt{-62})$	J^7	J^6	
			$(7, 1+\sqrt{-62})$	J^6	J^4	
			$(11, 2+\sqrt{-62})$	J^5	J^2	
			$(2, \sqrt{-62})$	J^4	J^7	$-\ -$
			$(11, 2-\sqrt{-62})$	J^3	J^5	
			$(7, 1-\sqrt{-62})$	J^2	J^3	
			$(3, 1+\sqrt{-62})$	J	J	
-65	$\sqrt{-65}$	$-2^2 \cdot 5 \cdot 13$	(1)	J^4	J^4	$+\ +\ +$
			$(3, 1-\sqrt{-65})$	J^3	J^2	
			$(9, 5-\sqrt{-65})$	J^2	AJ^2	$+\ -\ -$
			$(3, 1+\sqrt{-65})$	J	A	
			$(11, 1-\sqrt{-65})$	AJ^3	AJ^3	$-\ +\ -$
			$(2, 1+\sqrt{-65})$	AJ^2	AJ	
			$(11, 1+\sqrt{-65})$	AJ^1	J^3	$-\ -\ +$
			$(5, \sqrt{-65})$	A	J	
-66	$\sqrt{-66}$	$-2^3 \cdot 3 \cdot 11$	(1)	J^4	J^4	$+\ +\ +$
			$(5, 2-\sqrt{-66})$	J^3	J^2	
			$(3, \sqrt{-66})$	J^2	AJ^2	$+\ -\ -$
			$(5, 2+\sqrt{-66})$	J	A	
			$(7, 2+\sqrt{-66})$	AJ^3	AJ^3	$-\ +\ -$
			$(11, \sqrt{-66})$	AJ^2	AJ	
			$(7, 2-\sqrt{-66})$	AJ	J^3	$-\ -\ +$
			$(2, \sqrt{-66})$	A	J	

TABLE III, Part 1 (continued)

	ω	d	Ideal Structure		Genus Structure	
			Ideal‡	Class	Class	Character
-67	$\frac{1+\sqrt{-67}}{2}$	-67	(1)	I	I	$+$
-69	$\sqrt{-69}$	$-2^2 \cdot 3 \cdot 23$	(1)	J^4	J^4	$+\ +\ +$
			$(7, 1-\sqrt{-69})$	J^3	J^2	
			$(6, 3+\sqrt{-69})$	J^2	AJ^3	$+\ -\ -$
			$(7, 1+\sqrt{-69})$	J	AJ	
			$(5, 1+\sqrt{-69})$	AJ^3	J^3	$-\ +\ -$
			$(3, \sqrt{-69})$	AJ^2	J	
			$(5, 1-\sqrt{-69})$	AJ	AJ^2	$-\ -\ +$
			$(2, 1+\sqrt{-69})$	A	A	
-70	$\sqrt{-70}$	$-2^3 \cdot 5 \cdot 7$	(1)	$A^2A_1^2$	$A^2A_1^2$	$+\ +\ +$
			$(7, \sqrt{-70})$	AA_1	AA_1	$+\ -\ -$
			$(5, \sqrt{-70})$	A_1	A_1	$-\ +\ -$
			$(2, \sqrt{-70})$	A	A	$-\ -\ +$
-71	$\frac{1+\sqrt{-71}}{2}$	-71	(1)	J^7	J^7	$+$
			$(2, \omega')$	J^6	J^6	
			$(5, 1+\omega')$	J^5	J^5	
			$(3, 2+\omega)$	J^4	J^4	
			$(3, 2+\omega')$	J^3	J^3	
			$(5, 1+\omega)$	J^2	J^2	
			$(2, \omega)$	J	J	
-73	$\sqrt{-73}$	$-2^2 \cdot 73$	(1)	J^4	J^4	$+\ \ +$
			$(7, 2-\sqrt{-73})$	J^3	J^2	
			$(2, 1+\sqrt{-73})$	J^2	J^3	$-\ \ -$
			$(7, 2+\sqrt{-73})$	J	J	
-74	$\sqrt{-74}$	$-2^3 \cdot 37$	(1)	J^5	J^5	$+\ \ +$
			$(11, 5-\sqrt{-74})$	J^4	J^4	
			$(3, 1-\sqrt{-74})$	J^3	J^3	
			$(3, 1+\sqrt{-74})$	J^2	J^2	
			$(11, 5+\sqrt{-74})$	J	J	
			$(5, 1-\sqrt{-74})$	AJ^4	AJ^4	$-\ \ -$
			$(6, 2+\sqrt{-74})$	AJ^3	AJ^3	
			$(6, 2-\sqrt{-74})$	AJ^2	AJ^2	
			$(5, 1+\sqrt{-74})$	AJ	AJ	
			$(2, \sqrt{-74})$	A	A	

TABLE III, Part 1 (continued)

	ω	d	Ideal Structure		Genus Structure	
			Ideal‡	Class	Class	Character
-77	$\sqrt{-77}$	$-2^2 \cdot 7 \cdot 11$	(1)	J^4	J^4	$+\ +\ +$
			$(3, 1 - \sqrt{-77})$	J^3	J^2	
			$(14, 7 + \sqrt{-77})$	J^2	AJ^3	$+\ -\ -$
			$(3, 1 + \sqrt{-77})$	J	AJ	
			$(6, 1 - \sqrt{-77})$	AJ^3	AJ^2	$-\ +\ -$
			$(7, \sqrt{-77})$	AJ^2	A	
			$(6, 1 + \sqrt{-77})$	AJ	J^3	$-\ -\ +$
			$(2, 1 + \sqrt{-77})$	A	J	
-78	$\sqrt{-78}$	$-2^3 \cdot 3 \cdot 13$	(1)	$A^2A_1^2$	$A^2A_1^2$	$+\ +\ +$
			$(2, \sqrt{-78})$	AA_1	AA_1	$+\ -\ -$
			$(13, \sqrt{-78})$	A_1	A_1	$-\ +\ -$
			$(3, \sqrt{-78})$	A	A	$-\ -\ +$
-79	$\frac{1 + \sqrt{-79}}{2}$	-79	(1)	J^5	J^5	$+$
			$(2, 1 + \omega')$	J^4	J^4	
			$(5, \omega')$	J^3	J^3	
			$(5, \omega)$	J^2	J^2	
			$(2, 1 + \omega)$	J	J	
-82	$\sqrt{-82}$	$-2^3 \cdot 41$	(1)	J^4	J^4	$+\ +$
			$(7, 3 - \sqrt{-82})$	J^3	J^2	
			$(2, \sqrt{-82})$	J^2	J^3	$-\ -$
			$(7, 3 + \sqrt{-82})$	J	J	
-83	$\frac{1 + \sqrt{-83}}{2}$	-83	(1)	J^3	J^3	$+$
			$(3, \omega')$	J^2	J^2	
			$(3, \omega)$	J	J	
-85	$\sqrt{-85}$	$-2^2 \cdot 5 \cdot 17$	(1)	$A^2A_1^2$	$A^2A_1^2$	$+\ +\ +$
			$(5, \sqrt{-85})$	AA_1	AA_1	$+\ -\ -$
			$(10, 5 + \sqrt{-85})$	A_1	A_1	$-\ +\ -$
			$(2, 1 + \sqrt{-85})$	A	A	$-\ -\ +$
-86	$\sqrt{-86}$	$-2^3 \cdot 43$	(1)	J^{10}	J^{10}	$+\ +$
			$(3, 1 - \sqrt{-86})$	J^9	J^8	
			$(9, 2 + \sqrt{-86})$	J^8	J^6	
			$(5, 2 + \sqrt{-86})$	J^7	J^4	
			$(17, 4 - \sqrt{-86})$	J^6	J^2	
			$(2, \sqrt{-86})$	J^5	J^9	$-\ -$
			$(17, 4 + \sqrt{-86})$	J^4	J^7	
			$(5, 2 - \sqrt{-86})$	J^3	J^5	
			$(9, 2 - \sqrt{-86})$	J^2	J^3	
			$(3, 1 + \sqrt{-86})$	J	J	

TABLE III, Part 1 (continued)

	ω	d	Ideal Structure		Genus Structure	
			Ideal‡	Class	Class	Character
-87	$\frac{1+\sqrt{-87}}{2}$	$-3 \cdot 29$	(1)	J^6	J^6	
			$(2, \omega')$	J^5	J^4	$+ \quad +$
			$(7, 2+\omega)$	J^4	J^2	
			$(3, 1+\omega)$	J^3	J^5	
			$(7, 2+\omega')$	J^2	J^3	$- \quad -$
			$(2, \omega)$	J	J	
-89	$\sqrt{-89}$	$-2^2 \cdot 89$	(1)	J^{12}	J^{12}	
			$(3, 1-\sqrt{-89})$	J^{11}	J^{10}	
			$(17, 8-\sqrt{-89})$	J^{10}	J^8	$+ \quad +$
			$(7, 4-\sqrt{-89})$	J^9	J^6	
			$(5, 1-\sqrt{-89})$	J^8	J^4	
			$(6, 1+\sqrt{-89})$	J^7	J^2	
			$(2, 1+\sqrt{-89})$	J^6	J^{11}	
			$(6, 1-\sqrt{-89})$	J^5	J^9	
			$(5, 1+\sqrt{-89})$	J^4	J^7	
			$(7, 4+\sqrt{-89})$	J^3	J^5	$- \quad -$
			$(17, 8+\sqrt{-89})$	J^2	J^3	
			$(3, 1+\sqrt{-89})$	J	J	
-91	$\frac{1+\sqrt{-91}}{2}$	$-7 \cdot 13$	(1)	A^2	A^2	$+ \quad +$
			$(7, 3+\omega)$	A	A	$- \quad -$
-93	$\sqrt{-93}$	$-2^2 \cdot 3 \cdot 31$	(1)	$A^2A_1^2$	$A^2A_1^2$	$+ \quad + \quad +$
			$(6, 3+\sqrt{-93})$	AA_1	AA_1	$+ \quad - \quad -$
			$(3, \sqrt{-93})$	A_1	A_1	$- \quad + \quad -$
			$(2, 1+\sqrt{-93})$	A	A	$- \quad - \quad +$
-94	$\sqrt{-94}$	$-2^3 \cdot 47$	(1)	J^8	J^8	
			$(5, 1-\sqrt{-94})$	J^7	J^6	$+ \quad +$
			$(7, 2-\sqrt{-94})$	J^6	J^4	
			$(11, 4+\sqrt{-94})$	J^5	J^2	
			$(2, \sqrt{-94})$	J^4	J^7	
			$(11, 4-\sqrt{-94})$	J^3	J^5	$- \quad -$
			$(7, 2+\sqrt{-94})$	J^2	J^3	
			$(5, 1+\sqrt{-94})$	J	J	

TABLE III, Part 1 (continued)

	ω	d	Ideal Structure		Genus Structure	
			Ideal‡	Class	Class	Character
-95	$\frac{1+\sqrt{-95}}{2}$	$-5 \cdot 19$	(1)	J^8	J^8	$+\ +$
			$(2, 1+\omega')$	J^7	J^6	
			$(4, 1-\omega')$	J^6	J^4	
			$(3, \omega')$	J^5	J^2	
			$(5, 2+\omega)$	J^4	J^7	$-\ -$
			$(3, \omega)$	J^3	J^5	
			$(4, 1-\omega)$	J^2	J^3	
			$(2, 1+\omega)$	J	J	
-97	$\sqrt{-97}$	$-2^2 \cdot 97$	(1)	J^4	J^4	$+\ +$
			$(7, 1-\sqrt{-97})$	J^3	J^2	
			$(2, 1+\sqrt{-97})$	J^2	J^3	$-\ -$
			$(7, 1+\sqrt{-97})$	J	J	

TABLE III

Part 2. Real Fields*

	ω	d	η	$N(\eta)$	Ideal Structure		Genus Structure	
					Ideal†	Class	Class	Character
2	$\sqrt{2}$	2^3	$1+\sqrt{2}$	-1	(1) E	I	I	$+$
3	$\sqrt{3}$	$3\cdot 2^2$	$2+\sqrt{3}$	$+1$	(1) E	I	I	$+$
5	$\frac{1+\sqrt{5}}{2}$	5	ω	-1	(1) E	I	I	$+$
6	$\sqrt{6}$	$3\cdot 2^3$	$5+2\sqrt{6}$	$+1$	(1) E	I	I	$+$
7	$\sqrt{7}$	$7\cdot 2^2$	$8+3\sqrt{7}$	$+1$	(1) E	I	I	$+$
10	$\sqrt{10}$	$2^3\cdot 5$	$3+\sqrt{10}$	-1	(1) $(2, \sqrt{10})$	A^2 A	A^2 A	$++$ $--$
11	$\sqrt{11}$	$11\cdot 2^2$	$10+3\sqrt{11}$	$+1$	(1) E	I	I	$+$
13	$\frac{1+\sqrt{13}}{2}$	13	$1+\omega$	-1	(1) E	I	I	$+$
14	$\sqrt{14}$	$7\cdot 2^3$	$15+4\sqrt{14}$	$+1$	(1)	I	I	$+$
15	$\sqrt{15}$	$3\cdot 2^2\cdot 5$	$4+\sqrt{15}$	$+1$	(1) $(2, 1+\sqrt{15})$	A^2 A	A^2 A	$++$ $--$
17	$\frac{1+\sqrt{17}}{2}$	17	$3+2\omega$	-1	(1) E	I	I	$+$
19	$\sqrt{19}$	$19\cdot 2^2$	$170+39\sqrt{19}$	$+1$	(1) E	I	I	$+$
21	$\frac{1+\sqrt{21}}{2}$	$3\cdot 7$	$2+\omega$	$+1$	(1) E	I	I	$+$
22	$\sqrt{22}$	$11\cdot 2^3$	$197+42\sqrt{22}$	$+1$	(1)	I	I	$+$
23	$\sqrt{23}$	$23\cdot 2^2$	$24+5\sqrt{23}$	$+1$	(1)	I	I	$+$
26	$\sqrt{26}$	$2^3\cdot 13$	$5+\sqrt{26}$	-1	(1) $(2, \sqrt{26})$	A^2 A	A^2 A	$++$ $--$

* The fundamental units are designated by η.

† The Euclidean fields are designated by E; there is no other quadratic field beyond this table.

TABLE III, Part 2 (continued)

	ω	d	η	$N(\eta)$	Ideal Structure		Genus Structure	
					Ideal†	Class	Class	Character
29	$\frac{1+\sqrt{29}}{2}$	29	$2+\omega$	-1	(1) E	I	I	$+$
30	$\sqrt{30}$	$3 \cdot 2^3 \cdot 5$	$11+2\sqrt{30}$	$+1$	(1) $(2, \sqrt{30})$	A^2 A	A^2 A	$+\ +$ $-\ -$
31	$\sqrt{31}$	$31 \cdot 2^2$	$1520+273\sqrt{31}$	$+1$	(1)	I	I	$+$
33	$\frac{1+\sqrt{33}}{2}$	$3 \cdot 11$	$19+8\omega$	$+1$	(1) E	I	I	$+$
34	$\sqrt{34}$	$2^3 \cdot 17$	$35+6\sqrt{34}$	$+1$	(1) $(3, 1+\sqrt{34})$	A^2 A	A^2 A	$+\ +$ $-\ -$
35	$\sqrt{35}$	$7 \cdot 2^2 \cdot 5$	$6+\sqrt{35}$	$+1$	(1) $(2, 1+\sqrt{35})$	A^2 A	A^2 A	$+\ +$ $-\ -$
37	$\frac{1+\sqrt{37}}{2}$	37	$5+2\omega$	-1	(1) E	I	I	$+$
38	$\sqrt{38}$	$19 \cdot 2^3$	$37+6\sqrt{38}$	$+1$	(1)	I	I	$+$
39	$\sqrt{39}$	$3 \cdot 2^2 \cdot 13$	$25+4\sqrt{39}$	$+1$	(1) $(2, 1+\sqrt{39})$	A^2 A	A^2 A	$+\ +$ $-\ -$
41	$\frac{1+\sqrt{41}}{2}$	41	$27+10\omega$	-1	(1) E	I	I	$+$
42	$\sqrt{42}$	$3 \cdot 2^3 \cdot 7$	$13+2\sqrt{42}$	$+1$	(1) $(2, \sqrt{42})$	A^2 A	A^2 A	$+\ +$ $-\ -$
43	$\sqrt{43}$	$43 \cdot 2^2$	$3482+531\sqrt{43}$	$+1$	(1)	I	I	$+$
46	$\sqrt{46}$	$23 \cdot 2^3$	$24335+3588\sqrt{46}$	$+1$	(1)	I	I	$+$
47	$\sqrt{47}$	$47 \cdot 2^2$	$48+7\sqrt{47}$	$+1$	(1)	I	I	$+$
51	$\sqrt{51}$	$3 \cdot 2^2 \cdot 17$	$50+7\sqrt{51}$	$+1$	(1) $(3, \sqrt{51})$	A^2 A	A^2 A	$+\ +$ $-\ -$
53	$\frac{1+\sqrt{53}}{2}$	53	$3+\omega$	-1	(1)	I	I	$+$

TABLE III, Part 2 (continued)

	ω	d	η	$N(\eta)$	Ideal Structure		Genus Structure	
					Ideal†	Class	Class	Character
55	$\sqrt{55}$	$11 \cdot 2^2 \cdot 5$	$89 + 12\sqrt{55}$	+1	(1) $(2, 1 + \sqrt{55})$	A^2 A	A^2 A	+ + − −
57	$\frac{1 + \sqrt{57}}{2}$	$3 \cdot 19$	$131 + 40\omega$	+1	(1) E	I	I	+
58	$\sqrt{58}$	$2^3 \cdot 29$	$99 + 13\sqrt{58}$	−1	(1) $(2, \sqrt{58})$	A^2 A	A^2 A	+ + − −
59	$\sqrt{59}$	$59 \cdot 2^2$	$530 + 69\sqrt{59}$	+1	(1)	I	I	+
61	$\frac{1 + \sqrt{61}}{2}$	61	$17 + 5\omega$	−1	(1)	I	I	+
62	$\sqrt{62}$	$31 \cdot 2^3$	$63 + 8\sqrt{62}$	+1	(1)	I	I	+
65	$\frac{1 + \sqrt{65}}{2}$	$5 \cdot 13$	$7 + 2\omega$	−1	(1) $(5, 2 + \omega)$	A^2 A	A^2 A	+ + − −
66	$\sqrt{66}$	$3 \cdot 2^3 \cdot 11$	$65 + 8\sqrt{66}$	+1	(1) $(3, \sqrt{66})$	A^2 A	A^2 A	+ + − −
67	$\sqrt{67}$	$67 \cdot 2^2$	$48842 + 5967\sqrt{67}$	+1	(1)	I	I	+
69	$\frac{1 + \sqrt{69}}{2}$	$3 \cdot 23$	$11 + 3\omega$	+1	(1)	I	I	+
70	$\sqrt{70}$	$7 \cdot 2^3 \cdot 5$	$251 + 30\sqrt{70}$	+1	(1) $(2, \sqrt{70})$	A^2 A	A^2 A	+ + − −
71	$\sqrt{71}$	$71 \cdot 2^2$	$3480 + 413\sqrt{71}$	+1	(1)	I	I	+
73	$\frac{1 + \sqrt{73}}{2}$	73	$943 + 250\omega$	−1	(1) E	I	I	+
74	$\sqrt{74}$	$2^3 \cdot 37$	$43 + 5\sqrt{74}$	−1	(1) $(2, \sqrt{74})$	A^2 A	A^2 A	+ + − −

TABLE III, Part 2 (continued)

	ω	d	η	$N(\eta)$	Ideal Structure		Genus Structure	
					Ideal†	Class	Class	Character
77	$\frac{1+\sqrt{77}}{2}$	$7 \cdot 11$	$4+\omega$	$+1$	(1)	I	I	$+$
78	$\sqrt{78}$	$3 \cdot 2^3 \cdot 13$	$53+6\sqrt{78}$	$+1$	(1)	A^2	A^2	$++$
					$(2, \sqrt{78})$	A	A	$--$
79	$\sqrt{79}$	$79 \cdot 2^2$	$80+9\sqrt{79}$	$+1$	(1)	J^3	J^3	$+$
					$(3, 1-\sqrt{79})$	J^2	J^2	
					$(3, 1+\sqrt{79})$	J	J	
82	$\sqrt{82}$	$2^3 \cdot 41$	$9+\sqrt{82}$	-1	(1)	J^4	J^4	$++$
					$(3, 2-\sqrt{82})$	J^3	J^2	
					$(2, \sqrt{82})$	J^2	J^3	$--$
					$(3, 2+\sqrt{32})$	J	J	
83	$\sqrt{83}$	$83 \cdot 2^2$	$82+9\sqrt{83}$	$+1$	(1)	I	I	$+$
85	$\frac{1+\sqrt{85}}{2}$	$5 \cdot 17$	$4+\omega$	-1	(1)	A^2	A^2	$++$
					$(5, 2+\omega)$	A	A	$--$
86	$\sqrt{86}$	$43 \cdot 2^3$	$10405+1122\sqrt{86}$	$+1$	(1)	I	I	$+$
87	$\sqrt{87}$	$3 \cdot 2^2 \cdot 29$	$28+3\sqrt{87}$	$+1$	(1)	A^2	A^2	$++$
					$(2, 1+\sqrt{87})$	A	A	$--$
89	$\frac{1+\sqrt{89}}{2}$	89	$447+106\omega$	-1	(1)	I	I	$+$
91	$\sqrt{91}$	$7 \cdot 2^2 \cdot 13$	$1574+165\sqrt{91}$	$+1$	(1)	A^2	A^2	$++$
					$(2, 1+\sqrt{91})$	A	A	$--$
93	$\frac{1+\sqrt{93}}{2}$	$3 \cdot 31$	$13+3\omega$	$+1$	(1)	I	I	$+$
94	$\sqrt{94}$	$47 \cdot 2^3$	$2143295+$ $221064\sqrt{94}$	$+1$	(1)	I	I	$+$
95	$\sqrt{95}$	$19 \cdot 2^2 \cdot 5$	$39+4\sqrt{95}$	$+1$	(1)	A^2	A^2	$++$
					$(2, 1+\sqrt{95})$	A	A	$--$
97	$\frac{1+\sqrt{97}}{2}$	97	$5035+1138\omega$	-1	(1)	I	I	$+$

Index